Passive Optical Networks

Passive Optical Networks

Principles and Practice

CEDRIC F. LAM

ELSEVIER

AMSTERDAM • BOSTON • HEIDELBERG • LONDON
NEW YORK • OXFORD • PARIS • SAN DIEGO
SAN FRANCISCO • SINGAPORE • SYDNEY • TOKYO
Academic Press is an imprint of Elsevier

Academic Press in an imprint of Elsevier
30 Corporate Drive, Suite 400, Burlington, MA 01803, USA
525 B Street, Suite 1900, San Diego, California 92101-4495, USA
84 Theobald's Road, London WCIX 8RR, UK

This book is printed on acid-free paper. ♾

Library of Congress Cataloging-in Publication Data

Lam, Cedric F
Passive optical networks : principles and practice / Cedric F. Lam
p. cm.
Includes bibliographical references and index.
ISBN-13: 978-0-12-373853-0 (alk. paper)
ISBN-10: 0-12-373853-9
1. Optical communications. 2. Optoelectronics. I. Title.
TK5103.59.L36 2007
621.382′7—dc22 2007035594

British Library Cataloguing in Publication Data
A catalogue record for this book is available from the British Library

ISBN: 978-0-12-373853-0

For information on all Academic Press publications
visit our web site at www.books.elsevier.com

Printed in the United States of America
07 08 09 10 9 8 7 6 5 4 3 2 1

To my parents, for their unfailing love and encouragements

Cedric F. Lam

Contents

List of Acronyms xiii
List of Figures xxi
List of Tables xxxv
Preface xxxvii
Acknowledgments xxxix
List of Contributors xli

1 Introduction 1
Cedric F. Lam

1.1 History of Broadband Access Networks and PON 1
1.1.1 Digital Subscriber Line (DSL) 2
1.1.2 Cable Modem 3
1.1.3 Fiber Access Systems 7
1.1.4 Ethernet 8
1.1.5 WDM in Optical Access Networks 11
1.1.6 Killer Applications 11
1.2 Economic Considerations in PON Development 12
1.2.1 How Much Bandwidth Is Enough? 12
1.2.2 Policy and Regulation Influence 13
1.2.3 Standardization Efforts 14
1.2.4 Cost Considerations 14
1.3 Organization of the Book 15
References 16

2 PON Architectures Review 19
Cedric F. Lam

2.1 FTTx Overview 19
2.2 TDM-PON vs WDM-PON 20
2.3 Optical Transmission System 21
2.3.1 Optical Fiber 21
2.3.2 Chromatic Dispersion 23
2.3.3 Fiber Loss 25

2.3.4 Bidirectional Transmission 26
2.3.5 Wavelength Division Duplex 27
2.4 Power-Splitting Strategies in a TDM-PON 29
2.4.1 Splitting Architectures 29
2.4.2 Splitting Ratio 30
2.5 Standard Commercial TMD-PON Infrastructure 30
2.5.1 OLT and ONU Structures 31
2.5.2 Burst Mode Operation and Ranging 33
2.5.3 C-Band Analog CATV Signal Overlay 35
2.5.4 Security Concerns in Power-Splitting PON 36
2.6 APON/BPON and G-PON 37
2.6.1 ATM-PON and ITU-T G.983 37
2.6.2 Collision Resolution in APON/G-PON 42
2.6.3 Wavelength Overlay in APON/G-PON 42
2.6.4 G-PON and ITU-T G.984 43
2.6.5 APON/G-PON Protection Switching 54
2.7 EPON 54
2.7.1 Ethernet Layering Architecture and EPON 54
2.7.2 EPON PMD Layer 56
2.7.3 Burst Mode Operation and Loop Timing in EPON 56
2.7.4 PCS Layer and Forward Error Correction 59
2.7.5 Ethernet Framing 59
2.7.6 Multipoint Control Protocol (MPCP) 60
2.7.7 Point-to-Point Emulation (P2PE) in EPON 65
2.7.8 EPON Encryption and Protection 67
2.7.9 Ethernet OAM Sublayer 68
2.8 G-PON and EPON Comparison 69
2.9 Super PON 70
2.10 WDM-PON 71
2.10.1 Advantages and Challenges of WDM-PON 72
2.10.2 Arrayed Waveguide Grating (AWG) Router 72
2.10.3 Broadcast Emulation and Point-to-Point Operation 74
2.10.4 2-PONs-In-1 74
2.10.5 WDM-on-WDM 75
2.10.6 Hybrid WDM/TDM-PON 76
2.10.7 Multiply Fiber Plant Utility by WDM 80
2.11 Summary 84
References 84

3 Optical Technologies in Passive Optical Access Networks 87
Elaine Wong

3.1 Introduction 87
3.2 Optical Technologies in Passive Outside Plant 91

3.2.1 Planar Lightwave Circuit (PLC)-Based Optical Power Splitter 91
3.2.2 Arrayed Waveguide Grating 93
3.3 PON Technologies for Indoor Installation 108
3.3.1 Field Assembly and Indoor Connectors 108
3.3.2 Fiber for Indoor Installations 112
3.4 Transmitter Sources at Subscriber Premises 120
3.4.1 Introduction 120
3.4.2 Wavelength-Specific ONUs 122
3.4.3 Colorless ONUs 129
3.4.4 Source-Free ONUs Based on Wavelength Reuse Schemes 137
3.5 Summary 141
References 143

4 Transceivers for Passive Optical Networks 151
Yongmao Frank Chang and Badri Gomatam

4.1 Introduction 151
4.1.1 Historical Background 152
4.1.2 PON Transceiver Evolution 153
4.2 PON System Requirements 154
4.2.1 PON System Power Budgets and Specifications 154
4.2.2 Physical-Layer Specifications 155
4.2.3 PON Burst-Mode Timing Requirements 156
4.3 Transceiver Technologies 161
4.3.1 Transceiver Building Blocks 162
4.3.2 Optical Transmit and Receive Devices 163
4.3.3 Bidirectional Optical Subassembly (BOSA) 167
4.3.4 PON Transceiver Modules 172
4.4 Burst-Mode Electronics 174
4.4.1 Conventional vs Burst-Mode Data 174
4.4.2 Burst-Mode Transmitter (BM Tx) 175
4.4.3 Burst-Mode Receiver (BM Rx) 181
4.5 Transceiver Digital-Diagnostic Monitoring 187
4.5.1 Module Parameter Monitoring 188
4.5.2 Fiber OTDR Monitoring 192
4.6 PON Transceiver System Evaluation 196
4.6.1 G-PON Transceiver System Evaluation 197
4.6.2 EPON Transceivers 202

4.6.3 Impact of Analog CATV Overlay 205
4.7 Summary and Outlook 207
References 209

5 Ranging and Dynamic Bandwidth Allocation 215
Noriki Miki and Kiyomi Kumozaki

5.1 Ranging 215
5.1.1 Purpose of Ranging 215
5.1.2 Ranging Procedures Overview 216
5.1.3 Ranging Protocol of G-PON 218
5.1.4 Ranging Protocol of EPON 220
5.2 Dynamic Bandwidth Allocation (DBA) 221
5.2.1 DBA Overview 221
5.2.2 Target Service 222
5.2.3 Requirements of DBA 224
5.2.4 Traffic Control Fundamentals—Fair Queuing 226
5.2.5 IPACT and its Variants 228
5.2.6 Improved DBA 231
References 241

6 Protection Architectures for Passive Optical Networks 243
Calvin C. K. Chan

6.1 Introduction 243
6.2 Considerations of Protection in Passive Optical Networks 244
6.2.1 Protection or Restoration 245
6.2.2 Network Topology 245
6.2.3 Network Type 246
6.2.4 Resources To Be Protected 246
6.2.5 Single or Multiple Failures 247
6.2.6 Automatic Protection Switching 247
6.2.7 Operation, Administration, and Management 247
6.2.8 Traffic Restoration Time 247
6.2.9 Complexity 248
6.3 Protection Architectures 248
6.3.1 Conventional Passive Optical Networks 248
6.3.2 WDM Passive Optical Networks 253
6.4 Summary 261
Acknowledgments 264
References 264

7 Optical Characterization, Diagnosis, and Performance Monitoring for PON **267**
Alan E. Willner and Zhongqi Pan

7.1 Introduction 267
7.2 Network Testing, Characterization, and Monitoring Challenges for PON 269
7.2.1 Fiber's Key Characteristics 269
7.2.2 Characterization and Monitoring Challenges for PON 271
7.3 Methods for Characterization, Diagnosis, and Monitoring 275
7.3.1 Required Physical-Layer Measurement 275
7.3.2 Basic Test Equipment 275
7.3.3 Network Testing, Characterization, and Diagnosis Guidelines 277
7.3.4 Network Performance Monitoring 278
7.4 Conclusion 296
Acknowledgments 296
References 296

Appendix I G-PON PMD Characteristics **301**
Appendix II EPON MPCPDU Formats **311**

Index **315**

List of Acronyms

AAL	ATM Adaptation Layer
ADC	analog to digital converter
ADSL	asymmetric DSL
AGC	automatic Gain Control
AM-VSB	Amplitude Modulation Vestigial Side Band
AN	Access node
APC	angled physical contact
APC	automatic power control
APD	Avalanche Photo-Diode
APON	ATM-based passive optical network
APS	Automatic protection switching
ASE	amplified spontaneous emission
ASIC	Application Specific IC
ASK	amplitude shift keying
ATC	automatic threshold control
ATM	Asynchronous Transfer Mode
AWG	Arrayed Waveguide Grating
AWGR	arrayed-waveguide grating router
BER	bit error ratio
BER	Bit Error Rate
BiDi	bidirectional
BLS	broadband light source
BLSR	Bi-directional line-switched ring
BM	burst-mode
BOH	Burst Overhead
BOSA	bidirectional optical subassembly
BPF	band-pass filter
BPON	broadband passive optical network
BW	Bandwidth
CAPEX	Capital Expenditure
CATV	cable television
CATV	Community Antenna Television
CDMA	Code Division Multiple Access
CDR	clock and data recovery

CID	channel identification
CJPAT	continuous jitter test pattern
CMOS	complementary metal—oxide—semiconductor
CMTS	Cable Modem Termination System
CNR	Carrier to Noise Ratio
CO	central office
CRC	cyclic redundancy check
CRPAT	continuous random test pattern
CSMA/CD	Carrier Sense Multiple Access with Collision Detection
CW	continuous wave
CWDM	coarse wavelength-division-multiplexing
DA	Destination Address
DBA	Dynamic Bandwidth Allocation
DBR	distributed Bragg reflector
DBRu	Dynamic Bandwidth Report upstream
DFB	distributed feedback
DFB	distribute-feedback laser diode
DGD	differential group delay
DMT	Discrete Multi-Tone
DMUX	Demultiplexer
DOCSIS	Data Over Cable Service Interface Specification
DOP	degree of polarization
DPSK	differential phase shift keying
DRR	Deficit Round Robin Scheduling
DS	Downstream
DSL	digital subscriber line
DSLAM	DSL Access Multiplexer
DVB-ASI	Digital Video Broadcast—Asynchronous Serial Interface
DWDM	dense wavelength division multiplexed
EDFA	erbium doped fiber amplifier
EFM	Ethernet for the First Mile
EMEA	Europe, Middle East, and Africa
EOL	end-of-life
EPON	Ethernet PON
ER	extinction ratio
ETRI	Electronics and Telecommunications Research Institute
FA	field assembly
FBG	fiber Bragg grating
FCS	Frame Check Sequence
FDM	Frequency Division Multiplexing
FEC	forward error correction

FET	field effect transistor
FITL	Fiber in the Loop
FP	Fabry-Perot
F-P	Fabry-Perot laser
FP-LD	Fabry-Perot laser diode
FSAN	Full Service Access Network
FSK	frequency shift keying
FSR	free spectral range
FTTB	Fiber-to-the-building
FTTC	Fiber-to-the-Curb
FTTH	fiber to the home
FTTP	Fiber to the Premise
FTTx	fiber-to-the-x
FWM	four-wave-mixing
GBIC	gigabit interface converter
GCC	gain control circuit
GEM	G-PON encapsulation method
GEM	Gigabit Encapsulation Mode
GE-PON	gigabit Ethernet-PON
GFP	Generic Framing Procedure
GI-POF	graded index polymer optical fiber
GMII	Gigabit Media Independent Interface
G-PON	Gigabit-capable passive optical networks
GTC	GPON Transmission Convergence
HDSL	High data rate DSL
HDTV	high definition television
HEC	Header Error Control
HFC	hybrid fiber coaxial
IC	integrated circuit
IEEE	Institute of Electronic and Electrical Engineering
IETF	Internet Engineering Task Force
IFG	inter frame gap
IP	Internet Protocol
IPACT	Interleaved Polling with Adaptive Cycle Time
IPTV	internet protocol TV
ISDN	Integrated Services Digital Network
ITU	International Telecommunication Union
ITU-T	International Telecommunication Union Telecommunication Standardization Sector
L2	Layer-2
L3	Layer-3
LAN	local-area network

LD	laser diode
LDD	laser diode driver
LDS	laser driver stage
LED	Light Emitting Diode
LFD	live fiber detector
LLID	Logical Link ID
MAC	media access control
MEF	Metro-Ethernet Forum
MEMS	micro-electro-mechanical system
MFD	mode field diameter
MIB	Management Information Base
MII	Media Independent Interface
MLM	multi-longitudinal mode laser
MMF	Multi-mode Fiber
MP2MP	Multi-point to Multi-point
MPCP	Multi-point Control Protocol
MPCPDU	Multipoint Control Protocol Date Unit
MPEG	Motion Picture Expert Group
MPMC	Multi-point Media Access Control
MSA	multi-source agreement
MSB	Most Significant Bit
MSDSL	Multi-rate Symmetric DSL
MSO	Multiple Service Operator
MTW	Maximum Transmission Window
MUX	Multiplexer
NA	numerical aperture
NA	Not Applicable
NEXT	Near End Crosstalk
NMS	Network management system
NRZ	non-return to zero
nSR	non-Status Reporting
NT	network termination
NTSC	North American National Television System Committee
OADM	optical add/drop multiplexer
OAM	Operation, Administration, and Management
OCDM	optical code division multiple access
OCM	optical channel monitor
ODN	optical distribution network
OE	Optical to Electrical
OIL	optical injection locking
OLT	optical line termination
OLT	optical line terminal

OLTS	optical loss test set
OMA	optical modulation amplitude
OMCC	ONT Management and Control Channel
OMCI	ONT Management and Control Interface
ONT	optical network termination
ONU	optical network unit
OPM	optical performance monitoring
OPS	optical pulse suppressor
OPS	Operation System
ORL	optical return loss
OSA	optical spectrum analyzer
OSA	optical subassembly
OSI	Open System Architecture
OSNR	optical signal-to-noise ratio
OSW	Optical switch
OTDR	optical time-domain reflectometer/reflectometry
OWDR	optical wavelength domain reflectometry
P2MP	point-to-multipoint
P2P	point-to-point
P2P	Peer-to-Peer
P2PE	Point-to-Point Emulation
PAL	Phase Alternate Line
PCB	printed circuit board
PCBd	Physical Control Block downstream
PCS	Physical Coding Sub-layer
PD	photodiode
PDU	Protocol Data Unit
PECL	positive-emitter-coupled logic
PHY	Physical Layer
PIN	positive-insulator-negative type of detector
PLC	planar lightwave circuit
PLC-ECL	Planner Lightwave Circuit External Cavity Tunable Laser
PLM	power leveling mechanism
PLOAM	Physical Layer OAM
PLOu	Physical Layer Overhead upstream
PLP	physical layer preample
PLSu	Power Leveling Sequence upstream
PMD	polarization mode dispersion
PMD	physical medium-dependent sublayer
PMD	Physical Medium Dependent
PON	passive optical network
POTS	Plain Old Telephone Service

PRBS	pseudo-random bit sequence
PSP	principal states of polarization
PS-PON	Power Splitting PON
PTI	Payload Type Indicator
QAM	Quadrature Amplitude Modulation
QoS	quality of service
QPSK	Quadrature Phase Shift Keying
RADSL	Rate Adaptive DSL
RBOC	Regional Bell Operating Company
RF	radio frequency
RFP	Request for Proposal
RIN	relative intensity noise
RMS	Root Mean Square
RN	remote node
ROSA	receive optical subassembly
RSOA	reflective semiconductor optical amplifier
RT	Remote Terminal
RTD	Round Trip Delay
RTT	Round Trip Time
Rx	receiver
RZ	return to zero
SA	Source Address
SCB	Single Copy Broadcast
SDH	Synchronous Digital Hierarchy
SDSL	Symmetric DSL
SerDes	serializer/deserializer
SFF	small-form-factor
SFP+	small-form-factor pluggable plus
SHR	Self-healing ring
SLA	Service Level Agreement
SLD	Start LLID delimiter
SLED	superluminescent light emitting diode
SMF	single mode fiber
SMT	surface mount technology
SNR	signal-to-noise ratio
SOA	Semiconductor Optical Amplifier
SONET	Synchronous Optical Network
SPM	self-phase modulation
SR	Status Reporting
SRR	upstream signal to Rayleigh backscattering ratio
STB	Set-Top Box
STP	Spanning Tree Protocol

T-CONT	Transmission Container
TCP	Transmission Control Protocol
TDM	time-division multiplexing
TDMA	time division multiple access
TDM-PON	Time Division Multiplexing PON
TDP	transmit and dispersion penalty
TFF	thin-film filter
TIA	trans-impedance amplifier
TO-CAN	co-axial Transistor Outline "can"—type of packaging
TOSA	transmit optical subassembly
Tx	transmitter
UNI	User Network Interface
UPSR	Uni-directional path-switched ring
US	Upstream
VC	Virtual Circuit
VCI	Virtual Circuit Identifier
VCSEL	vertical cavity surface-emitting laser
VDSL	Very high data rate DSL
VFL	visual fault locator
VOD	Video On Demand
VoIP	Voice over IP
VP	Virtual Path
VPI	Virtual Path Identifier
VSB	vestigial-sideband
WCT	wavelength coded tag
WDM	wavelength division multiplexing
WDMA	Wavelength Division Multiple Access
WDM-PON	wavelength division multiplexed passive optical network
WE-PON	WDM-Ethernet PON
WFQ	Weighted Fair Queuing
WGR	Waveguide Grating Router
XPM	cross-phase modulation

List of Figures

Figure 1.1	Development trend of Ethernet technologies.	9
Figure 2.1	Generic structure of a modern telecommunication network.	20
Figure 2.2	FTTx alternatives (from [1]).	21
Figure 2.3	Architecture of (a) TDM-PON and (b) WDM-PON.	22
Figure 2.4	Single-mode fiber (SMF) vs multimode fiber (MMF).	23
Figure 2.5	Dispersion coefficient as a function of wavelengths for various types of optical fibers.	24
Figure 2.6	Loss in an optical fiber at different wavelengths.	25
Figure 2.7	(a) One-fiber single-wavelength bidirectional transmission. (b) Near-end cross talk (NEXT).	27
Figure 2.8	Wavelength division duplex uses 1.3/1.5-μm coarse WDM coupler (diplexer) to separate upstream and downstream signals.	28
Figure 2.9	Splitting strategies in a TDM-PON: (a) one-stage splitting, (b) multistage splitting, and (c) optical bus.	29
Figure 2.10	Standard commercial TMD-PON architecture.	31
Figure 2.11	Generic structure of a standard TDM-PON OLT. This diagram represents a chassis with multiple OLT cards, which are interconnected through a back plane switch. Each OLT card with its own MAC and PMD layer serves a separate PON.	32
Figure 2.12	Generic structure of a standard TDM-PON ONU.	33
Figure 2.13	Overlaying analog broadcast TV services on a TDM-PON using the 1.55-μm wavelength.	36
Figure 2.14	Downstream and upstream APON frame formats at 155.52-Mbps speed. For 622.08-Mbps and 1244.16-Mbps speed, the numbers of time slots are simply multiplied by 4 and 8 to the numbers shown in the above diagrams.	39
Figure 2.15	ATM switching examples: (a) VP and VC switching and (b) VC switching.	41
Figure 2.16	Wavelength allocation plan in ITU-T G.983.3.	43
Figure 2.17	A T-CONT represent a logical link between the OLT and an ONU.	45

Figure 2.18 ATM-based T-CONT vs GEM-based T-CONT. 46
Figure 2.19 Protocol stack of the G-PON transmission convergence (GTC) layer (from [24]). 46
Figure 2.20 GTC downstream signal consists of 125-μs frames with a PCBd header and a payload section. 48
Figure 2.21 GTC downstream frame and media access control concept (from [24]). 48
Figure 2.22 GTC downstream frame formats. 49
Figure 2.23 GTC upstream framing: each ONT starts the upstream transmission with PLOu. An ONT assigned with two Alloc-ID in two consecutive upstream allocations only needs to transmit PLOu once. 50
Figure 2.24 GTC upstream frame format. 51
Figure 2.25 GEM frames in upstream payload (from [24]). 52
Figure 2.26 GEM encapsulation formats. 52
Figure 2.27 Mapping and fragmentation of user data frames into GEM payload (from [24]). 53
Figure 2.28 Multiplexing of urgent data using GEM fragmentation process (from [24]). 54
Figure 2.29 Point-to-point (P2P) Ethernet and point-to-multipoint (P2MP) EPON layering architecture. 55
Figure 2.30 Standard Ethernet frame format. 60
Figure 2.31 EPON ranging process. 61
Figure 2.32 EPON Gate operation. 62
Figure 2.33 EPON Report operation. 63
Figure 2.34 Generic format of MPCPDU. 63
Figure 2.35 Autodiscovery process (from [9]). 64
Figure 2.36 Although all the ONU traffic arrives at the same physical port at the OLT, because of the directional power-splitting coupler used at the remote node, ONUs cannot see each other's traffic without the forwarding aid of OLT. 66
Figure 2.37 Point-to-point emulation in EPON. 66
Figure 2.38 Modified preamble with LLID for point-to-point emulation in EPON. 67
Figure 2.39 EPON point-to-point emulation operation. 68
Figure 2.40 Point-to-point and single-copy broadcast (SCB) MACs in an EPON model. 68
Figure 2.41 Conventional WDM coupler vs arrayed waveguide grating. 73
Figure 2.42 Emulation of broadcast services on a WDM-PON with a broadband source. 74

Figure 2.43	A modified AWG for 2-PONs-in-1.	75
Figure 2.44	By using CWDM devices to combine and separate optical signals in multiple FSRs of an AWG device, a highly flexible WDM-on-WDM system can be achieved.	76
Figure 2.45	Split-net test-bed demonstrating the WDM-on-WDM concept (from [50]).	77
Figure 2.46	Demonstration of a hybrid WDM/TDM-PON with wavelength-selection-free transmitters: (a) downstream link and (b) upstream link (from [52]).	78
Figure 2.47	WE-PON optical layer block diagram. (Courtesy of ETRI and Korea Telecom.)	79
Figure 2.48	RSOA module used in ETRI WE-PON prototype. (Courtesy of ETRI and Korea Telecom.)	80
Figure 2.49	Feed-forward current injection to RSOA in WE-PON. (Courtesy of ETRI and Korea Telecom.)	81
Figure 2.50	PLC-ECL used in WE-PON prototype. (Courtesy of ETRI and Korea Telecom.)	82
Figure 2.51	A band sorter interleaves the optical spectrum into bands (top) which can be further separated with diplexers (bottom) into bands for upstream and downstream PON connections.	83
Figure 2.52	Overlaying APON and EPON on the same fiber plant using "band sorters."	83
Figure 3.1	A PLC splitter is manufactured using two fiber arrays and one PLC chip all aligned within one package. (From Ref [1])	91
Figure 3.2	Measured insertion loss of a 1×32 PLC splitter as a function of wavelength, showing uniform insertion loss ($\leq$1.25-dB variation) and low excess insertion less ($\leq$1.25 dB) above the theoretical value. (From Ref [3])	92
Figure 3.3	Schematic illustration of arrayed waveguide grating comprising two free-propagating regions and a waveguide array. (From Ref [7])	94
Figure 3.4	Schematic layout of $N\times N$ arrayed waveguide grating router. (From Ref [12])	95
Figure 3.5	Schematic illustration of the cyclic property of an arrayed waveguide grating router. (From Ref [12])	95
Figure 3.6	Schematic illustration of dual-function reflective arrayed waveguide grating. (From Ref [13])	96

Figure 3.7 (a) Fiber-to-fiber insertion loss of 1×14 power splitter function; (b) transmission spectra of 14-channel wavelength-router function. (From Ref [13]) 97

Figure 3.8 Schematic illustration of 1×8 two-PONs-in-one (2P1) device. For clarity, the vertical scale has been expanded by four times. (From Ref [14]) 98

Figure 3.9 Spectra from the output ports of a 1 × 8 2P1 device measured around the (a) 1.3-μm waveband and (b) 1.55-μm waveband. (From Ref [14]) 98

Figure 3.10 Schematic diagram of athermal AWG with silicone resin-filled triangular groove. (Adapted from Ref [15]) 100

Figure 3.11 Temperature-dependent transmission spectrum of Channel 4 over the temperature range of 0–85 °C for (a) conventional silica-based AWG and (b) athermal AWG with silicon resin-filled triangular groove. (From Ref [16]) 100

Figure 3.12 Temperature-dependent wavelength shift of Channel 4 in the temperature range of 0–85 °C. (From Ref [16]) 101

Figure 3.13 Reduction of excess loss by replacing triangular groove of wider width with several grooves of smaller widths. (From Ref [17]) 101

Figure 3.14 Schematic diagram of modified low-loss groove and array structure for a 1.5%-Δ athermal AWG. (From Ref [21]) 102

Figure 3.15 Schematic diagram of silicon resin-filled trenches in the slab region of 1.5%-Δ athermal AWG. (From Ref [20]) 102

Figure 3.16 Schematic diagram of silicon resin-filled trenches with spot size converters based on segmented core in the slab region of 2.5%-Δ athermal AWG. (From Ref [19]) 103

Figure 3.17 (a) Cross-sectional view of silica/polymer hybrid waveguide; (b) comparison of measured TE/TM polarization shift of the central wavelength, Channel 9, with polymer and silica overcladding; (c) measured temperature-dependent wavelength shift of Channel 9 with polymer and silica cladding. (Adapted from Ref [22]) 104

Figure 3.18 All-polymer 8×8 AWG with 200 GHz channel spacing: (a) measured spectra; (b) measured TE/TM polarization shift of Channel 4; (c) measured temperature-dependent transmission spectrum of Channel 4 for the temperature range of 25–65 °C. (Adapted from Ref [23]) 105

Figure 3.19 Schematic diagram of an athermal AWG with temperature compensating plate. (From Ref [24]) 105
Figure 3.20 Optical spectra of the athermal AWG with temperature compensating plate shown in Fig. 3.19. (From Ref [24]) 106
Figure 3.21 Schematic diagram of an athermal AWG chip with temperature-compensating rod. (From Ref [25]) 106
Figure 3.22 (a) Optical spectra in C- and L-bands of athermal AWG using temperature-compensating rod; (b) temperature-dependent wavelength shift of four channels in the C-band from −30 °C to +70 °C. (Adapted from Ref [25]) 107
Figure 3.23 Schematic diagram of AWG with bimetal plate. (From Ref [26]) 108
Figure 3.24 Center wavelength of AWG as a function of temperature. (From Ref [26]) 109
Figure 3.25 Field assembly (FA) connector: (a) structure of plug and socket; (b) in optical cabinet at customer premises. (From Refs [32] and [33]) 110
Figure 3.26 (a) Connection loss trials and (b) return loss trials. (From Ref [32]) 111
Figure 3.27 Bendable SC connector. (From Ref [33]) 111
Figure 3.28 Cross-sectional view of angled physical contact (APC) connector. The 8-degree fiber end face directs reflected light into the cladding. (From Ref [34]) 112
Figure 3.29 (a) Bend loss characteristics of bend resistant fiber; (b) comparison of conventional optical fiber with minimum bending radius of 30 mm and bend-resistant with minimum bending radius of 15 mm. (From Ref [32]) 113
Figure 3.30 (a) Typical spectral attenuation of PureAccess-Ultra; (b) bending loss properties as a function of wavelength. (From Ref [37]) 114
Figure 3.31 Schematic diagram of optical curl cord. (From Ref [39]) 115
Figure 3.32 Connection of laptop terminal to optical outlet in the wall using optical curl cord. (From Ref [39]) 115
Figure 3.33 Cross section and refractive index profile of hole-assisted fiber. (From Ref [38]) 116
Figure 3.34 Bending loss characteristics of hole-assisted fibers: (a) calculated bending loss performance as a function of normalized hole distance; (b) wavelength dependence of measured bending loss. (From Ref [38]) 117

Figure 3.35 (a) Calculated and experimental mode field diameter mismatch as a function of hole arrangement; (b) splice loss induced by MFD mismatch as a function of refractive index difference between N_{hole} and N_{clad}. (From Ref [38]) 118
Figure 3.36 Comparison of three GI-POFs with differing core diameter and numerical aperture (NA): (a) refractive index profile; and (b) bending loss as a function of radius. (From Ref [47]) 120
Figure 3.37 (a) Core diameter and (b) NA dependence loss at bending radius of 5 mm. (From Ref [47]) 121
Figure 3.38 (a) Bending loss of GI POF with a core diameter = 200 μm and NA = 0.24; (b) optimum waveguide parameters of GI POF to suppress the bending loss. (From Ref [47]) 122
Figure 3.39 Schematic diagram of an index-guided Fabry–Perot laser diode. Optical spectra showing multiple longitudinal lasing modes. 123
Figure 3.40 Schematic diagram of a distributed feedback laser. Optical spectrum showing excellent single-mode behavior. 124
Figure 3.41 Schematic diagram of a vertical-cavity surface-emitting laser (VCSEL) with circular output beam. 125
Figure 3.42 WDM-PON using directly modulated OIL-VCSELs as ONU transmitters. Each slave VCSEL, located at the ONU is injection-locked by modulated downstream signal transmitted from a master DFB laser located at the CO. (From Ref [56]) 126
Figure 3.43 Upstream BER curves for different values of (a) upstream signals to Rayleigh backscattering ratio (SRR) at a constant downstream extinction ratio (ER) = 4.5 dB; (b) downstream ER at a constant SRR = 13.4 dB. 127
Figure 3.44 Received optical power at BER = 10^{-9} as a function of downstream extinction ratio. Solid line: bidirectional 25-km transmission with Rayleigh backscattering effects. Dash line: unidirectional back-to-back transmission. 128
Figure 3.45 Schematic diagram of WDM-PON with broadband optical source (LED) in ONU. (From Ref [62]) 130

Figure 3.46 Experimental setup of WDM-PON with centralized supercontinuum broadband light source. (From Ref [67]) 131
Figure 3.47 Schematic diagram of colorless bidirectional (BiDi) transceivers based on FP-LDs injection-locked by the spectrally sliced light from a broadband super-luminesent LED (SLED). (From Ref [72]) 132
Figure 3.48 Optical spectra of total injection light and spectrum-sliced injection lights measured at the input and output ports of the (a) remote node AWG and (b) central office AWG. (From Ref [72]) 133
Figure 3.49 Schematic diagram of colorless reflective semiconductor optical amplifier (RSOA) wavelength seeded by spectrally sliced light from a broadband (SLED). (From Ref [81]) 134
Figure 3.50 Monolithic indium phosphide-based RSOA features a curved waveguide architecture. (From Ref [82]) 135
Figure 3.51 WDM-PON with self-injection-locked Fabry–Perot laser diodes (FP-LDs). (From Ref [81]) 136
Figure 3.52 Optical spectra of self-injection-locked FP-LD. (From Ref [81]) 136
Figure 3.53 Schematic diagram of WDM-PON with self-seeding reflective SOAs. (From Ref [85]) 137
Figure 3.54 Overlapped optical spectra of 1.25 Gbps self-seeded upstream channels. (From Ref [85]) 138
Figure 3.55 RITE-net architecture. The CW portion of each downstream packet is remodulated with upstream data at the optical network unit. (From Ref [86]) 138
Figure 3.56 WDM-PON using reflective semiconductor optical amplifiers to amplify and remodulate downstream wavelengths carrying downstream signals (ASK format) for upstream transmission. (From Ref [90]) 139
Figure 3.57 Optical spectrum of adiabatic-chirped DFB laser source directly modulated with 1.25 Gbps NRZ signal. Superimposed on the spectrum is the passband characteristic of negatively detuned (–18.7 GHz) demultiplexer. (From Ref [92]) 140
Figure 3.58 Wavelength reuse scheme for WDM-PON using multichannel fiber Bragg grating to filter out downstream carriers for remodulation with upstream data. (From Ref [93]) 141

Figure 4.1 The general PON network architecture for the FTTx scenario. 152
Figure 4.2 Physical layer burst-mode timing definition. T_{ON} – Tx turn-on time; T_{OFF} – Tx turn-off time; T_{DSR} – Rx dynamic sensitivity recovery time; T_{LR} – Rx level recovery time; T_{CR} – Rx clock recovery time; T_{DL} – Rx delimiter time. 157
Figure 4.3 Various PON diplexer and triplexer transceivers for optical access systems. 161
Figure 4.4 Feature function blocks of optical transceivers showing both downstream and upstream directions. In the upstream direction, an ONU transmitter consists of a burst-mode laser driver (BM-LD), and an F-P laser in the form of transmit optical subassembly (TOSA); the OLT receiver consists of a PIN or avalanche photodiode (PIN or APD), BM-trans-impedance (BM-TIA) ROSA (receive optical subassembly), a limiting post-amplifier (post-amp) and a burst-mode clock/data recovery (BM-CDR) unit. 162
Figure 4.5 Typical L-I curves of (a) DFB-LD and (b) F-P LD. 164
Figure 4.6 APD V-I curve and multiplication factor. 165
Figure 4.7 Schematics structures of TO-CAN packaging for DFB (a) and APD (b). Their photographs are shown on the right. 166
Figure 4.8 Transmission spectrum characteristic (left) and photograph (right) of the WDM filter formed on a fiber facet [29]. The two curves represent two different light polarizations. 166
Figure 4.9 BOSA schematics for three wavelengths based on thin-film technology. 168
Figure 4.10 The BOSA in TO-CAN package with driving circuit [32]. 169
Figure 4.11 Optical circuit configurations with external filters. 170
Figure 4.12 The feasibility of grating assisted WDM filter on a PLC platform (top) and conceptual schematic (bottom) of a triplexer using this structure [43]. 171
Figure 4.13 Bidirectional ONU diplexer transceiver module. 173
Figure 4.14 Bidirectional ONU triplexer transceiver module. 173
Figure 4.15 Data formats in digital communication: (a) continuous-mode data, (b) burst-mode data, (c) burst packet data [47]. 175

Figure 4.16 Temperature characteristics of a typical F-P laser as L-I curves. 176
Figure 4.17 Block diagram of a typical burst-mode laser driver IC [49]. 177
Figure 4.18 Comparison of (a) continuous and (b) burst-mode laser driver stage. 178
Figure 4.19 Block diagrams of two typical APC circuits. 179
Figure 4.20 Optical and timing performance of an EPON BM-Tx: (top left) Optical eye diagram using 4th-order Thompson filter at 1.25 Gbps; numbers inside box are GbE eye-mask margin. (Bottom left) Optical bursty signal with Laser bias on/off. The measured laser burst-on and burst-off times are shown on the right side [22]. 180
Figure 4.21 Optical eye diagrams at −40 °C and 80 °C case temperature with optical power above 0 dBm and ER above 10 dB. The numbers inside box are GbE eye-mask margin. 181
Figure 4.22 Feed-forward and feedback implementations of optical burst-mode receivers. 182
Figure 4.23 Block diagram of a burst-mode preamplifier IC employing ATC (left) and the response of ATC circuit (right) [56]. 183
Figure 4.24 The comparison of (a) a conventional AGC and (b) a cell-AGC for burst-mode inputs with low extinction ratio [60]. 184
Figure 4.25 Configuration and operation principles of cell-AGC based preamplifier IC. 185
Figure 4.26 Comparison of AC-coupled (left) versus DC-coupled (right) burst-mode receivers. 186
Figure 4.27 An example of 1.25 Gbps BM APD/TIA output eye diagram (left) and settling time measurements (right) [22]. 187
Figure 4.28 Examining temperature error at different ADC resolutions. The *y*-axis shows the absolute value of the maximum error. The overall error for the temperature monitoring is $\pm y$ °C. The blue, pink, and green lines represent the 8-, 10-, and 12-bit ADCs, respectively. The solid lines represent the cases in which the accuracy level of both the temperature sensor and the voltage reference is within $\pm 1\%$, and the dashed lines represent a $\pm 0.5\%$ accuracy level. The red solid line is a 10-bit ADC without oversampling [70]. 189

Figure 4.29 A Laser output power and monitoring current versus driving current. 191
Figure 4.30 A tunable OTDR example for in-service monitoring of the fiber fault in WDM-PON [74]. 192
Figure 4.31 Fiber fault detection results for fault locations at (a) 3 km, (b) 3.4 km, (c) 4 km, and (d) 5.2 km from the remote node using the setup from Fig. 4.30. 194
Figure 4.32 Embedded OTDR into ONU burst-mode transmitter. OTDR analog front-end bandwidth is limited to 5 MHz [76]. 194
Figure 4.33 ONU block diagram with OTDR functionality integrated [77]. 195
Figure 4.34 Typical burst-mode test setup configuration consisting of two ONUs. A weak packet from BM-Tx as ONU#1 followed by a strong packet from commercial SFF module Tx as ONU#2 emulates the worst-case condition [22]. 198
Figure 4.35 A typical G-PON upstream burst-mode test pattern when strong packets from BM-Tx as ONU#1 are followed by the weak packets from ONU#2. The overhead guard time consists of the mandatory 32 bits, preamble 44 bits, and delimiter time 16–20 bits. 199
Figure 4.36 Typical measurements of burst-mode BER (top) and sensitivity penalty (bottom) performed for a DC-coupled G-PON OLT Rx at 1.244 Gbps. 200
Figure 4.37 1.25-Gbps G-PON uplink performance with PLM. 201
Figure 4.38 (a) Measured continuous-mode (P2P) versus burst-mode BER. (b) Input and output for burst-mode receiver. 203
Figure 4.39 Burst-mode BER curves for the weak packet as a function of the power levels of the strong packet at 1.25 Gbps, indicating the degradation of OLT Rx sensitivity due to the influence of the strong packet. 204
Figure 4.40 (a) Impact of CATV signal on BER of the digital receiver. (b) Impact of digital signal cross talk on the CNR of CATV receiver. 206
Figure 5.1 Time division multiple access. 216
Figure 5.2 Ranging window. 217
Figure 5.3 G-PON ranging phase 1: serial number process. 219
Figure 5.4 GPON ranging phase 2: delay measurements. 220
Figure 5.5 Dynamic bandwidth allocation. 222
Figure 5.6 An example network configuration which offers Layer-2 link services to ISPs. 224

Figure 5.7 The principle of Fair Queuing. 226
Figure 5.8 Operation of limited service IPACT: (a) when only ONU#3 has upstream traffic; and (b) when all ONUs have upstream traffic. Here, R1, R2, and R3 represent request messages from each ONU respectively; data#3 represents user data transmitted from ONU#3, G3 represents grant for ONU#3. In (a), the values of R1 and R2 are equal to 0 and the value of R3 indicates amount of upstream user data stored in ONU#3. 230
Figure 5.9 Deficit round-robin (DRR) scheduling. 234
Figure 5.10 Maintaining short round-robin cycle using multiple queue request. 235
Figure 5.11 An example with two types of requests. 235
Figure 5.12 An IEEE802.3ah REPORT frame with multiple requests. 236
Figure 5.13 Burst allocation example of DRR using multiple queue report set. 237
Figure 5.14 Comparison of bandwidth efficiency (Emax) among the three DBAs covered in this chapter. Here we have chosen the number of ONU $N = 32$, Burst overhead $= 1.4\,\mu$sec, Max RTD $= 100\,\mu$sec, $w_i = w$, $\text{Th}_i = w/N$,OLT processing time $= 16\,\mu$sec, and ONU processing time $= 16\,\mu$sec. 241
Figure 6.1 Protection switching architectures suggested by ITU-T G.983.1. 249
Figure 6.2 A 1:N protection scheme at OLT [14]. LT: Line Terminal. 250
Figure 6.3 A PON architecture with protection of distribution fibers by interconnecting ONUs with protection switching [17]. OSW: optical switch, FBG: fiber Bragg gratings, λ_d: downstream wavelength, λ_u: upstream wavelength. 251
Figure 6.4 A PON architecture with protection of both feeder and distribution fibers by an additional loopback distribution fiber with protection switching [18]. λ_d: downstream wavelength, λ_u: upstream wavelength. 252
Figure 6.5 Protection of feeder fibers in two PONs using CWDM technology [19]. OSW: optical switch, λ_d: downstream wavelength, λ_u: upstream wavelength. 252
Figure 6.6 An example of a four-fiber shared rings in (a) normal operation; (b) span switching; and (c) ring switching [25]. 254

Figure 6.7 A modified star-ring PON with protection capability [26]. Inset shows the structure of the RN to illustrate the protection mode. OS: optical switch. 255

Figure 6.8 The structure of (a) the OLT and (b) RN of a WDM-PON with duplicated fiber feeders for protection [27]. AN: access node, CN: central node, OSC: optical supervisory channel, PD: photodiode, OSW: optical switch. 256

Figure 6.9 Self-protected network architecture for WDM-PON. LD1–4: laser diode; PD1–4: PIN photodiode; WC: WDM coupler; M1&M2: optical power monitors; {A*i*, C*i*} for $i \in \{1, \ldots, N\}$: upstream wavelengths; {B*i*, D*i*} for $i \in \{1, \ldots, N\}$: downstream wavelengths. Inset shows the spectral response of one of the output ports of the AWG. FSR: free-spectral range of the AWG [28]. 258

Figure 6.10 (a) A WDM-PON survivable architecture with eight ONUs and centralized protection switching control; (b) OLT configuration under normal operation; (c) wavelength assignment plan. B/R: blue/red filter; OC: 3-dB fiber coupler; LD: laser diode; PD: photodiode. Note: FSR1 stands for free-spectral range of the $N \times 2$ ($N = 8$) AWG at the OLT while FSR2 stands for that of both AWG1 and AWG2 at the RN. The wavelengths quoted in boxes are the working upstream wavelengths. The wavelengths in blue band are underlined but those in red band are not [31]. 259

Figure 6.11 A WDM-PON architecture with self-protection capability against both feeder fiber and distribution fiber failures [32]. WC: wavelength coupler, B/R: blue/red filter, OS: optical switch, M: power monitoring modules. 260

Figure 6.12 (a) A single-fiber CWDM optical access ring network; (b) & (c) the structure of AN2 in (b) normal state and (c) protection state when there was a fiber cut between AN2 and AN3 [45]. AN: access node, Tx: transmitter, Rx: receiver. 262

Figure 6.13 (a) A protected optical star-shaped ring network; (b) lightpath diagram, dotted lines were the designated protection paths; (c) the protection lightpath was adopted when node 1 failed [47]. FBG: fiber Bragg gratings. 263

Figure 7.1 The architecture of a passive optical-access network. 268

Figure 7.2 Points where the test required for a WDM, point-to-multipoint, and bidirectional PON. 272
Figure 7.3 Three layers of optical performance monitoring: transport monitoring, signal quality monitoring, and protocol monitoring. 279
Figure 7.4 Example on the detection and localization of fiber failures in a bidirectional WDM-PON [15]. 283
Figure 7.5 Fiber-link loss monitoring example in a bidirectional WDM-PON using ASE-injected FP-LD [17]. 284
Figure 7.6 Schematic of an OWDR-embedded FTTH network from OLT to ONT [18]. 285
Figure 7.7 (a) Undermined noise level: different noise on adjacent signal channels; (b) "Missing" noise between channels: optical filters or MUX/DEMUX may remove the out of band noise. 287
Figure 7.8 Operating principle of the polarization technique: the polarized noise (i.e. half of the total noises) can be measured by using the second linear polarizer, which is aligned to be orthogonal to the signal's polarization [25]. 287
Figure 7.9 RF pilot tone added to the channel bandwidth as the signal quality/degradation monitor. 288
Figure 7.10 (a) Typical measured data for the logarithm of the BER versus the decision threshold [37]; (b) the BER as a function of the received optical SNR [41]. 290
Figure 7.11 OCDM-based PON-monitoring system where every network leg is assigned an encoder. One tunable decoder is employed at the CO [42]. 291
Figure 7.12 System configuration of optical fiber line testing system using a 1650-nm testing window [43]. 292
Figure 7.13 Conceptual diagram for monitoring chromatic dispersion using optical vestigial-sideband (VSB) filtering: the recovered bits from either part of the spectrum arrive at slightly different times depending on the chromatic dispersion [50]. 294
Figure 7.14 The use of the degree of polarization (DOP) to monitor the effects of PMD: (a) DOP measurements as a function of instantaneous DGD. Note that the DOP is pulse-width-dependent [52]; and (b) measured DOP for a 40-Gbit/s RZ signal with concatenation of 6-ps and 4-ps DGD sections [58]. Note that higher-orders

of PMD decrease the signal's maximum DOP at the receiver to less than unity. 295
Figure A1.1 Generic physical configuration of the optical distribution network (reproduced from Figure 5/G.983.1). 301
Figure A2.1 Gate frame format. 311
Figure A2.2 Report frame format. 312
Figure A2.3 Register request frame format. 312
Figure A2.4 Register frame format. 313
Figure A2.5 Register acknowledgment frame format. 313

List of Tables

Table 1.1	Summary of different DSL technology performances	4
Table 1.2	Summary of DOCIS data rates	6
Table 1.3	Bandwidth requirements for different IP services	13
Table 2.1	APON downstream/upstream bit-rate combinations	38
Table 2.2	Selected properties of IEEE802.3ah EPON transmitters	57
Table 2.3	Selected properties of IEEE 802.3ah receiver characteristics	58
Table 2.4	G-PON and EPON comparison	69
Table 4.1	PON power budgets	154
Table 4.2	ITU-T G.983.3 BPON standards	156
Table 4.3	Physical-layer requirements of ITU-T G-PON and IEEE EPON standards	157
Table 4.4	G-PON and EPON burst-mode timing comparison	158
Table 4.5	Key PMD parameters of G-PON class B 1.244 Gbps upstream [10, 20]	159
Table 4.6	Key PMD parameters of IEEE 802.3ah EPON 20-km 1.25 Gbps upstream [9, 22]	160
Table 4.7	State-of-the-art G-PON transceiver performance parameters for the upstream link at 1.244 Gbps, in comparison with the ITU-T G.984.2 specifications [10]	201
Table 4.8	A state-of-the-art EPON transceiver uplink performance summary in comparison with the IEEE 802.3ah standard	205
Table A1.1	G.984.2—Physical medium dependant layer parameters of ODN (reproduced from Table 2a/G.984.2)	302
Table A1.2	G.984.2—Optical interface parameters of 1244 Mbit/s downstream direction (reproduced from Table 2b/G.984.2)	303
Table A1.3	G.984.2—Optical interface parameters of 2488 Mbit/s downstream direction (reproduced from Table 2c/G.984.2)	305
Table A1.4	G.984.2—Optical interface parameters of 1244 Mbit/s upstream direction (reproduced from Table 2f-1/G.984.2)	307
Table A1.5	G.984.2—Optical interface parameters of 1244 Mbit/s upstream direction using power-levelling mechanism at ONU Transmitter (reproduced from Table 2f-2/G.984.2)	309

Preface

Passive optical network (PON) technologies have been researched for over 20 years. It is not until recently that PON has seriously appeared on service providers' radar screens as an important broadband access infrastructure option, as a result of the fast-growing and bandwidth-hungry video-on-demand (VOD) applications, and continued deregulation and fierce competition between traditional telecommunication companies and cable TV service providers.

Yet, there is only a limited number of books in this area that give a coherent and comprehensive review of PON technologies. Most of the PON-related research materials are scattered around in journals, periodicals, conference proceedings, and a number of technical standards. Therefore, we feel that it is timely to publish a book that covers the different aspects of the PON technology.

There are many different flavors and nomenclatures of PON technologies (EPON, BPON, G-PON, APON, WDM-PON, TDM-PON, etc.) Although some of the standards are gaining more popularity over the others, none of them has become a clear winner yet. As demand picks up, the research and development in PON technologies also accelerates due to the heavy involvement of startup and incumbent vendors, system manufacturers, as well as component and chip companies.

Unlike other access technologies, which use the conventional copper media for communications, PON makes use of optical fiber as the transmission medium, lasers and photodiodes as transmitters and detectors. Fiber optics are traditionally used in long-haul transmission systems. In the mean time, to understand PON as an access technology requires a good background of networking knowledge and media access control (MAC) protocols. Most people working on fiber communications are more focused on transmission properties of fiber, laser physics, etc. It is one of the purposes of this book to provide a balanced coverage of networking technologies, fiber-optic transmission technologies, and the electronics involved in PON system development.

Very often, researchers in new technologies such as PON will be driven by the desire to demonstrate technical smartness and overlook practicality. This tends to generate a lot of literature which eventually becomes irrelevant due to economic reasons. As an access technology, PON equipment will be deployed at the end-user premises. Unlike backbone fiber systems, where cost is shared by all the users on the same network, end users will eventually bear the cost of PON equipment directly. So PON system design needs to be very cost-conscious.

Therefore, when selecting the materials for this book, we tried to balance research interests with practical economic and engineering considerations.

This book is intended as a general reference for researchers, senior and graduate-level college students working in the field of broadband optical access networks. It can also be used by engineers and managers to obtain a working knowledge of passive optical networks. The book provides the breadth, for people who need to have general understanding of the field, and the depth, for those who would like to dig deeper into PON technologies and relevant areas.

Cedric F. Lam
Silicon Valley, 2007

Acknowledgments

First, I would really like to thank all the authors who have made tremendous efforts to contribute to this book out of their busy schedule. I am indebted to Professor Nick Frigo of the US Naval Academy, who brought me into this field when I worked with him at AT&T Labs—Research several years ago. My sincere gratitude is extended to Dr. Tingye Li for his unfailing friendship, love, encouragement, and fatherly mentorship, both in my life and in my career.

I would like to thank Academic Press for giving me the opportunity to start this project and the valuable support from its staff, especially Tim Pitts, who has been extremely patient with my constant slipping of the schedule and Ganesan Murugesan for his excellent project management. I also want to thank Dr. Katsunari Okamoto for his generosity in letting us reproduce many figures from his past presentations.

Cedric F. Lam

List of Contributors

Cedric F. Lam obtained his B.Eng. with First Class Honors from the University of Hong Kong, Hong Kong and Ph.D. degree from University of California, Los Angeles (UCLA), both in Electrical Engineering. He then joined AT&T Labs—Research as senior technical staff member of the Broadband Access Research Department. His researches covered fiber to the home (FTTH), hybrid fiber coax (HFC) systems, optical metropolitan/regional area networks, optical-signal modulation techniques, etc. He received the AT&T Research Excellence Award for his contribution to the Metro-DWDM project in 2000. In 2002, Dr. Lam joined OpVista Inc., where he is now Chief System Architect, responsible for the development of metro WDM and Ethernet systems for video and high speed Internet transport. He has served the technical program committee of many technical conferences including OFC and Asian Pacific Optical Communication Conference (APOC). Dr. Lam is a senior member of IEEE, a member of OSA, and a member of SCTE. He has 14 patents in the field of optical communications and has published over 60 technical papers in international journals and conferences.

Dr. Elaine Wong received her Ph.D. in December 2002 from the University of Melbourne. Shortly, she joined the Photonics Research Laboratory at The University of Melbourne to carry out research in the areas of optical local area networks, optical access networks, and optical performance monitoring. From 2003 to 2005, she was managing the signal impairments and mitigation challenge project for the Australian Photonics Cooperative Research Centre. In 2006, Elaine spent the year at the University of California, Berkeley, working on the applications of optical injection-locked VCSELs in broadband access networks with the CCH Optoelectronics Group. Currently, Elaine is a senior lecturer at the Department of Electrical and Electronic Engineering, The University of Melbourne where she is also affiliated with the ARC Special Research Centre for Ultra-Broadband Information Networks, and the Victoria Research Laboratory of the National ICT Australia. Elaine is the Associate Editor for the *OSA Journal of Optical Networking*, and reviewer for numerous IEEE and OSA journals. She is also currently serving as a member in the Optical Networks and Systems Committee of the IEEE Lasers and Electro-Optics Society (LEOS) and the Technical Program Committee of the SPIE Asia Pacific Optical Communications Conference. She has co-authored more than 80 refereed international journal and conference publications, and holds numerous patents on access network technologies. She is a member of IEEE LEOS and OSA.

Yongmao Frank Chang obtained his Ph.D. in Optoelectronics from the University of Montreal, Canada for his research thesis on ultrashort optical pulse generation of 1550m tunable solid-state lasers. He has been principally specializing in optical system engineering and IC product specifications issues for telecom, datacom, and PON markets at Vitesse since 2002. Prior to Vitesse, Dr. Chang worked as individual contributor and project manager roles at JDS Uniphase, Cisco/Pirelli, and Mahi Networks for the development of WDM subsystem, optical transceivers and fiberoptics components. From 1992–1995, he was employed by City University of HK, Hong Kong and Concordia University, Canada as research associates to perform optoelectronics materials research. Dr. Chang has authored or co-authored over 50 peer-reviewed journal and conference articles, and represents Vitesse to actively contribute to standard-setting bodies including OIF/ITU, IEEE 802.3, FSAN etc for the definition of various optical specifications. Dr. Chang has chaired the OFC/NFOEC technical sessions for several years in a row since 2004, and serves as the active reviewer for *Optical Express, Optical Letters, Applied Optics, Journal of Optical Networking, Applied Physics Letters, IEEE Photonics Technology Letters,* and *Journal of the Optical Society of America B.* He is a member of IEEE/LEOS and OSA.

Badri Gomatam received his MS in 1989 and Ph.D. in 1993 from the University of Massachusetts, Amherst, for work on diode laser dynamics and strained quantum well electroabsorption modulators. From 1993 to 1996, he performed research in several optoelectronics technologies at Foster-Miller Inc., Waltham, Massachusetts, notably on plastic optoelectronic packaging, semiconductor spatial light modulators and novel holographic optical switch elements. Dr. Gomatam joined Vitesse Semiconductor Corp. in 1996 in the Optoelectronics/Analog Circuits Division and has served in several capacities including Product Engineering, Applications Engineering, and Product Marketing for Optoelectronics IC products. He represents Vitesse at standards organizations including the ITU-T and FSAN. He currently serves as Director, Systems Engineering at Vitesse with a focus on strategic marketing and is a member of the IEEE-LEOS Technical Committee on Optoelectronic Packaging, Manufacturing, and Reliability (OPMR).

Calvin C.K. Chan received the B.Eng., M.Phil., and Ph.D. degrees from The Chinese University of Hong Kong, Hong Kong, all in information engineering. In 1997, he joined the Department of Electronic Engineering, City University of Hong Kong, as a Research Assistant Professor. In 1999, he joined Bell Laboratories, Lucent Technologies, Holmdel, NJ, as a Member of Technical Staff, where he worked on control of widely tunable semiconductor lasers and realization of an optical-packet switch fabric with terabit-per-second capacity. In August 2001, he joined the Department of Information Engineering, CUHK, where he is currently an Associate Professor. He has served as the member of the technical program committee for OFC/NFOEC from 2005–2007 and a number of other international conferences. Currently, he serves as an Associate Editor of *OSA Journal of Optical Networking*. He has published more than 150 technical papers in refereed international journals and conference proceedings, and holds one issued US patent. His main research interests include optical-access networks, optical-packet switching, and optical-performance monitoring.

Noriki Miki received the B.S. and M.S. degrees in electronics engineering from the Shinshu University, Nagano, Japan, in 1982 and 1984, respectively. In 1984, he joined the Yokosuka Electrical Communication Laboratory, Nippon Telegraph and Telephone Public Corporation, Kanagawa, Japan, where he was engaged in research on digital subscriber loop transmission systems using digital signal processing. From 1991 to 1994, he was at the NTT Transmission Systems Laboratories, Kanagawa, Japan, where he carried out research on passive optical networks for providing data communication and telephony. From 1994 to 1997, he was at the NTT Network Systems Development Department, Chiba, Japan, where he worked on development of fiber-optic access systems. Currently, he is working on research and development of higher-speed passive optical networks at the NTT Access Network Service Systems Laboratories. In 1999, he received the Achievement Award from the Institute of Electronics, Information, and Communication Engineers.

Kiyomi Kumozaki received the B.E. and M.E. degrees in electrical and electronics engineering from Toyohashi University of Technology, Toyohashi, Japan, in 1980 and 1982, respectively, and Ph.D. degree from Tohoku University, Sendai, Japan, in 2000. In 1982, He joined the NTT Yokosuka Electrical Communication Laboratory, Kanagawa, Japan, where he was engaged in research on metallic digital transmission, and development of metallic digital access systems. From 1987 to 1990, he worked on the development of ISDN access systems at the NTT Network Systems Development Center, Kanagawa, Japan. From 1991 to 1997, he was at the NTT Transmission Systems Laboratories, Kanagawa, Japan, where he carried out the research on passive-optical access network for providing data communication, telephony, and video distribution services. From 1998 to 2000, he worked for NTT Communications Corporations, Tokyo, Japan, where he was engaged in building large-scale LAN, WAN, and global network infrastructure for business enterprises. Since 2001, he has been with the NTT Access Network Service Systems Laboratories, Chiba, Japan, and working on research and development of next-generation access network including high-speed passive optical network.

Alan E. Willner received the Ph.D. from Columbia University, has worked at AT&T Bell Labs and Bellcore, and is Professor of Electrical Engineering at University of Southern California (USC). He has received the NSF Presidential Faculty Fellows Award from the White House, Packard Foundation Fellowship, NSF National Young Investigator Award, Fulbright Foundation Senior Scholars Award, IEEE Lasers and Electro-Optics Society (LEOS) Distinguished Traveling Lecturer Award, USC University-Wide Award for Excellence in Teaching, IEEE Fellow, Optical Society of America (OSA) Fellow, and Eddy Paper Award from Pennwell Publications for the Best Contributed Technical Article. Prof. Willner's professional activities have included: President of the IEEE LEOS, Editor-in-Chief of the *IEEE/OSA Journal of Lightwave Technology*, Editor-in-Chief of the *IEEE Journal of Selected Topics in Quantum Electronics*, Co-Chair of the OSA Science and Engineering Council, General Co-Chair of the Conference on Lasers and Electro-Optics (CLEO), General Chair of the LEOS Annual Meeting Program, Program Co-Chair of the OSA Annual Meeting, Chair of the IEEE TAB Ethics Committee, and Steering and Program Committee Member of the Conference on Optical Fiber Communications (OFC). Professor Willner has 630 publications, including one book.

Zhongqi Pan received his Ph.D. in electrical engineering from the University of Southern California (USC). He is currently an Assistant Professor of Electrical and Computer Engineering and Bell South/LEQSF Regents Professor in Telecommunications at the University of Louisiana at Lafayette. Dr. Pan's research is in the area of optical-fiber communication systems, wavelength-division multiplexing, optical amplification, and optical networks. He has over 100 publications, including three book chapters. Dr. Pan has served as the reviewer for *Applied Physics B—Lasers and Optics, Applied Physics Letters, Optics Letters, Optics Express, IEEE/OSA Journal of Lightwave Technology, IEEE Photonic Technology Letters, IEEE Selected Topics of Quantum Electronics*, and *Optics Communications.* Dr. Pan is a member of the Institute of Electrical and Electronics Engineers (IEEE) and the Optical Society of America (OSA).

Chapter 1

Introduction

Cedric F. Lam

OpVista Inc.

The recent surge in bandwidth demand, driven by fast-growing video-on-demand (VOD) services and emerging applications such as network gaming, peer-to-peer downloading etc. has revitalized the optical communication industry. After more than 20 years of active research, passive optical network (PON)-based broadband optical access systems are finally seeing wide-scale deployments in Asia and North America. In Europe, carriers and service providers are also actively looking into PONs as the next-generation broadband access solutions.

1.1 HISTORY OF BROADBAND ACCESS NETWORKS AND PON

Access networks have been traditionally called last-mile networks as they comprise the last segment connection from service providers' central office (CO) to end users. They are also called first-mile networks in recent years as they are the first segment of the broader network seen by users of telecommunication services. Example of access networks are twisted copper pairs connecting to each individual household (also called local loops) and residential coaxial cable drops from community antenna TV (CATV) service providers. Wi-Max is another type of access technology which uses radio waves for last-mile connectivity. Traditionally, optical fibers have been widely used in backbone networks because of their huge available bandwidth and very low loss. Although fiber has also been touted as the next-generation access technology for a long time, it is not until the beginning of this century that fiber has finally seen its growing commercial importance as the technology of last-mile connection.

Traditional telecommunication networks were developed for analog voice services. For a long time, 4 kHz was the bandwidth required to connect to end users for voice services. A ubiquitous twisted copper network has been deployed by telephone companies (the Bell System in the United States and PTTs in other countries) in industrialized countries for decades. It is not difficult to imagine that such networks were optimized for analog voice frequency transmissions. As a matter of fact, in order to achieve better economy and allow longer local loop drops, inductors called loading coils have been installed in many old twisted copper pair plants to enhance the voice frequency band performance. Loading coils, however, significantly attenuate high frequency signals outside the voice frequency band and make them unsuitable for broadband digital subscriber line (DSL) services.

Although voice signal transmission has been digitized into 64 kbps digital channel (DS0) for TDM switching long time ago, digital voice signals have been converted back to analog format to be backward-compatible with analog telephone sets before they are delivered to end users. The Internet is the driving force for digital local loops, also called DSLs [1]. The Internet was first invented in the 1960s. For a long time after its invention, the Internet was mainly used by research and academia for data sharing. The first popular Internet application was e-mail, which was invented in the early 1970s. It was not until the early 1990s, when the World Wide Web and its graphical user interface MOSAIC came out, that the Internet started to become an important part of people's lives. Because of the voice bandwidth limitation and transmission line noise, the best data rate available from an analog modem working on voice grade twisted copper lines is 56 kbps.

1.1.1 Digital Subscriber Line (DSL)

The first broadband DSL standard was the integrated services digital network (ISDN) system developed in the 1980s by CCITT which was the predecessor of ITU-T. The ISDN (also called IDSL) standard offers the so-called 2B + 1D encoding scheme on a single twisted pair. It includes two 64-kbps (2B) channels for voice and data, and one optional 16-kbps (1D) digital channel. This gives a total of 144 kbps data rate in both directions. ISDN services were never popular because of the high cost and lack of killer applications.

As web pages incorporate more and more multimedia data, the demand for bandwidth is slated to grow in order to satisfy customer expectations. Various flavors of DSL technologies (collectively called xDSL) have been invented for broadband data delivery on twisted copper pairs. DSL services make use of the higher frequency range on twisted pairs for data transmission. The 0- to 4-kHz band carries the traditional plain old telephone service (POTS) line. Typically, the 25- to 160-kHz band carries the upstream (user to carrier) direction data and the 240-kHz to 1.5-MHz band carries the downstream (carrier to user) direction data.

DSL data rates and transmission distances are limited by signal impairments inside twisted copper pairs [2]. The attenuation of electromagnetic waves on copper wires increases as the square root of the signal frequency. Therefore, higher frequency signals attenuate faster on twisted pairs. The aforementioned loading coils on old twisted pair plants must be removed in order to offer DSL services. Moreover, copper wire quality, bridge taps on twisted pairs, and cross talk between neighboring twisted pairs all degrade the signal quality.

Most of the DSL technologies today use a modulation technique called discrete multitone modulation (DMT), which divides the whole frequency bands into 247 channels of 4-kHz slots. Signal quality in each slot is constantly monitored and the signals are shifted from bad slots to good ones in an adaptive manner.

Asymmetric DSL (ADSL) and very high data rate DSL (VDSL) are the two most common DSL services. ADSL provides a downstream data rate of up to 8 Mpbs and upstream data rate up to 800 kbps, over a maximum transmission distance of 18000 ft or 5500 m. VDSL services are usually supported with a fiber deep infrastructure such as fiber-to-the-curb (FTTC), which has a short distance of twisted copper loop. For a 4000-ft (1200-m) twisted pair drop distance, VDSL can support up to 52 Mbps and 16 Mbps downstream and upstream data rates respectively. Table 1.1 gives a summary of the different DSL technology performances.

To receive DSL services, typically, users need to install a filter to separate the voice signal from data signals. A DSL modem is employed at the user end to connect the user to the service provider through a DSL access multiplexer (DSLAM), which is located at a remote node or CO. The DSLAM provides point-to-point dedicated bandwidth between the service provider and each end user. This scenario is different from the cable modem and PON scenarios described later.

1.1.2 Cable Modem

Traditional CATV networks were one-way broadcast systems. CATV programs are transmitted as analog signals in amplitude modulation–vestigial side band (AM-VSB) format [3]. Each CATV channel occupies a 6-MHz frequency division multiplexed (FDM) frequency slot in the North American National Television System Committee (NTSC) standard or 8-MHz slot in the phase alternate line (PAL) standard used in other parts of the world. Broadcast TV signals normally occupy a frequency band from 50 MHz to 500 MHz or 750 MHz.

Table 1.1

Summary of different DSL technology performances

DSL type	Max Speed: Upstream	Max Speed: Downstream	Max transmission distance	Number of lines required	POTS support
ADSL (asymmetric DSL)	800 kbps	8 Mbps	18,000 ft (5500 m)	1	Yes
HDSL (high data rate DSL)	1.54 Mbps	1.54 Mbps	12,000 ft (3650 m)	2	No
IDSL (ISDN)	144 kbps	144 kbps	35,000 ft (10700 m)	1	No
MSDSL (multirate symmetric DSL)	2 Mbps	2 Mbps	29,000 ft (8800 m)	1	No
RADSL (rate adaptive DSL)	1 Mbps	7 Mbps	18,000 ft (5500 m)	1	Yes
SDSL (symmetric DSL)	2.3 Mbps	2.3 Mbps	22,000 ft (6700 m)	1	No
VDSL (very high data rate DSL)	16 Mbps	52 Mbps	4000 ft (1200 m)	1	Yes

Linear optical transmission system has been used for analog TV signal transmission. CATV signals are transmitted from service providers' headend[1] offices to remote fiber nodes (called hub nodes) where they are converted back to the radio frequency (RF) domain and transmitted through coaxial cables to end users. Therefore, CATV systems are also called hybrid fiber coax (HFC) systems. Coaxial cable plants are laid out in a tree-and-branch architecture with cascaded RF amplifiers. Compared to twisted pairs, the coaxial cable is a very good

[1] CATV service providers have different terminologies than traditional telecommunication service providers. Some of the terminologies could be confusing between CATV and telecommunication service providers. A cable headend is called a central office or core node by telecommunication service providers. It is usually located directly on the backbone. Telecommunication service providers also call their core backbone nodes hubs, and small distribution nodes edge node. However, a hub node for CATV providers is a local signal distribution node, which parallels telecom service providers' edge nodes.

broadband medium with excellent frequency responses. It has a usable frequency range up to 1 GHz.

In North America, CATV service providers also call themselves multiple service operators (MSOs). In the mid-1990s, MSOs started to reform their traditional one-way analog broadcast systems by putting in bidirectional RF amplifiers and return optical links so that they can provide data and other services such as VOD. The 1996 Telecom Act [4] in the United States liberalized the telecommunication service market. MSOs took advantage of their coaxial cable to offer broadband data services through cable modems.

Unlike the rest of the world, where DSL services are far more popular than cable modem, in the United States, cable modem was the dominating form of broadband access for several reasons. Firstly, the coaxial CATV plant coverage in the United States is far more complete than in other countries. The majority of households are connected to MSO providers through coaxial cable drops. Secondly, being the first nation with ubiquitous telephone service deployments, most of the twisted pairs in the United States were too old, too long, and too poor in quality to enable DSL services. Thirdly, the data over cable service interface specification (DOCSIS) [5] 1.0 released in March 1997 made significant contribution to the success of cable modem in the United States by offering a common specification for multivendor interoperability and therefore helped to lower the equipment and service costs.

DOCSIS standards are developed by Cable Labs. In a cable modem system, cable modems at individual households are connected to a cable modem termination system (CMTS) at a headend office. The tree-and-branch coaxial cable plant forms a shared medium among the cable modem users. Customer data are multiplexed using the time division multiplexing (TDM) scheme. Downstream data signals from headends are broadcast to individual cable modems through the coaxial cable plant. Each individual cable modem recognizes its data by the ID embedded in downstream data. The CMTS acts a medium access control (MAC) master which assigns upstream time slots for each cable modem. This is very different from DSL systems where a dedicated line is provided between a DSL modem and the DSLAM.

Cable modems use the 0- to 45-MHz frequency band in the coaxial cable for upstream transmission. This frequency range usually has poorer channel characteristics due to the coupling from sources such as home electrical appliances. Another source of degradation in the upstream band is due to the noise funneling effect from all the upstream users.

DOCSIS specified a modulation technique called quadrature amplitude modulation (QAM) [6], which encodes multiple bits of information on each symbol. For example, on a 256-QAM channel, each symbol represents 8 bits ($\log_2 256$) of information, so that a 5-MSps (mega symbol per second) channel can carry a 40-Mbps data stream. DOCSIS 1.0 specified channel widths between

200 kHz and 3.2 MHz. 64-QAM and 256-QAM are used for downstream modulation. For upstream connections, quadrature phase shift keying (QPSK, which is equivalent to 4-QAM) and 16-QAM are used. This gives an upstream and downstream throughput of 10 Mbps and 38 Mbps respectively. DOCSIS 2.0 [5] increased the upstream capacity by employing 32-, 64-, and 128-QAM.

To cope with increasing bandwidth demand from customers, DOCSIS 3.0 [5] further increased the available data bandwidth to above 100 Mbps in both directions through a technique called channel bonding. Significant performance improvement is achieved by bonding 4 RF channels as a single logical data channel. Table 1.2 summarizes the data rates of different DOCSIS versions. Different versions of the DOCSIS protocols have been made backward compatible.

Although the maximum cable modem throughput can reach 38 Mbps downstream and 10 Mbps upstream in DOCSIS 1.0, most MSOs throttle the maximum data rate from users from 3 Mbps to 8 Mbps downstream and from 200 kbps to 800 kbps upstream because of the shared nature of the cable modem bandwidth. Higher rates are also available to users who are willing to pay premium prices. Per user available bandwidth can be improved by shrinking the sharing group size, i.e. decreasing the size of each coaxial tree by pushing fiber deeper into the field. In modern HFC systems, the share group size is usually between 50 and 100 households per fiber node [7].

The cable modem architecture has a lot of similarities to the most commonly seen power-splitting PON architecture, although they use completely different media for transmission. Both cable modem system and power-splitting PON use a point-to-multipoint (P2MP) tree-and-branch distribution plant as shared transmission medium among all the end nodes. They both use TDM for MAC. In a cable modem access system, the CMTS and cable modem form a master–slave relationship for medium control. The CMTS controls the bandwidth allocation to cable modems. In a power-splitting PON, the optical line terminal (OLT) and optical network unit (ONU) form a master–slave relationship for medium control. The OLT controls bandwidth allocation to the ONUs. Dynamic bandwidth allocation (DBA) is required in both cable modem systems and power-splitting PONs.

Table 1.2

Summary of DOCIS data rates

	Max data rate	
DOCSIS version	Downstream	Upstream
1.0	38 Mbps	10 Mbps
2.0	40 Mbps	30 Mbps
3.0	160 Mbps	120 Mbps

1.1.3 Fiber Access Systems

Local loops using optical fiber for access connections are called fiber-in-the-loop (FITL) systems [8–9]. Optical fiber has the advantage of high bandwidth, low loss, and low noise. Compared to the coaxial cable plant, which usually requires many cascaded RF amplifiers, fiber plants are in general much cleaner and require very little maintenance.

Studies for FITL started in the 1980s [10–11]. Fiber access systems are also referred to as fiber-to-the-x (FTTx) system, where "x" can be "home," "curb," "premises," "neighborhood," etc., depending on how deep in the field fiber is deployed or how close it is to the user. In a fiber-to-the-home (FTTH) system, fiber is connected all the way from the service provider to household users. In an FTTC system, fiber is connected to the curb of a community where the optical signal is converted into the electrical domain and distributed to end users through twisted pairs. Therefore, an FTTC system can also be regarded as a hybrid fiber twisted pair system.

Nowadays, most people think of FTTx as the P2MP power-splitting PONs (PS-PONs). In reality, fiber access systems can be point-to-point (P2P) or P2MP. Moreover, they can use an active remote distribution node such as an Ethernet switch or a simple passive splitter as the remote distribution node used in PS-PONs. In fact, NTT adopted P2P architectures in some early FTTH trials [12]. Another type of PON called WDM-PON uses a wavelength multiplexer as the remote distribution node [11]. PON architectures will be described in detail in Chap. 2.

Although FITL was in trial for a long time since its proposal, the high cost of fiber-optic components and lack of killer applications for the high bandwidth offered by optical fibers have been barriers to its real applications. The PON architecture was proposed as a way to share the large fiber bandwidth among many users through a passive splitter, and hence improve the per user cost of FITL.

PON standardization work began in the 1990s when carriers anticipated fast growth in bandwidth demands. In 1995, the full service access network (FSAN) [13] consortium was formed by seven global telecommunication operators including British Telecom, NTT, and Bell South to standardize common requirements and services for a passive optical access network system. One of the goals of FSAN was to create the economy of scale and lower the cost of fiber-optic access systems by promoting common standards.

FSAN recommendations were later adopted by the International Telecommunication Union (ITU) as the ITU-T G.983 BPON (i.e. broadband PON) standards [14–16]. G.983 specified 622 Mbps downstream and 155 Mbps or 622 Mbps aggregate upstream data rate. Each OLT is shared by up to 32 ONUs for a maximum separation of 20 km between the OLT and ONU.

BPONs use TDM for multiple access and asynchronous transfer mode (ATM) cells for data framing. Therefore, a BPON is also called an ATM PON or APON for short.

The G.983.3 standard [16] specified wavelength division duplex on a single fiber with 1.3-μm wavelength for upstream transmission and 1.49-μm wavelength for downstream transmission. The 1.55-μm wavelength window was reserved for analog TV signal overlay. Early BPON standard defined the reference architecture model and the physical medium dependent (PMD) layer. But it also left many of the control and management message formats unspecified for a considerable while.

BPONs only had limited trials and deployments. In the past few years, Ethernet emerged as the dominating framing technology for packetized IP data transmission. In March 2001, the IEEE 802.3 standard group started the 802.3ah Ethernet in the First Mile (EFM) project [17]. One of the charters of the 802.3ah work group was to standardize the transport of Ethernet frames on P2MP PONs or EPON. The IEEE802.3ah Standard was ratified in June 2004. It specifies an upstream and downstream throughput of 1 Gbps and a transmission distance of 10 km or 20 km with 16 ONUs per OLT.

EPON has gained tremendous popularity in East Asian countries, especially Japan and Korea. NTT has selected EPON as the standard for its large-scale FTTH rollout [12]. Nevertheless, EPON did not achieve much commercial success in the United States.

At the same time that EPON was developed by IEEE, the ITU-T Study Group 15 (SG15) was also working on the next-generation PON called Gigabit-capable PON (G-PON). G-PON specifications are captured in the G.984 series recommendations [18–20]. G-PON increased the transmission speed to 2.5 Gbps downstream and 1.25 Gbps or 2.5 Gbps upstream. Besides, it uses a new framing mechanism called G-PON encapsulation mode (GEM), which is based on the original idea of generic framing procedure (GFP).

G-PON was selected as the standard by Verizon, SBC (now AT&T), and Bell South in January 2003 when the three incumbent telecommunication operators issued a joint request for proposal (RFP) for fiber-to-the-premise (FTTP). These companies will use G-PON to compete with MSOs in delivering the so-called triple-play (video, voice, and data) services.

1.1.4 Ethernet

Ethernet was invented in the 1970s [21]. It was originally invented as a local area network technology for interconnecting desktop computers. Ethernet has been firmly established as the standard user interface of choice for connecting IP devices. More than 90% of the IP packets are generated and terminated as

Ethernet frames.[2] Ethernet covers the bottom two layers (i.e. physical layer and data link layer) of the OSI reference model. It is standardized by the IEEE802.3 standard group. Figure 1.1 shows the development trend of the Ethernet technology and local area networks.

Because of its high bandwidth, low cost, and ease of use and installation, Ethernet has become the most popular technology for data networking. After the Internet boom, the amount of IP data traffic in telecommunication networks has greatly surpassed that of the traditional TDM voice traffic. Such traffic begins and terminates as Ethernet frames. Moreover, video streaming services are also moving on to IP-based digital platforms. New video servers stream both broadcast and on-demand TV signals as Ethernet packets. Efficient handling of Ethernet packets is therefore very important in next-generation networks.

The first-generation Ethernet used MAC protocol called Carrier Sense Multiple Access with Collision Detection (CSMA/CD) for local area networking [22]. All the hosts are connected in a multipoint-to-multipoint (MP2MP) coaxial

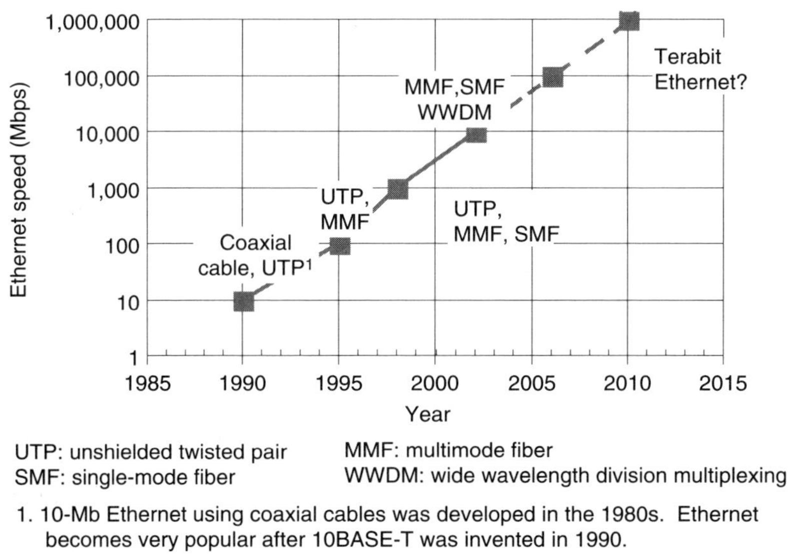

Figure 1.1 Development trend of Ethernet technologies.

[2] In switched networks, protocol data units (PDUs) on Layer 3 of the ISO reference model or the IP layer are usually referred to packets. Layer 2 or Ethernet protocol data units are usually called frames. These two terms are often used interchangeably.

bus line. Each station can directly communicate with another station in a peer-to-peer fashion. The CSMA/CD protocol is completely distributed. It does not require a master controller for bandwidth arbitration on the transmission medium. However, the CSMA/CD protocol also limits the transmission speed and distance. As the data rate increases, the network size has to be scaled down accordingly.

The transmission distance limit imposed by the CSMA/CD protocol was removed when full-duplex Ethernet was introduced. In a modern full-duplex Ethernet, all the stations in the network communicate with each other through a P2P link to a switch (also called bridge). Switches perform medium arbitration among connected stations and relay packets from station to station in a transparent fashion (i.e. connected stations have no knowledge of the existence of switches).

Ethernet switch operations are specified in the IEEE802.1 Spanning Tree Protocol (STP) [23]. STP is very important to full-duplex mode Ethernet operation, which removes the CSMA/CD protocol-imposed distance limitation. It allows Ethernet frames to be transmitted to distances only limited by the physical channel impairments such as noise and attenuation. To understand the principle of modern Ethernet, one really needs to have knowledge of both IEEE802.3 and 802.1 standards.

EPON created a new P2MP Ethernet architecture which is different from the original MP2MP and P2P models. For this reason, a P2P emulation function has been introduced in EPON in order to operate with the 802.1-based Ethernet switching.

Whether a PON system uses Ethernet (as in EPON), ATM (as in BPON), or GEM (as in G-PON) for data encapsulation between the OLT and ONU, there is no doubt that Ethernet is a must-support user network interface (UNI) that an ONU has to provide for connecting to customers' network equipment.

In the last few years, Ethernet has quickly moved out from the local area networks into backbone networks. End-to-end private-line Ethernet services are becoming more and more popular. Traditional Ethernet delivers best-effort services and lacks the capabilities of their TDM rivals such as SONET for management. The IEEE802.3ah EFM standard added an operation, administration, and management (OAM) sublayer to Ethernet. However, the OAM sublayer is mainly focused on a limited subset of performance monitoring functions of the physical and link layers.

To actually provide Ethernet services, the carrier must have the capability to control and manage Ethernet connections. Metro Ethernet Forum (MEF) [24], an industry consortium formed to promote Ethernet services, has been carrying out the effort of specifying Service Level Agreement (SLA) and OAM for carrier Ethernet. Other standard organizations such as ITU and Internet Engineering

Task Force (IETF) have also invested many efforts to make Ethernet more manageable and suitable for end-to-end service delivery.

1.1.5 WDM in Optical Access Networks

Wavelength division multiplexing (WDM) increases system capacity by transmitting multiple wavelengths on a single fiber. Coarse WDM techniques have already been applied in PON systems to separate upstream and downstream signals, and provide analog video overlay [16]. An important advantage of the optical fiber is its virtually unlimited bandwidth from an access viewpoint. Coarse WDM overlay on a power-splitting PON is an obvious way to provide different services and increase system capacity. For example, one can segregate the optical spectrum into different coarse WDM bands and engineer a G-PON system in one wavelength band and an EPON in a different band, doubling the value of the costly fiber plant [25].

All the PON systems mentioned so far use a power coupler to distribute the signal from OLT to users at different ONUs. In a WDM-PON system, a WDM coupler is used to distribute signals to different users. Each ONU is allocated with its own wavelengths. Such a system has the advantage of high capacity, privacy, and protocol transparency. The idea of WDM-PON was first proposed by Wagner and is now in field trial in Korea by Korea Telecom.

By using a wavelength cyclic (or so-called colorless) AWG device, one can spawn or realize multiple WDM-PON networks on a single physical fiber plant. The challenge of WDM-PON is wavelength stability control and low-cost colorless optical sources for the ONUs. A lot of research has been done in these areas to produce practical colorless AWG devices which are temperature-compensated, and colorless sources using ideas of injection locking a Fabry–Perot laser diode or reflective semiconductor optical amplifiers (RSOAs). Chapter 3 will describe these technologies in more detail. It is a matter of time before these devices will eventually become readily available when bandwidth requirement reaches the point at which WDM-PON systems will be necessary to satisfy customer demands. WDM-PON should enable broadband optical access network to stay passive for a considerable while before optical access networks eventually become active.

1.1.6 Killer Applications

The field of telecommunications took a significant dive in the beginning of this century because of the exuberance in capacity deployment. After some years of stagnancy, demands for bandwidths are growing again, and have become the driving force for the recent enthusiasm in PONs and FTTx developments.

It would be interesting to know that e-mail and the World Wide Web were the first and second killer applications in the history of the Internet since it was proposed in the 1960s. The Internet has been touted as the bandwidth driver in modern telecommunication networks. A slew of new applications have emerged in the past few years. Examples of these include peer-to-peer networking, sharing of music and video clips, network gaming, and voice-over IP.

Among all the emerging applications, the most important is the widespread adoption of digital TV and VOD. Digital TV signals are much easier to transport than analog ones because of the much lower linearity and signal-to-noise ratio requirements. From a technical viewpoint, VOD services have now become feasible because: (1) new mpeg video compression technologies have tremendously reduced the bandwidth and capacity required for digitized video transmission and storage; (2) electronic memory, storage, and processing technologies have made it possible to store and switch thousands of movies in practical-size video servers; and (3) low-cost WDM, Gigabit, and 10Gigabit Ethernet transmission technologies have enabled economical transport of high bandwidth video signals. For example, a Gigabit Ethernet link is capable of carrying 240 streams of standard resolution video signals in mpeg-2 format [26], each of which requires 3.5 Mbps bandwidth compared to the 200 Mbps per channel used in the last-generation digital video broadcast–asynchronous serious interface (DVB-ASI)-based digital TV systems.

From a business perspective, benefited from continual improvements of access bandwidths, the Internet has become an important new form of media in people's lives. The billion-dollar acquisition of YouTube by Google in 2006 is a clear reflection of such developments. In order to compete with the Internet, cable companies are busy reviving their services with VOD programs using IPTV technologies. In addition, in order to reduce both the capital and operational expenditures, carriers are merging their video and data delivery platform into a unified platform based on IP technologies. VOD has become the killer application for broadband access network development.

1.2 ECONOMIC CONSIDERATIONS IN PON DEVELOPMENT

1.2.1 How Much Bandwidth Is Enough?

An old question often asked is: "how much bandwidth will be enough for an end user?" The demand for bandwidth is driven by the contents available on the network, which is only limited by imagination. Capable technologies will find

Table 1.3

Bandwidth requirements for different IP services

Application	Bandwidth	QoS
Video (SDTV)	3.5 Mbps	Low loss, low jitter, constant bit rate
Video (HDTV)	15 Mbps	Same as above
Telecommuting	10 Mbps	Best effort, bursty
Video gaming	10 Mbps	Low loss, low jitter, bursty
Voice	64 kbps	Low loss, low latency, constant bit rate
Peer-to-peer downloading	100 kbps–100 Mbps	Best effort

their applications through people's creativity. The Internet was nothing new when the World Wide Web was invented in the early 1990s. Since then it has changed the way people live within a few years.

We also saw that the general availability of applications such as VOD happens naturally when a number of techno-economical factors come together. Without affordable broadband networks, even when compression and storage technologies are perfected, there will be no VOD services. Instead of speculating how much bandwidth is enough, we just summarize the bandwidth requirements for different applications in a typical household in Table 1.3.

A sustained connection of 100 Mbps will allow an 8-GB DVD movie to finish downloading in about 10 minutes, which will be comparable to the time taken to go to the video rental store at the next block. It is not difficult to imagine the requirement of 100 Mbps per broadband household. In fact, capacity is a major consideration when the RBOCs issued their joint G-PON RFP for FTTP applications in January 2003.

1.2.2 Policy and Regulation Influence

Communication network deployment requires significant upfront capitals. Whenever possible, companies would always like to evolutionarily upgrade their legacy infrastructure and maximize the value of existing investment.

Unlike in the United States, in most of the Asian countries government restrictions exist for telecommunication service providers to offer content services such as streaming video [27–28]. Nevertheless, countries such as South Korea and Japan have made significant progress in broadband access networks as a result of government incentives for new technology development. Percentage wise, Korea has the most broadband coverage in the world, whereas Japan accounts for two-thirds of global FTTH users.

In North America, continual deregulation of the telecommunication industry introduces fierce competition between MSOs and incumbent Regional Bell Operation Companies (RBOCs). New regulations allow MSOs and RBOCs to enter each other's traditional markets.

MSOs have not only finished converting their legacy one-way broadcast coaxial cable network into a bidirectional network, but are also taking the advantage of their broadband HFC network to provide high-speed data, digital TV, and voice services. Faced with the competition from MSOs and the rapid price erosion of their legacy voice services, RBOCs need to upgrade their 100-year-old twisted copper access plant in order to match or surpass services offered by MSOs. A fiber deep access network architecture or FTTx is the natural choice to future-proof their new investment.

In the United States, one of the incentives for RBOCs to build up fiber deep infrastructure is to bypass the requirement to share their access loops with competitors. In order to introduce competitions into the telecommunication market, the US government requires traditional incumbent carriers to unbundle (open) their twisted pair access loops to their competitors. Such requirements, however, do not apply to new access infrastructure investment such as FTTx.

1.2.3 Standardization Efforts

The DOCSIS [5] standards jointly developed by the cable industry had been a key factor for the success of the cable modem market in the United States by creating common multivendor interoperable specifications, and hence the necessary economy of scale.

The joint RFP from the US RBOCs in 2003 was aiming at recreating the DOCSIS story in the PON field. FSAN and ITU are the standard organizations responsible for the G.983 series BPON and G.984 series G-PON standards. These two organizations are also hosting interoperability test events among vendors manufacturing BPON and G-PON equipment.

The IEEE 802.3 standard group is responsible for EPON developments. This group traditionally had a good track record of keeping track of the details in technical requirements to ensure interoperability.

1.2.4 Cost Considerations

As mentioned in the beginning, although PON has been invented for over 20 years, it has not been commercially successful until recently. An access system connects end users to COs through local loops in two ways. The straightforward approach is to run a separate pair of wires (called home run) from each end user to the assigned CO. Alternatively, local loops first connect end users to a remote terminal (RT), which multiplexes the individual signals on a feeder line. The feeder line in turn connects to a local CO. Feeder lines provide pair gain by reducing the amount of wires required to connect each user to the CO.

There is trade-off between the cost of electronics in the RT and the savings achieved from the pair gain. If the cost of RT is high, in general, more users are required to share the cost, and the length of local loops will tend to be longer. The local loop distance where the cost of RT starts to make economic sense is called prove-in distance. Prove-in distance reduces as electronic technologies improve. In a PON system, the RT is simply a passive optical power splitter or WDM coupler.

It should be realized that lack of killer applications was not the reason for the slow adoption of the PON technology. In order for PON to become commercially viable, the cost of running a fiber local loop (labor + capital) needs to be in par with that of running a twisted pair loop, which is about $1000 in the United States. For a considerable period of time, fiber-optic components accounted for a significant portion of PON deployment cost. The high cost of optical components was the main barrier for FTTH deployments.

In order to improve the economic model of fiber access networks, instead of pulling fiber all the way to the home, PON systems have been proposed for various FTTx applications where the ONU is placed at a curb or building basement so that it can be shared by a group of users through a short drop of twisted pairs. The cost of optical components has come down significantly in recent years and FTTH is now making a lot of sense for high bandwidth broadband access.

PON belongs to access networks. Access equipment is usually deployed in large volumes. They are therefore very cost-sensitive. In a PON system, the cost of ONU needs to be multiplied by the number of users. This cost is either borne by the service provider or the end user. Therefore, low cost is the most important consideration in ONU designs.

1.3 ORGANIZATION OF THE BOOK

This book consists of seven chapters, covering different aspects of PON technologies.

Chapter 1 begins with the history of broadband access network and PON developments. It also discusses the economical and policy forces behind broadband infrastructure developments.

Chapter 2 reviews the various PON architectures and technologies. It paves the way to understand the reasons and philosophies behind the PON technology development, which will be covered in the ensuing chapters.

Chapter 3 thoroughly covers PON-related optical technologies and their state of the art.

Chapter 4 focuses on the properties and characteristics of PON transceivers, which are quite different from those used with conventional two-fiber, continuous-mode optical transmission systems.

Chapter 5 reviews the ranging process and dynamic bandwidth allocation protocols, which are essential to the operation and performance of power-splitting TDM PON systems.

As the PON speed and service group size increases, reliability and availability becomes more and more important. Chapter 6 discusses protection switching and traffic restoration schemes for PON systems.

A PON system differs from the traditional fiber-optic system in its P2MP and one-fiber bidirectional transmission architecture. Chapter 7 studies the challenges to characterize, monitor, and diagnose optical links in a PON system.

It is our wish to present this book as a comprehensive reference of PON technologies for those interested in developing and understanding this fast-growing area.

REFERENCES

[1] K.G. Coffman and A.M. Odlyzko, "Growth of the Internet," in *Optical Fiber Telecommuncations IV B*, edited by I. Kaminow and T. Li, pp17–59, Academic Press, 2002.

[2] J.A.C. Bingham, *ADSL, VDSL and Multicarrier Modulation*, John Wiley & Sons, 2000.

[3] W.I. Way, *Broadband Hybrid Fiber/Coax Access System Technologies*, Academic Press, 1999.

[4] Federal Communications Commission, *Telecommunications Act of 1996*, available from http://www.fcc.gov/Reports/tcom1996.pdf and http://www.fcc.gov/telecom.html

[5] Cable Labs, Data Over Cable Service Interface Specification (DOCSIS), available from http://www.cablemodem.com/specifications/

[6] J.G. Proakis, *Digital Communications*, 3/e (Chap. 4), McGraw-Hill, 1995.

[7] O.J. Sniezko, T. Warner, D. Combs, E. Sandino, X. Lu, T. Darcie, A. Gnauck, S. Woodward, and B. Desai, "HFC Architecture in the Making," *NCTA Technical Papers*, 1999.

[8] N.J. Frigo, "A survey of fiber optics in local access architectures," in *Optical Fiber Telecommunications, III A*, edited by I. Kaminow and T.L. Koch, pp461–522, Academic Press, 1997.

[9] E. Harstead and P.H. van Heyningen, "Optical Access Networks," in *Optical Fiber Teclecommunications, IV B*, edited by I. Kaminow and T. Li, pp438–513, Academic Press, 2002.

[10] S.S. Wagner, H. Kobrinski, T.J. Robe, H.L. Lemberg, and L.S. Smmot, "Experimental demonstration of a passive optical subscriber loop architecture," *Elect. Lett.*, vol.24, pp344–346, 1988.

[11] S.S. Wagner and H.L. Lemberg, "Technology and system issues for a WDM-based fiber loop architecture," *IEEE J. Lightwave Tech.*, vol.7, no.11, pp1759–1768, 1989.

[12] H. Shinohara, "Broadband access in Japan: rapidly growing FTTH market," *IEEE Commun. Mag.*, pp72–78, Sept. 2005.

[13] FSAN website: http://www.fsanweb.org/

[14] ITU-T, 1998, Recommendation G.983.1, *Broadband optical access systems based on Passive Optical Networks (PON).*

[15] ITU-T, 2000, Recommendation G.983.2, *ONT management and control interface specifications for ATM PON.*

[16] ITU-T, 2001, Recommendation G.983.3, *A broadband optical access system with increased service capability by wavelength allocation.*

[17] IEEE 802.3ah Ethernet for the First Mile website, available from http://www.ieee802.org/3/efm/

[18] ITU-T, 2003, Recommendation G.984.1, *Giagbit-capable Passive Optical Networks (G-PON): General characteristics.*

[19] ITU-T, 2003, Recommendation G.984.2, *Gigabit-capable Passive Optical Networks (G-PON): Physical Media Dependent (PMD) layer specification.*

[20] ITU-T, 2004, Recommendation G.984.3, *Gigabit-capable Passive Optical Networks (G-PON): Transmission convergence layer specification.*

[21] R.M. Metcalfe, "Ethernet: Distributed packet switching for local computer networks," *Communications of the ACM*, vol.19, no.9, pp395–404, July 1976.

[22] IEEE Standard 802.3, *Carrier sense multiple access with collision detection (CSMA/CD) access method and physical layer specifications*, 2005 Edition.

[23] IEEE Standard 802.1D, *Media access control (MAC) bridges*, 1993.

[24] Website of MEF: http://www.metroethernetforum.org/

[25] N. Frigo and C.F. Lam, "WDM overlay of APON with EPON—a carrier's perspective," available from http://grouper.ieee.org/groups/802/3/efm/public/sep01/lam_1_0901.pdf

[26] ISO/IEC 13818–1:2000, *Information technology—Generic coding of moving pictures and associated audio information: Systems*, 2000.

[27] K. Song and Yong-Kyung Lee, "Broadband access networks in and services in Korea," in *Broadband Optical Access Networks and Fiber-to-the-Home*, edited by C. Lin, pp69–106, John Wiley & Sons, 2006.

[28] J. Wang and Z. Zhao, "FTTH system, strategies, and deployment plans in China," in *Broadband Optical Access Networks and Fiber-to-the-Home*, edited by C. Lin, pp233–249, John Wiley & Sons, 2006.

Chapter 2

PON Architectures Review

Cedric F. Lam

OpVista Inc.

2.1 FTTx OVERVIEW

The general structure of a modern telecommunication network consists of three main portions: backbone (or core) network, metro/regional network, and access network (Fig. 2.1).

On a very high level, core backbone networks are used for long-distance transport and metro/regional networks are responsible for traffic grooming and multiplexing functions. Structures of backbone and metro networks are usually more uniform than access networks and their costs are shared among large numbers of users. These networks are built with state-of-the-art fiber optics and wavelength division multiplexing (WDM) technologies to provide high-capacity connections.

Access networks provide end-user connectivity. They are placed in close proximity to end users and deployed in large volumes. As can be seen from Fig. 2.1, access networks exist in many different forms for various practical reasons. In an environment where legacy systems already exist, carriers tend to minimize their capital investment by retrofitting existing infrastructure with incremental changes, whereas in a green-field environment, it often makes more sense to deploy future-proof new technologies which might be revolutionary and disruptive.

Compared to traditional copper-based access loops, optical fiber has virtually unlimited bandwidth (in the range of tera-hertz or THz of usable bandwidth). Deploying fiber all the way to the home therefore serves the purpose of future-proofing capital investment. A passive optical network (PON) is a form of fiber-optic access network. Most people nowadays use PON as a synonym of FTTx, despite the fact that the latter carries a much broader sense.

Figure 2.2 shows the alternatives of FTTx [1]. As seen from the figure, in the simplest case, individual optical fibers can be run directly from the central office (CO) to end users in a single star architecture. Alternatively, an active or passive remote

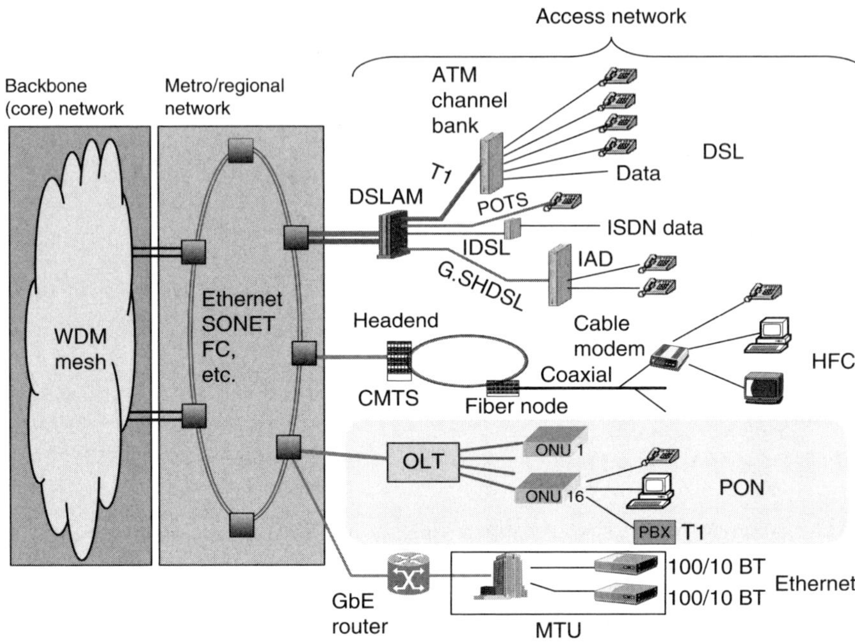

Figure 2.1 Generic structure of a modern telecommunication network.

terminal[1] (RT) with multiplexing functions may be placed in the field to reduce the total fiber mileage in the field. A PON network is characterized by a passive RT.

In an optical access network, the final drop to customers can be fiber (FTTH), coaxial cable (as in an HFC system), twisted pairs or radio (FTTC). In fact, a PON system can be used for FTTH or FTTC/FTTP depending on whether the optical fiber termination (or the ONU location) is at the user, or in a neighborhood and extended through copper or radio links to the user. In this book, we do not make a distinction between FTTH and FTTC/FTTP.

2.2 TDM-PON VS WDM-PON

Figure 2.3 shows the architecture of a time division multiplexing PON (TDM-PON) and a wavelength division multiplexing (WDM) PON [2]. In both structures, the fiber plant from the optical line terminal (OLT) at a CO to the optical network units (ONUs) at customer sites is completely passive.

[1] Another term for RT is remote node (RN). In this book, these two terminologies will be used interchangeably from place to place.

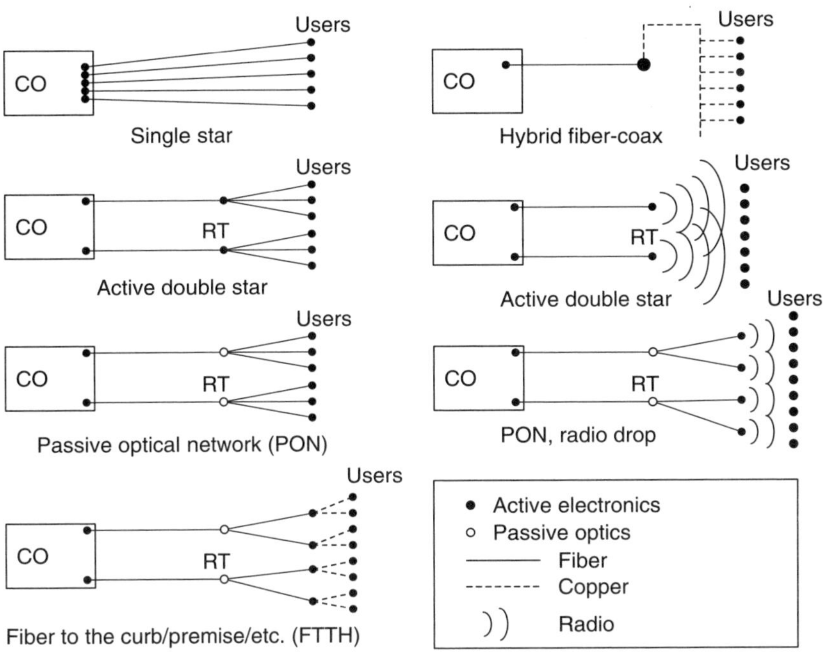

Figure 2.2 FTTx alternatives (from [1] copyright [2004] by IEEE, reprinted with permission).

A TDM-PON uses a passive power splitter as the remote terminal. The same signal from the OLT is broadcast to different ONUs by the power splitter. Signals for different ONUs are multiplexed in the time domain. ONUs recognize their own data through the address labels embedded in the signal. Most of the commercial PONs (including BPON, G-PON, and EPON) fall into this category.

A WDM-PON uses a passive WDM coupler as the remote terminal. Signals for different ONUs are carried on different wavelengths and routed by the WDM coupler to the proper ONU. Since each ONU only receives its own wavelength, WDM-PON has better privacy and better scalability. However, WDM devices are significantly more expensive, which makes WDM-PONs economically less attractive at this moment.

2.3 OPTICAL TRANSMISSION SYSTEM

2.3.1 Optical Fiber

The low loss, low noise, and exceptionally large bandwidth of optical fiber makes it ideal for long-distance backbone network transmission. Recently, the

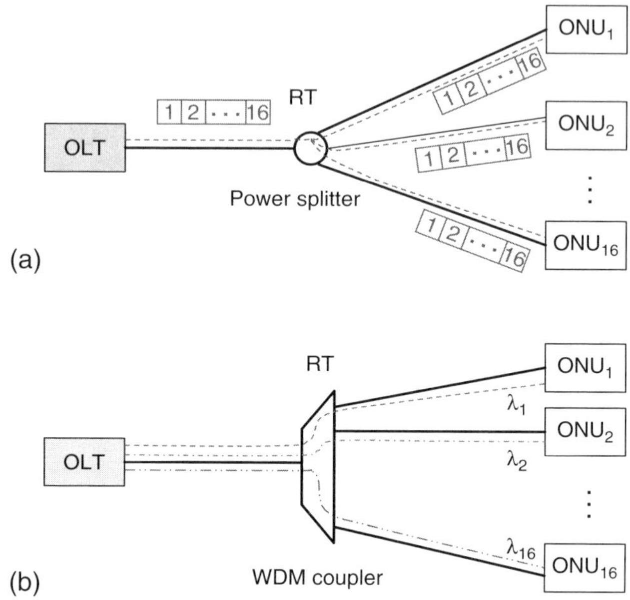

Figure 2.3 Architecture of (a) TDM-PON and (b) WDM-PON.

field of fiber-optic communications has experienced tremendous growth, thanks to the development of WDM technologies. As a result of this development, costs of fiber-optic components have dramatically decreased to the point that it is now commercially viable to apply fiber-optic technologies in access networks.

Optical fibers are waveguides made of high-purity glasses [3]. The cylindrical core of a fiber has a slightly lower refractive index than the cladding surrounding it. Optical fibers can be classified as single mode or multimode. Standard single-mode fiber (SMF) has a small core diameter of about 10 μm and requires high mechanical precision for signal coupling. On the other hand, multimode fibers (MMFs) have large core diameters for easy alignment and coupling. There are two commonly seen MMFs with core diameters of 50 μm and 62.5 μm respectively.

As shown in Fig. 2.4, light can only propagate in one mode in an SMF whereas there are multiple modes that light signals can propagate in an MMF because of the large core size [3]. These modes propagate at different speeds and result in modal dispersion in MMF. Modal dispersion causes signal pulses to broaden, thus limiting the signal bandwidth and transmission distance. The bandwidth–distance product of MMF is measured in mega-hertz kilometer (MHz km).

Traditionally, SMF has been used for long-distance backbone transmissions and MMF for local building connections. As we have seen from Chap. 1 that

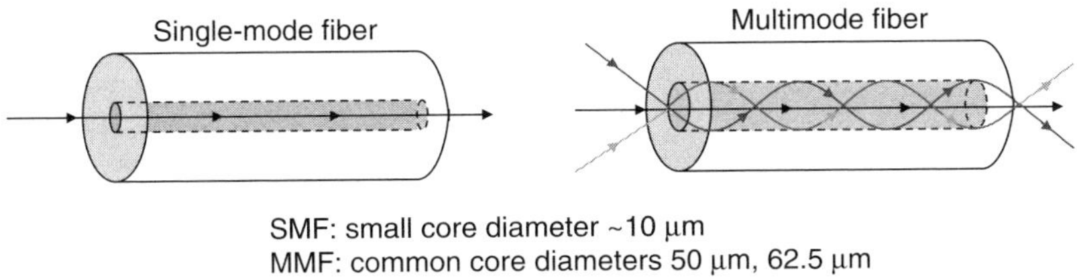

Figure 2.4 Single-mode fiber (SMF) vs multimode fiber (MMF).

with the existing twisted copper pair infrastructure, one can already achieve tens of megabit per second data rate. Therefore, in order for PON, or FTTx system to make economic sense and future-proof the investment, it has to offer unparalleled capabilities that traditional copper plant cannot provide, i.e. gigabit per second data rate at kilometer distances. For this reason, high-speed optical access networks use SMF as opposed to MMF. In this book, we mainly deal with SMF.

2.3.2 Chromatic Dispersion

Modal dispersion is not an issue in SMF. However, chromatic dispersion exists in SMF. It is caused by the different propagation speeds of light signals of different wavelengths or frequencies. As data symbols occupy finite frequency spans, chromatic dispersion broadens optical signal pulses as they propagate through optical fibers and produces power penalties at the receiver.

Chromatic dispersion is characterized by the dispersion parameter D, which is measured in units of ps/nm/km. It gives the broadening ΔT (in pico-second) of a pulse with bandwidth $\delta\lambda$ of 1 nm on the optical spectrum, after the pulse propagates through 1-km distance of fiber. For an arbitrary optical pulse propagating in a fiber network, the total broadening is given by:

$$\Delta T = \mathrm{D} \cdot \delta\lambda \cdot L \tag{2.1}$$

where L is the transmission distance given in kilometers.

The modulation bandwidth $\delta\lambda$ of a transform limited optical signal with NRZ (nonreturn zero) modulation, is roughly related to the data rate R as:

$$\delta\lambda \cdot = (\lambda^2/\mathrm{c}) \cdot \mathrm{R} \tag{2.2}$$

where λ is the wavelength of the optical carrier. Such a signal is usually produced with a high-quality single-wavelength laser through external modulation. In many cases, the output from the optical source spreads over a finite spectral region much wider than the modulation bandwidth of the data, which dominates δλ and hence the chromatic dispersion.

The dispersion coefficient is a function of optical wavelength [3]. Figure 2.5 shows the dispersion coefficients as a function of wavelengths for different types of optical fibers. Standard single-mode fiber is usually the only type of fiber used in a PON system. As can be seen from Fig. 2.5, standard SMF has nearly 0 dispersion around the 1.3-μm wavelength region and 17-ps/nm/km dispersion around 1.55 μm, which is where erbium doped fiber amplifier (EDFA)—the most mature optical amplification technology works the best. Chromatic dispersion affects the choice of optical wavelength and transmitter technologies for upstream and downstream connections, which will be discussed later.

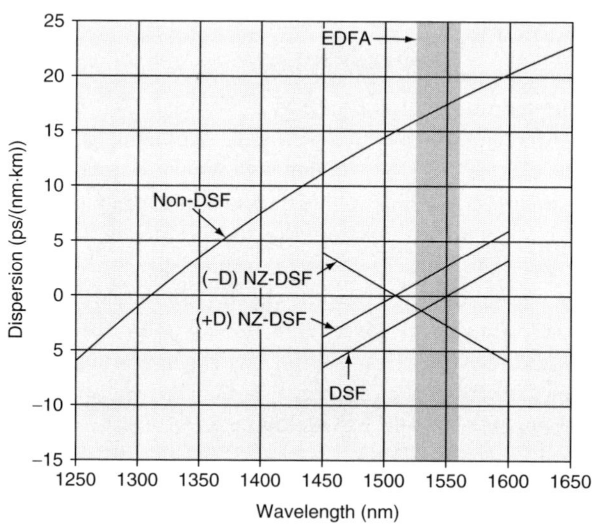

Non-DSF: Nondispersion shifted fiber, which is also called standard single-mode fiber or ITU-T G.652 fiber
(−D) NZ-DSF: negative nonzero dispersion shifted fiber
(+D) NZ-DSF: positive nonzero dispersion shifted fiber
DSF: dispersion shifted fiber with nearly zero dispersion in C-band

Figure 2.5 Dispersion coefficient as a function of wavelengths for various types of optical fibers.

2.3.3 Fiber Loss

Optical fiber loss affects the power budget, which limits the physical distance and splitting ratio that can be achieved in a PON system. Standard fibers are made of Silica (SiO_2), the very same material of glass and sand. Figure 2.6 shows the loss in a Silica optical fiber at different wavelengths.

In the short-wavelength region, optical signal loss is limited by Rayleigh scattering whereas in the long-wavelength region, optical scattering due to lattice vibrations limits signal loss. Figure 2.6 also shows the names of different optical bands and their spectral locations. It indicates that the C-band (stands for conventional band) wavelengths experience the minimum signal loss. As mentioned earlier, this spectral region is the region where optical amplification can be easily achieved with EDFAs [4]. This makes it the most suitable wavelength band of choice for long-haul WDM transmissions and for analog CATV transmissions [5] where high power is required to achieve the stringent carrier-to-noise ratio (CNR) requirement.

In conventional fibers, the attenuation peaks locally around the 1.38 μm wavelength (inside the E-band). This local peak (also called the water peak) is due to the absorption by the OH^- impurities left from fiber manufacturing. Better purification techniques have removed the water peak in new fibers such as the All-Wave fiber from OFS-Fitel Corporation and the SMF-28e fiber from Corning Inc. This makes the E-band spectrum available for coarse WDM applications.

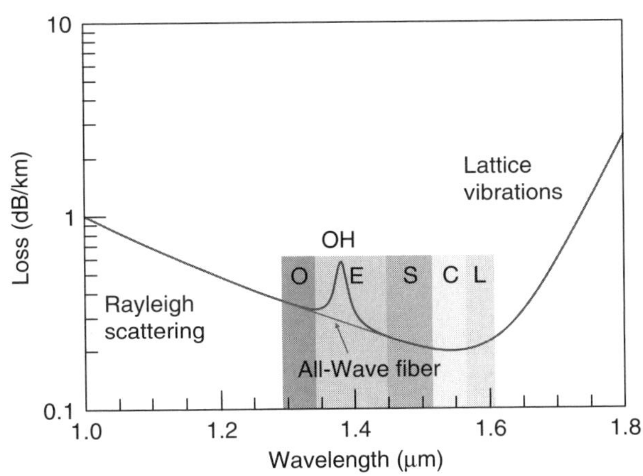

Figure 2.6 Loss in an optical fiber at different wavelengths.

PON systems are mostly built with 1.3 μm (O-band) for upstream signal transmission, 1.49 μm (S-band) for downstream transmission, and 1.55-μm (C-band) wavelengths for an optional analog CATV signal overlay. The use of these bands will be explained in more details in a latter section. The loss of new fibers around the 1.55-μm wavelength region can be as low as 0.19 dB/km [6]. It should also be noticed that around the 1.3-μm wavelength region used for upstream transmission, optical fiber loss is around 0.33–0.35 dB/km, significantly higher than that in the 1.5-μm wavelength region.

2.3.4 Bidirectional Transmission

2.3.4.1 Two-Fiber vs One-Fiber

Conventional fiber-optic communication systems use two separate fibers for bidirectional communications. This is also called space division duplex, or two-fiber approach. This straightforward method does not require separation of the upstream (ONU to OLT) and downstream (OLT to ONU) signals in time, frequency, or wavelength domains, and is simple to implement. In a TDM-PON system, the two-fiber approach requires two optical power splitters in field whereas in a WDM-PON system, one or two WDM multiplexers may be used in the field (Chap. 3). Our main focus in this section will be power-splitting-based TDM-PON. WDM-PONs will be discussed in a separate section.

Usually, for power-splitting PONs with two fibers, the 1.3-μm wavelength is used for both upstream and downstream transmissions because low-cost Fabry–Perot (FP) lasers are readily available at this wavelength, and they can be used without much worry about fiber dispersion effects.

Despite its simplicity, because of the extra fiber required and the necessity to terminate and manage a second splitter, the two-fiber solution is more costly than one-fiber solutions from both the capital and operational standpoints.

2.3.4.2 One-Fiber Single-Wavelength Full Duplex

In this approach, only one optical fiber is used for both upstream and downstream connections. A simple 3-dB 1:2 directional coupler is used at the OLT and ONU to separate the upstream and downstream optical signal. Such a system is illustrated in Fig. 2.7 (a).

The problem of this approach is that the 3-dB couplers introduce about 3.5-dB signal loss on each end of the transmission link and hurt the system power budget. Moreover, the transmitted signal can be scattered into the local receiver as near-end cross talk (NEXT), which is illustrated in Fig. 2.7 (b). Given that the

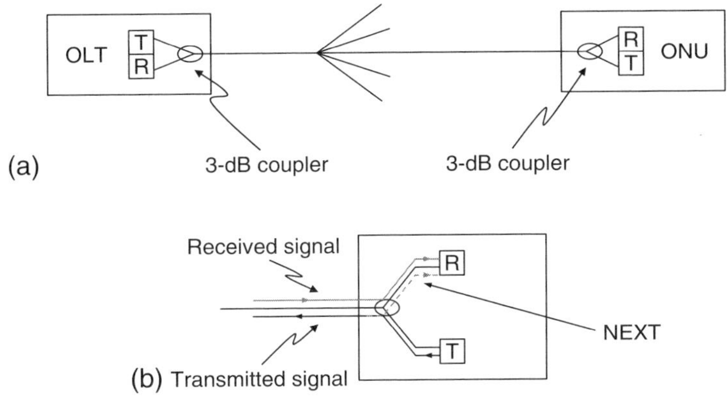

Figure 2.7 (a) One-fiber single-wavelength bidirectional transmission. (b) Near-end cross talk (NEXT).

received downstream signal strength is reduced by the optical coupler at the RT and local transmitter power is usually high, NEXT puts a stringent requirement on reflection controls [7]. Furthermore, if the ONU and OLT wavelengths happen to be very close, NEXT will produce coherent cross talk, which is even more detrimental.

2.3.4.3 Time Division Duplex

In the time division duplex approach, the OLT and ONU take turns to use the fiber in a ping-pong fashion for upstream and downstream transmissions. Similar to the one-fiber single-wavelength full duplex approach, directional couplers are used at OLT and ONUs to separate upstream and downstream optical signals. The NEXT effect is avoided by separating upstream and downstream signals in the time domain, at the cost of reducing the overall system throughput by about 50%. The OLT coordinates the time slots assigned for upstream and downstream transmissions. Burst mode receivers are required at both the OLT and ONU.

2.3.5 Wavelength Division Duplex

The wavelength division duplex method separates upstream and downstream transmission signals using different wavelengths. To ease wavelength control, a coarse 1.3/1.5-μm wavelength duplexing scheme[2] is chosen to separate upstream

[2] As mentioned earlier, the exact downstream wavelength used in industry standard PONs is actually 1.49 μm.

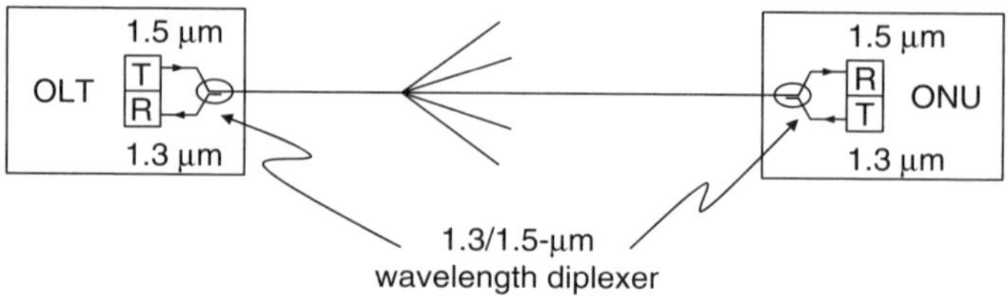

Figure 2.8 Wavelength division duplex uses 1.3/1.5-μm coarse WDM coupler (diplexer) to separate upstream and downstream signals.

and downstream signals (Fig. 2.8) [8]. The window at each wavelength is made sufficiently large so that no temperature control is needed to stabilize laser output wavelengths.

To lower the overall system cost, it is important to employ low-cost optical components. F-P lasers are simple to manufacture and have good output power and reliability. However, these lasers emit more than one wavelength defined by repetition frequency of the cavity (Chap. 3). As a result of chromatic dispersions, these different longitudinal modes will propagate at different speeds, leading to pulse-broadening and intersymbol interference.

At high speed, dispersion effects due to the multiple F-P laser emission modes can limit the transmission distance. As mentioned earlier, such effect is minimum for standard SMF near the 1.3-μm wavelength region where the dispersion coefficient is nearly zero. F-P lasers at 1.5-μm wavelength region ($D = 17$ ps/nm/km) has a typical spectral width of $\delta\lambda = 2.5$ nm. Therefore, for a transmission distance of 10 km, the pulse-broadening $\Delta T = 425$ ps according to Eq. (2.1). As an example, in the IEEE802.3ah EPON standard [9, Clause 60], the symbol rate is 1.25 Gbaud/s after 8B10B physical layer encoding. The pulse-broadening due to typical 1.5-μm F-P laser at 10-km transmission distance is more than half of the symbol period (800 ps) (assuming NRZ modulation). Therefore, without dispersion compensation, F-P lasers are unsuitable at 1.5-μm transmission wavelength.

The more expensive single-mode DFB (distributed feedback) lasers [10] with narrow-output spectrum are needed for 1.5-μm transmission. Therefore, low-cost 1.3-μm F-P lasers are used at ONUs for upstream transmission and 1.5-μm DFB lasers are used at the OLT for downstream transmission where its cost can be shared by the multiple ONUs connected to the OLT.

One of the advantages of wavelength division duplex is that the reflected downstream light from unterminated splitter ports is also flittered by the diplexer at the OLT. This reduces the connector reflectivity requirements at the remote node power splitter. During the initial deployment stage of a PON system, the take-rate would be low and most of the remote node splitter fan-out ports may be unused. Without the wavelength diplexer, this could produce a lot of unwanted interference to the upstream-received signal at the OLT if those connectors are not properly terminated. Another source of unwanted reflected light is from breaks of distribution fibers in the field.

2.4 POWER-SPLITTING STRATEGIES IN A TDM-PON

2.4.1 Splitting Architectures

The purposes of power splitting include: (1) sharing the cost and bandwidth of OLT among ONUs and (2) reducing the fiber mileage in the field. Apart from the simple one-stage splitting strategy (Fig 2.9 (a)), splitters may also be cascaded in the field as shown in Fig. 2.9 (b). In the most extreme case, the feeder fiber forms an optical bus and ONUs are connected to it at various locations along its path through 1:2 optical tap splitters as shown in Fig. 2.9 (c).

The actual splitting architecture depends on the demography of users and the cost to manage multiple splitters. From a management point of view, it is usually simpler to have a single splitter for distribution in the field, which makes splicing easier and minimizes connector and splicing losses.

In a bus or tree architecture like Fig. 2.9 (c), if all the splitters have the same power splitting ratio, the furthest ONU will suffer the most transmission and

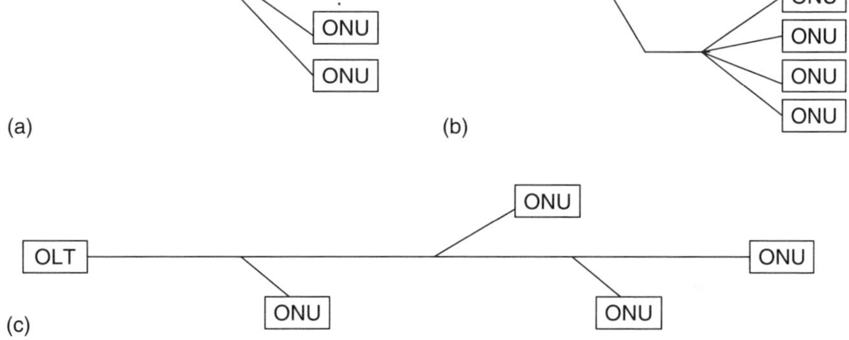

Figure 2.9 Splitting strategies in a TDM-PON: (a) one-stage splitting, (b) multistage splitting, and (c) optical bus.

splitting loss and become the system bottleneck. Splitters with uneven splitting ratios may be used to improve the overall power margin. However, such optimization requires stocking nonuniform splitters and is hence difficult to manage.

2.4.2 Splitting Ratio

Most of the commercial PON systems have a splitting ratio of 1:16 or 1:32. A higher splitting ratio means that the cost of the PON OLT is better shared among ONUs. However, the splitting ratio directly affects the system power budget and transmission loss. The ideal splitting loss for a 1:N splitter is $10 \times \log(N)$ dB. To support large splitting ratio, high-power transmitters, high-sensitivity receivers, and low-loss optical components are required. Higher splitting ratio also means less power left for transmission fiber loss and smaller margin reserved for other system degradations and variations. Therefore, up to a certain point, higher splitting ratio will create diminishing returns. Studies showed that economically the most optimal splitting ratio is somewhere around 1:40 [11].

A high splitting ratio also means the OLT bandwidth is shared among more ONUs and will lead to less bandwidth per user. To achieve a certain bit error rate (BER) performance, a minimum energy per bit is required to overcome the system noise. Therefore, increasing the bit rate at the OLT will also increase the power (which is the product of bit rate and bit energy) required for transmission. The transmission power is constrained by available laser technology (communication lasers normally have about 0–10-dBm output power) and safety requirements issued by regulatory authorities [12].

2.5 STANDARD COMMERCIAL TMD-PON INFRASTRUCTURE

Figure 2.10 shows the architecture of a standard commercial TDM-PON structure. This general architecture applies to APON, EPON, and G-PON.

In this architecture, an OLT is connected to the ONUs via a 1:32 splitter.[3] The maximum transmission distance covered is usually 10–20 km. Upstream direction traffic from ONUs is carried on 1.3-μm wavelength and downstream traffic from OLTs on 1.49-μm wavelength.

An ONU offers one or more ports for voice connection and client data connections (10/100BASE-T Ethernets). The voice connections can be T1/E1

[3] In some cases, smaller splitting ratios such as 1:8 or 1:16 may be used.

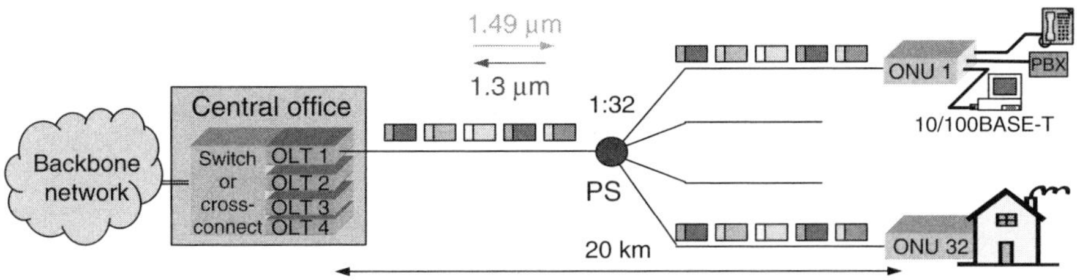

Figure 2.10 Standard commercial TMD-PON architecture.

ports for commercial users and plain old telephone service (POTS) for residential users.

Multiple OLTs in the CO are interconnected with a backbone switch or cross-connect, which also connects them to the backbone network. In a carrier environment, OLTs are usually constructed as line cards that are inserted into a chassis. The chassis can also host the backbone switch/cross-connect and provide the interconnect to the OLTs through a high-speed back plane.

The connection between the OLT and ONU is called the PON section. The signals transported in this section can be encoded and multiplexed in different formats and schemes depending on the PON standard implemented. Nevertheless, beyond the PON section, standard format signals are used for client interface hand-off, switching, and cross-connect. As mentioned in Chap. 1, the most common standard interface used today is the Ethernet interface.

In the PON section, signals from and to different ONUs are frame interleaved. Each frame is identified with a unique ONU ID in the frame header. The directional property of 1:N optical splitter made the downstream link a one-to-many broadcast connection. On the other hand, the upstream direction is a many-to-one connection, i.e. frames sent from all ONUs will arrive at the OLT, but no two ONUs can directly send signals toward each other on the optical layer. This implies that communications between ONUs need to be forwarded to the CO and relayed with the help of the OLT.

2.5.1 OLT and ONU Structures

Figures 2.11 and 2.12 show the generic structure of a TDM-PON OLT and ONU respectively [8]. The PMD (physical layer dependent) layer defines the optical transceiver and the wavelength diplexer at an OLT or ONU.

The medium access control (MAC) layer schedules the right to use the physical medium so that contention for the shared fiber link is avoided among different

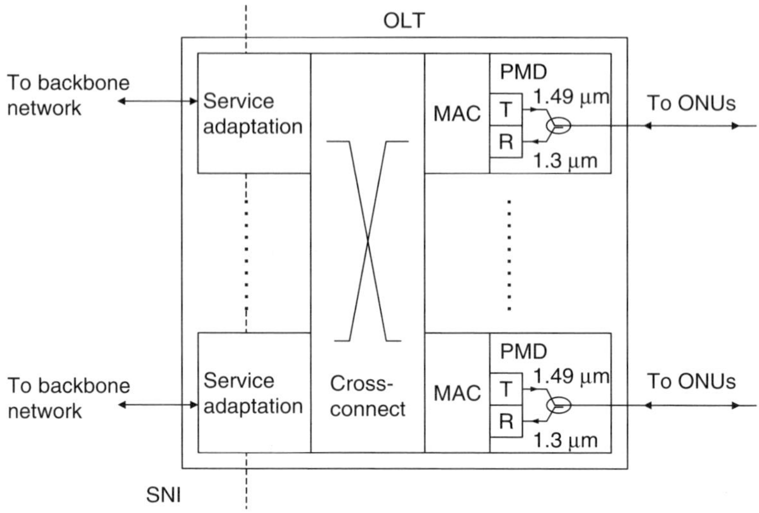

Figure 2.11 Generic structure of a standard TDM-PON OLT. This diagram represents a chassis with multiple OLT cards, which are interconnected through a back plane switch. Each OLT card with its own MAC and PMD layer serves a separate PON.

ONUs. In a PON system, the MAC layer at the OLT serves as the master and the MAC layer at an ONU as a client. The OLT specifies the starting and ending time that a particular ONU is allowed to transmit.

As shown in Fig. 2.11, an OLT may contain multiple MAC and PMD instances so that it may be connected to multiple PON systems. A cross-connect at the OLT provides the interconnection and switching among different PON systems, ONUs, and the backbone network. The service adaptation layer in an OLT provides the translation between the backbone signal formats and PON section signals. The interface from an OLT to the backbone network is called service network interface (SNI).

An ONU provides the connection to the OLT in the PON section through the ONU MAC and PMD (Fig. 2.12). The service adaptation layer in the ONU provides the translation between the signal format required for client equipment connection and the PON signal format. The interface from an ONU to client network equipment is called user network interface (UNI). The Service MUX/DMUX section provides multiplexing function for different client interfaces. Usually, multiple UNIs are available from an ONU for different type of services (e.g. data, voice). Each UNI may support a different signal format and require its own different adaptation service.

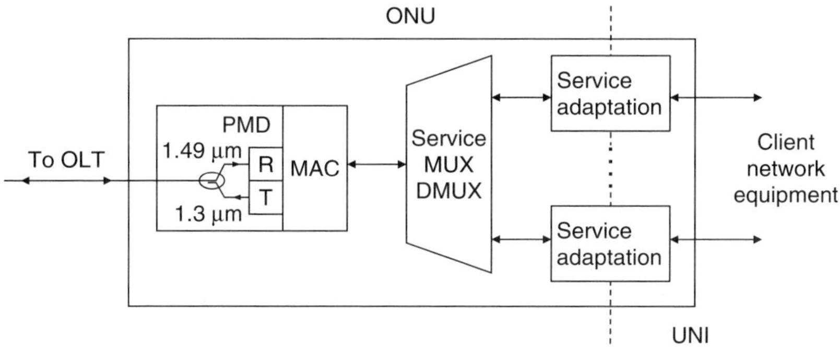

Figure 2.12 Generic structure of a standard TDM-PON ONU.

2.5.2 Burst Mode Operation and Ranging

In the downstream direction, the OLT interleaves frames destined for different ONUs as a continuous stream and broadcast to all ONUs. Each ONU extracts its own frame based on the header address.

In the upstream direction, each ONU has its own optical transmitter to communicate with the OLT. Since there is only one optical receiver at the OLT, ONUs need to take turns to send their data to the OLT. In a conventional transmission system, the transmitter and its receiver always maintain synchronization by transmitting an idle pattern when there is no data to send. In a PON system, when an ONU is not sending upstream data, it has to turn off its transmitter to avoid interfering with other ONUs' upstream transmission. Therefore, in a TDM-PON system, burst mode transmission is used in the upstream direction. Burst mode transmission was also used in early generation coaxial-cable Ethernet LAN systems such as 10BASE-2 and 10BASE-5.

Every time an ONU transmits a signal burst to the OLT, it needs to first send a preamble sequence to the OLT. The OLT uses the preamble as a training sequence to adjust its decision threshold and perform synchronization with the ONU. Moreover, a guard time is reserved between bursts from different ONUs to allow the OLT receiver to recover to its initial state before the next burst. The guard time between bursts adds overheads to bandwidth efficiency and should be minimized in system designs.

To avoid collision between bursts from different ONUs, it is necessary to coordinate or schedule upstream transmissions. Early generations of Ethernet uses a distributed control mechanism called carrier sense multiple access with collision detection (CSMA/CD) to avoid scheduling [9]. However, the CSMA/CD mechanism cannot be used in a PON system for the following reasons: (1) directional power splitters makes carrier sense and collision detection impossible because without using special tricks no ONU can monitor the optical transmission from other ONUs on the same PON; and (2) the data rate and distance covered by a PON system greatly exceeds the limits imposed by the CSMA/CD protocol. The CSMA/CD protocol becomes very inefficient under high bandwidth and long transmission distance [13, Chap. 4].

Scheduling is performed by the OLT MAC layer in a PON system. ONUs are located at different distances from the OLT. Therefore, signals from different ONUs will experience different delays before reaching the OLT. It is therefore important to establish a timing reference between the OLT and an ONU so that after accounting for the fiber delay, when the ONU signal arrives at the OLT, it arrives at precisely the same moment that the OLT intends for the ONU to transmit. The timing reference between the OLT and ONUs is established through the *ranging* process.

Ranging measures the round-trip delay between an ONU and OLT. In order to perform ranging, the OLT sends out a *ranging request* to ONU(s) to be ranged. An ONU participating in ranging then replies with a *ranging response* to the OLT. The OLT measures the round-trip time (RTT) from the ranging response and updates the ONU with this delay. The RTT is stored in the OLT or ONU which uses it to adjust the time that data frames from the ONU should be transmitted. All the ONUs are aligned to a common logical time reference after ranging so that collision does not occur in a PON system.

When ranging request is sent out, the OLT must reserve a time period called ranging window for unranged ONUs to respond. The size of the ranging window depends on the maximum differential delays between the closest ONU and the furthest ONU. Optical signal delay in 1 km of fiber is 5 μs. Therefore, for 20 km of differential distances between ONUs, an RTT difference of 200 μs needs to be reserved in the ranging window. Most PON standards specify a maximum physical distance of 20 km from an ONU to the OLT [8, 9, Clause 60, 14]. It should be realized that if the upper bound and lower bound of ONU separations are known to the OLT (e.g. through management provision), then instead of reserving a ranging window covering the maximum allowed separation between an ONU and an OLT, the size of ranging window can be reduced to cover only the maximum differential distance among ONUs. This reduces the ranging overhead and improves the overall bandwidth efficiency.

Ranging is usually done at the time an ONU joins a PON. The OLT periodically broadcasts ranging requests for ONU discovery. A new ONU detects the ranging request and responses to the OLT within the reserved ranging window after the ranging request. If multiple ONUs attempt to join the PON at the same time, collision may occur. Collisions in discovery are resolved by ONUs backing off with a random delay.

During operation, the ONU and OLT may continuously monitor the fluctuation of RTT due to changes such as temperature fluctuation, and perform fine adjustment by updating the RTT register value.

2.5.3 C-Band Analog CATV Signal Overlay

In addition to broadband data delivery, TV program distribution was envisioned as a primary PON service in the very beginning. As pointed out in Chap. 1, analog TV services have been distributed using the 1550-μm wavelength in an HFC system for a long time [5]. Despite its limitations, one way analog broadcast TV is a simple and efficient method to stream video services to a large group of users.

A quick and easy method to offer TV services on a TDM-PON is to directly broadcast analog TV signal to end users on the 1.55-μm wavelength (Fig. 2.13) using a wavelength coupler [16]. As indicated in Fig. 2.13, the 1.55-μm wavelength can be amplified by an EDFA at the CO for broadcasting to multiple PONs to achieve better sharing of the analog TV resources.

The analog TV signal is peeled off at the ONU and converted into the regular RF signal on a coaxial cable using a set-top box (STB), which provides a simple OE (Optical to Electrical) conversion function. In reality, the analog conversion function is built into the ONU which offers a standard 75-Ω coaxial cable client interface. Furthermore, a 1.3/1.49/1.55-μm wavelength triplexer is used instead of two cascaded diplexers as shown in Fig. 2.13.

One potential issue in such configuration is the degradation of the subcarrier analog TV signals on the 1.55-μm wavelength by the 1.49-μm wavelength downstream digital signal, which acts a Raman pump to the 1.55-μm analog signal [17]. Since the downstream wavelength is modulated, it degrades the carrier-to-noise ratio (CNR) of the analog signals especially for low-frequency subcarrier channels.

In the past few years, digital and on-demand IPTV services are fast growing and replacing traditional analog TV services. The current trend is to take the advantage of a converged IP network and low-cost Ethernet technology to encode TV signals as MPEG/IP/Ethernet packets. Such a trend speeds up the unification of TV distribution network with data networks, which not only reduces the total capital expenditure (CAPEX) but also simplifies network management.

In the United States, MSOs are busy retiring analog services. With proper QoS (quality of service) control, digital IPTV services can be more cost-effectively delivered through the data channel. The author expects the 1.55-μm analog overlay technique to be less and less popular because of the following reasons:

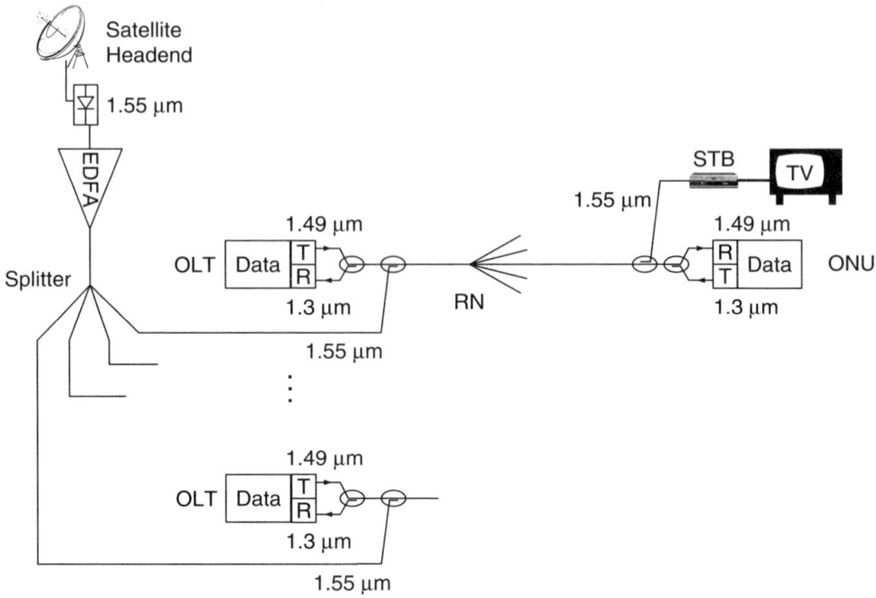

Figure 2.13 Overlaying analog broadcast TV services on a TDM-PON using the 1.55-μm wavelength.

1. The triplexer and analog RF receiver increase the cost of ONU.
2. The overlaid analog service creates additional management complexities. MSOs may be able to take the advantage of their existing legacy analog HFC system and the know-how they have developed for that. However, for conventional telecom operators entering the triple-play market, it is much better to leapfrog with digital IPTV technology as they do not have the legacy burden of analog TV distribution systems.
3. Regulatory bodies around the world are now phasing out analog TV broadcast and replacing them with digital services.

2.5.4 Security Concerns in Power-Splitting PON

One important concern of power-splitting PON is the broadcast nature of the downstream channel, which makes it easy to eavesdrop downstream communication signals by malicious users. It should be realized that because of the directional nature of the optical star coupler at the remote node, an ONU cannot listen to the upstream transmission from another ONU.

The biggest security exposure is in the ranging process when the OLT broadcasts the serial number and ID of the ranged ONU. A malicious user can make use of this information for spoofing. This problem can be avoided through an authentication process during which the ONU is verified by a password known only to the OLT (e.g. through management provision).

To improve security, the ITU-T G.983.1 standard (Sect. 8.3.5.6, [8]) defines a churning procedure to scramble the data for downstream connections with a key established between the ONU and OLT. The encryption key is sent from an ONU to the OLT with a defined protocol. (Note that the encryption key is only sent in the secure upstream link.) To further enhance the security, the encryption key can be periodically updated. Churning provides some level of security in the physical layer. When security concern becomes very important, encryption should be used at the application layer. The IEEE 802.3ah EPON standard does not include transmission encryption. Nevertheless, many ASIC vendors provide their own encryption mechanisms on EPON chip sets. Depending on the security and interoperability requirements, users can decide to enable or disable these vendor-specific encryption mechanisms.

2.6 APON/BPON AND G-PON

APON/BPON and G-PON are standardized by the ITU-T Study Group 15 (SG15). APON (ATM-PON) and BPON (Broadband PON) are different aliases of the TDM-PON architecture based on the ITU-T G.983 series standards [8, 16, 18]. While the name BPON serves its marketing purpose, APON clearly conveys that ATM frames are used for transport in the ITU-T G.983 standards. For this reason, we will simply use APON to refer to this class of PON designs. G-PON stands for gigabit-capable PON and is covered by the ITU-T G.984 series standards [14, 15]. It is the next generation PON technology developed by ITU-T after APON.

Both APON and G-PON defined line rates as multiples of 8 kHz [8, 14], the basic SONET/SDH frame repetition rate. As a matter of fact, the OLT distribute the 8-kHz clock timing from OLT to ONUs. This makes it easier to support TDM services on APON and GPON.

2.6.1 ATM-PON and ITU-T G.983

2.6.1.1 APON System Description

The work on APON was started by the full service access network (FSAN) consortium [19] and later transferred to ITU-T SG15 [20] as the G.983 standards.

APON systems were mostly deployed in North America by RBOCs for their FTTP projects. Many ideas covered in the G.983 standards were carried over to the G.984 G-PON standards.

The original G.983.1 standard published in 1998 defined 155.52-Mbps and 622.08-Mbps data rates. A newer version of the standard published in 2005 [8] added 1244.16-Mbps downstream transmission rate. APON vendors can choose to implement symmetric or asymmetric downstream and upstream transmission rates. Table 2.1 shows the possible combinations of downstream and upstream data rates for an APON system.

In addition to the one-fiber wavelength diplex solution explained earlier, both the G.983.1 and G.984.1 standards specify a two-fiber solution with dedicated upstream and downstream transmission fibers. The 1.3-μm wavelength is used in both directions in the two-fiber solution. However, to the knowledge of the author, no system has been deployed with the two-fiber solution.

All ITU PON standards feature three classes of optical transmission layer designs with different ODN (Optical Distribution Network) attenuations between ONU and OLT. The three classes are specified in ITU-T G.982 [21] as:

- Class A: 5–20 dB
- Class B: 10–25 dB
- Class C: 15–30 dB

Class C design is a very demanding power budget requirement for a passive fiber plant. For practical implementation yield and cost reasons, Class B+ with 28-dB attenuation was later introduced by most PON transceiver vendors (Chap. 4).

ITU-T G.983.1 specifies the reference architecture, transceiver characteristics, transport frame structures, and ranging functions in APON [8]. APON signals are transported in time slots. Each time slot contains either an ATM cell or a PLOAM (physical layer OAM) cell. PLOAM cells are used to carry physical layer management information such as protocol messages for ranging, churning

Table 2.1

APON downstream/upstream bit-rate combinations

Downstream	Upstream
1. 155.52 Mbps	155.52 Mbps
2. 622.08 Mbps	155.52 Mbps
3. 622.08 Mbps	622.08 Mbps
4. 1,244.16 Mbps	155.52 Mbps
5. 1,244.16 Mbps	622.08 Mbps

key request and update, time slot requests from ONUs to OLT, assignments of upstream time slots to various ONUs by the OLT, system error, and performance monitoring reports, etc. The definition of PLOAM cell contents can be found in Sect. 8 of G.983.1.

The downstream time slots are exact 53-octet (byte) long ATM or PLOAM cells. A PLOAM cell is inserted every 28 time slots (or 27 ATM cells). At 155.52-Mbps speed, APON designates 56 time slots (with 54 ATM cells and 2 PLOAM cells) as a downstream frame. Each 155.52-Mbps upstream frame contains 53 time slots of 56 octets. An upstream time slots contains a 3-octet overhead in addition to an ATM or PLOAM cell. Besides, each upstream time slot may be optionally divided into multiple mini-slots at the request of OLT if necessary. Details of mini-slot definitions can be found in ITU-T G.983.4 [22], which covers the dynamic bandwidth allocation (DBA) mechanisms for APON systems. Figure 2.14 illustrates APON frames at 155.52 Mbps.

Frame structures at 622.08 Mbps and 1244.16 Mbps are similar except that the numbers of time slots per frame are multiplied by 4 and 8

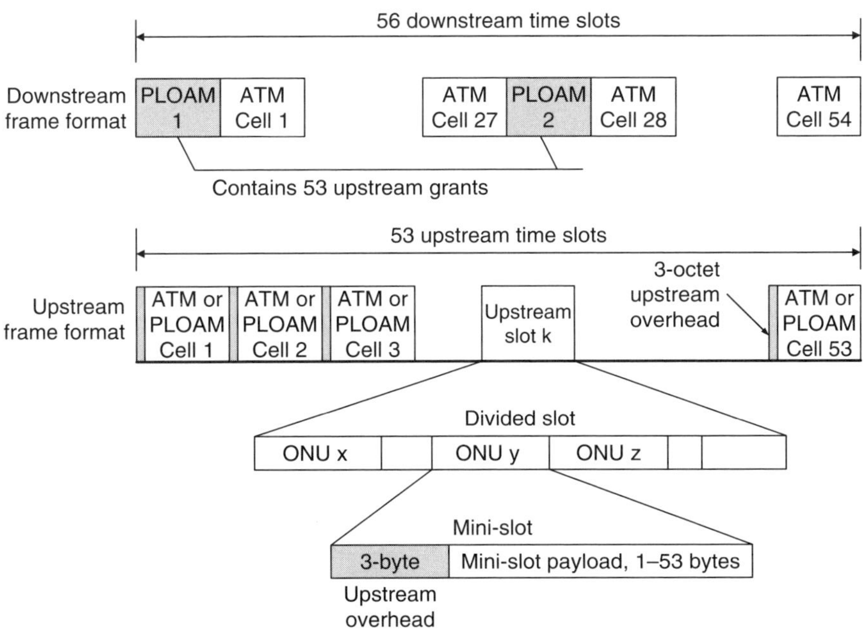

Figure 2.14 Downstream and upstream APON frame formats at 155.52-Mbps speed. For 622.08-Mbps and 1244.16-Mbps speed, the numbers of time slots are simply multiplied by 4 and 8 to the numbers shown in the above diagrams.

respectively. Transmission of an ATM cell, PLOAM cell, or divided slot in the upstream direction is controlled by OLT using downstream PLOAM cells. G.983 requires a minimum of one PLOAM cell per ONU in every 100 ms [8].

The 3-octet upstream overhead contains the guard time, preamble, and the cell-start delimiter. APON specifies a minimum guard time of 4 bits. The contents of these fields are programmable and defined by OLT in the "Upstream_overhead" message, which is broadcast through downstream PLOAM cells.

2.6.1.2 Services in APON

Connections between ONU and OLT are established as ATM virtual circuits in an APON system. ATM services are connection-oriented. Each virtual circuit is identified by a virtual path identifier (VPI) and a virtual channel identifier (VCI) which are embedded in the cells comprising its data flow. VPI and VCI are indices providing different levels of ATM signal multiplexing and switching granularity. Multiple virtual circuits (VCs) can exist within a single virtual path (VP). An ATM connection is identified by its VPI/VCI pair. Figure 2.15 illustrates the idea of ATM signal switching.

Services in APON are mapped to ATM virtual circuits through the ATM adaptation layer (AAL). ATM cells include information cells, signaling cells, OAM cells, unassigned cells, and cells used for cell-rate decoupling [23]. AAL implements different levels of quality of service (QoS). ITU-T G.983.2 includes three different ATM adaptations for APON [18]: AAL-1, AAL-2, and AAL-5. AAL-1 provides adaptation functions for time-sensitive, constant bit-rate, and connection-oriented services such as T1 and E1 circuits. AAL-2 is used for variable-bit-rate connection-oriented services such as streaming audio and video signals and AAL-5 is used for connectionless data services such as TCP/IP applications.

The OLT and ONU[4] as a whole can function as a VP or VC switch. Depending on the implementation, an ONU can cross-connect traffic at the VP level or VC level.

2.6.1.3 APON Control and Management

APON control and management functions are specified in the ITU-T G.983.2 standard [18]. An OLT manages its ONUs through a master–slave interface

[4] The G.983.2 standard refers to ONU as optical network terminator (ONT). The G.983.1 standard defines the ONU as the PON optical termination unit at the customer end in an FTTH system and ONT as the remote termination unit for FTTB systems. However, the use of these two terms seems to be quite arbitrary and interchangeable in the G.983 series standards. In this book, we do not distinguish between ONU and ONT.

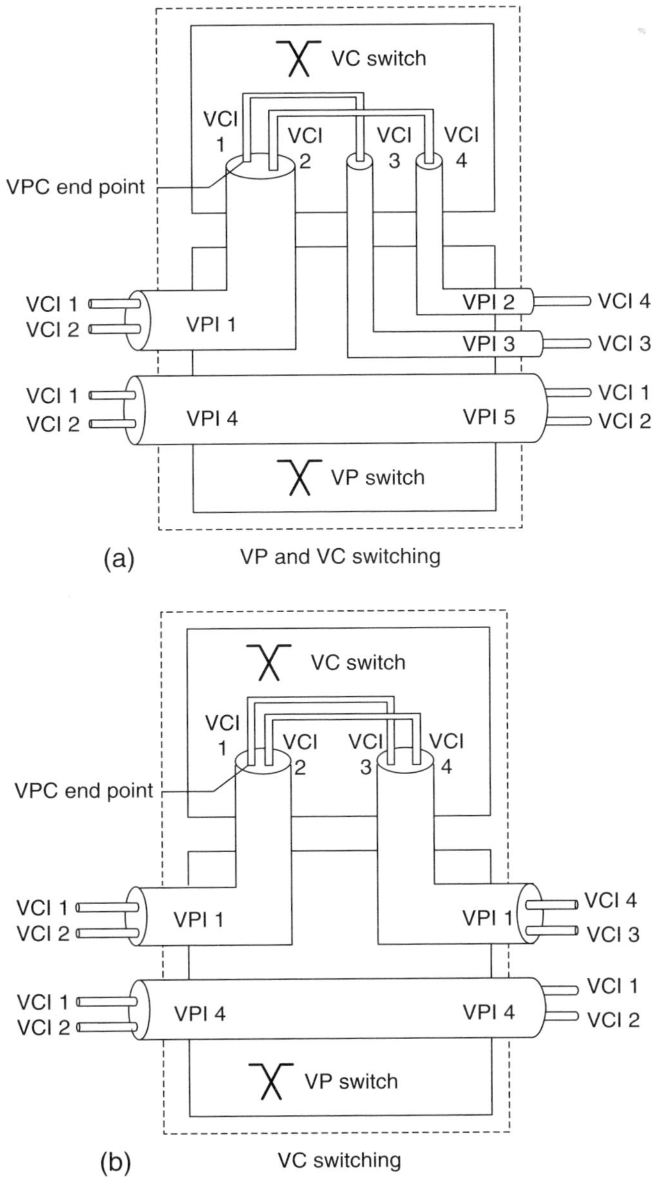

Figure 2.15 ATM switching examples: (a) VP and VC switching and (b) VC switching.

called ONT management and control interface (OMCI), with the OLT function as the master and ONU as slaves.

An ATM connection called ONT management and control channel (OMCC) provides the communication channel for OMCI. An OMCC is established by activating a pair of VPI/VCI using the particular PLOAM messages specified in ITU-T G.983.1. ITU-T G.983.2 requires each ONU to use a different VPI for the OMCC channel.

The G.983.2 standard specifies the management information bases (MIBs) with different parameters required for managing various APON objects and their behaviors. The OAM functions and parameters described in this standard are also applicable to the APON successor G-PON, except that different implementations of OMCI and OMCC are adopted.

2.6.2 Collision Resolution in APON/G-PON

In normal operation, a TDM-PON performs scheduled upstream transmission coordinated by the OLT and is therefore collision-free. The only exception is during the ranging process when multiple ONUs may respond to the broadcast ranging request at the same time. A collision will occur.

To resolve this contention, an APON system uses a binary tree mechanism specified in G.983.1 (Sect. 8.4.4, [18]). After detecting a ranging cell collision, an OLT will send a *Serial_number_mask* message followed by a ranging grant to allow ONUs whose serial number matches the mask to transmit a ranging cell. The size of the *Serial_number_mask* is increased by one bit at a time until only one ONU is transmitting a ranging cell. This resolves the contention and allows ONUs to be ranged one at a time. This mechanism may also help to avoid overloading the optical input of the OLT during ONU power setup.

In a G-PON system, ONUs use a random delay method to avoid collision during ONU initial registration process. The OLT may or may not enable the *Serial_number_mask* mechanism in G-PON systems as specified in G.984.3 (Sect. 10, [24]).

2.6.3 Wavelength Overlay in APON/G-PON

The original APON system only specified 1.3-μm wavelength for upstream and 1.5-μm wavelength for downstream transmissions. To leverage the vast bandwidth in optical fiber and make the value out of capital-intensive fiber plants, ITU-T G.983.3 [16] specified coarse WDM overlay on the original

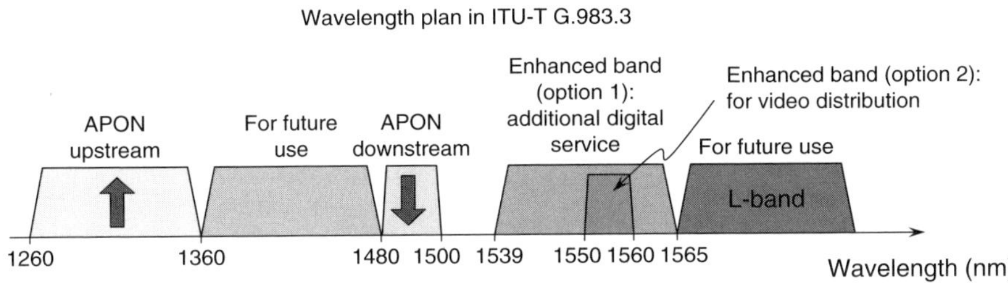

Figure 2.16 Wavelength allocation plan in ITU-T G.983.3.

APON system. This wavelength plan has been adopted by most PON manufactures to transport different services on the same physical infrastructure.

Figure 2.16 illustrates the wavelength allocation plan specified in ITU-T G.983.3. This ITU recommendation specifies the C-band as enhanced band for carrying either analog TV signals or additional digital services such as SONET/SDH links. Two more bands, (1) 1360–1480 nm and (2) 1565 nm and beyond, are reserved for future studies. Additional wavelength multiplexers are required at ONU and OLT in order to make use of the enhancement bands specified in G.983.3. These additional wavelength multiplexers add losses and cross talks to the system. The G.983.3 standard also defines the loss and isolation requirements of these additional wavelength multiplexers.

2.6.4 G-PON and ITU-T G.984

As explained before, ITU-T G.983 was based on the ATM technology. Unfortunately, ATM did not live up to the expectation of becoming the universal network protocol to carry different applications. Instead, Ethernet and IP successfully evolved into that role. In ITU-T G.983, the OLT and ONU as a whole function as VP and VC switches. This requires APON ONU and OLT to implement ATM switching capabilities. The complicated adaptation model and QoS support made ATM switches costly. Moreover, it is necessary to implement translations between ATM and the protocols used at UNI and SNI. These requirements increase the system cost and complexity, and hinder the growth of APON systems in a fast-evolving broadband communication world.

To better cope with the changes in communication technologies and meet fast-growing demand, ITU-T created the G.984 series standards for PONs with Gigabit capabilities, or G-PON [14, 15, 24].

2.6.4.1 G-PON Architecture

ITU-T G.984.1 gives a high-level overview of G-PON components and reference structure. G-PON PMD layer or transceiver requirements are covered by the ITU-TG.984.2 standard. Similar to APON, G-PON also defined single-fiber and dual-fiber PMDs. The bit rates defined in G.984 are:

- Downstream: 1244.16 Mbps/2488.32 Mbps
- Upstream: 155.52 Mbps/622.08 Mbps/1244.16 Mbps/2488.32 Mbps

Appendix I lists the optical layer characteristics for gigabit speed interfaces specified in G.984.2. At the time of writing, characteristics of 2488.32 Mbps upstream transmission link are yet to be studied and finalized.

As the bit rate advances into the gigabit regime, PON optical layer starts to become challenging. First, to cover the full 20-km transmission distance, multi-longitudinal-mode (MLM) lasers cannot be used at ONU any more in order to avoid excessive dispersion penalty. Second, to cover the loss budget requirements for Class B (10–25 dB) and Class C (15–30 dB) fiber plants, more sensitive avalanche photo-diodes (APDs) are required instead of the lower cost PIN receivers [25]. Without proper protection circuits, APDs are susceptible to damages due to avalanche breakdown if the inputs optical power becomes too high.

With the same fiber plant loss budget as in APON, to support the high bit rates, higher power transmitters are used in G-PON to meet the power budget requirements. This also implies that G-PON receivers need to handle higher receiver overload powers and therefore larger dynamic ranges. To ease the requirements and implementation of the upstream OLT burst mode receiver, G-PON has specified a power-leveling mechanism for "dynamic" power control (Sect. 8.3, [15]).

In the power-leveling mechanism, the OLT tries to balance the power it received from different ONUs by instructing ONUs to increase or decrease the launched power. Consequently, an ONU which is closer to the OLT and seeing less loss, will launch at a smaller power than an ONU which is further apart and experiencing more loss. Such concepts of power-leveling or power control have long existed in cellular networks to deal with the near–far cross talk effect and save cellular device battery power.

2.6.4.2 G-PON Transmission Convergence Layer

The main function of the G-PON transmission convergence (GTC) layer is to provide transport multiplexing between the OLT and ONUs [24]. Other functions provided by the GTC layer include:

- Adaptation of client layer signal protocols
- Physical layer OAM (PLOAM) functions
- Interface for dynamic bandwidth allocation (DBA)
- ONU ranging and registration
- Forward error correction (optional)
- Downstream data encryption (optional)
- Communication channel for the OMCI

GTC functions are realized through transmission containers or T-CONTs. Each T-CONT, which is identified by an allocation ID (Alloc-ID) assigned by the OLT, represents a logical communication link between the OLT and an ONU. A single ONU may be assigned with one or more T-CONTs as shown in Fig. 2.17. Five different types of T-CONTs with different QoS attributes have been defined in the ITU-T G.983.4 standards. G-PON standards defined two different operation modes, ATM and GEM (G-PON encapsulation mode). The GEM mode encapsulation is similar to generic framing procedure (GFP) [26]. A T-CONT can be either ATM- or GEM-based. An ATM-based T-CONT multiplexes virtual circuits identified by VPI and VCI, whereas a GEM-based T-CONT contains connections identified by 12-bit port numbers. These are illustrated in Fig. 2.18.

The protocol stack of GTC is shown in Fig. 2.19. Although the physical reach and splitting ratio are defined as 20 km and 1:32 respectively in G.984.2, the GTC layer has

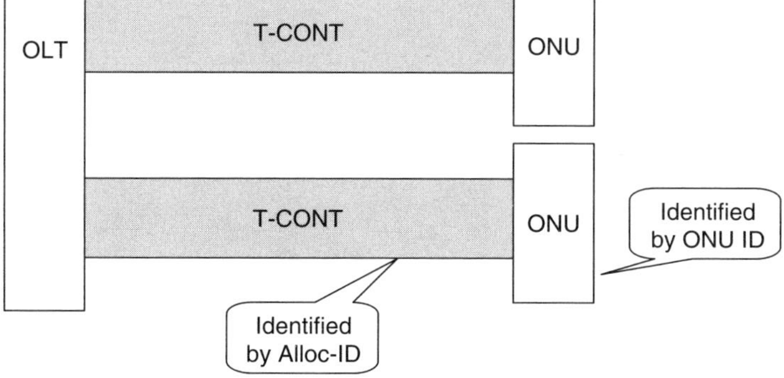

Figure 2.17 A T-CONT represent a logical link between the OLT and an ONU.

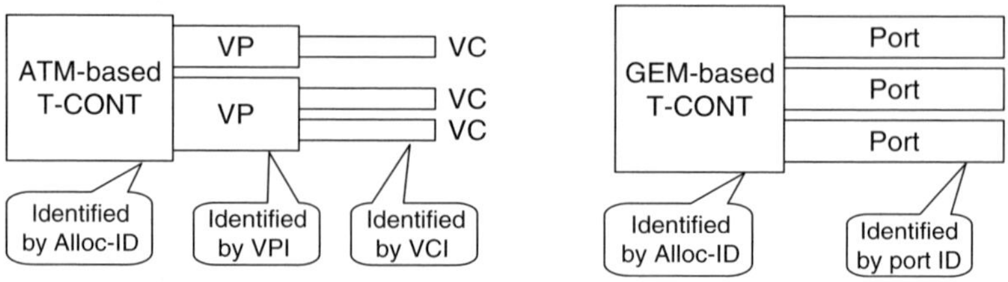

Figure 2.18 ATM-based T-CONT vs GEM-based T-CONT.

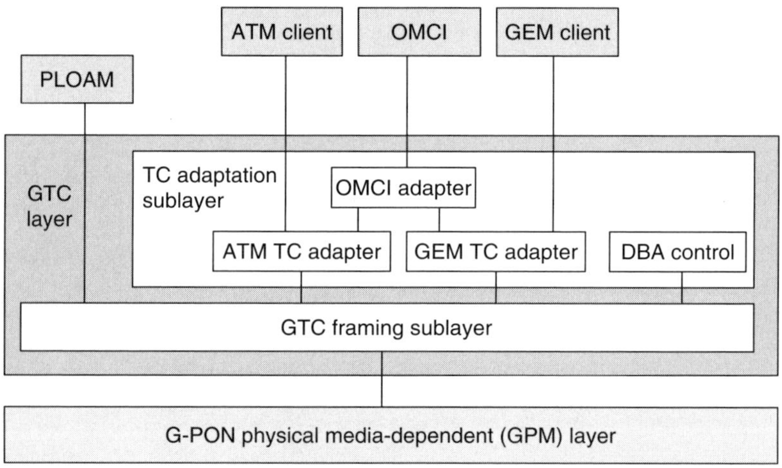

Figure 2.19 Protocol stack of the G-PON transmission convergence (GTC) layer (from [24], reproduced with kind permission from ITU).

specified a maximum logical reach of 60 km and splitting ratio support of 128, in anticipation for future technology developments. Such high splitting ratio and long-distance coverage are being studied under a series of super PON projects around the world. ONUs and OLTs may support either ATM-based or GEM-based T-CONT or a mix of both (dual-mode). A G-PON using ATM-based T-CONT is however not backward compatible with APONs. Therefore, in reality, most of the G-PONs only implement GEM-based T-CONT. In the rest of the discussion, we will mainly focus on GEM.

The GTC layer supports the transports of 8-kHz clock from the OLT to ONUs and an additional 1-kHz reference signal from the OLT to ONUs using a control channel.

2.6.4.2.1 GTC Framing

The GTC framing sublayer offers multiplexing capabilities and embedded OAM functions for upstream time slot grants and dynamic bandwidth allocation (DBA). Embedded OAM is implemented in the GTC frame header.

2.6.4.2.1.1 GTC Downstream Framing

Each GTC downstream frame is 125 μs long, which contains a downstream physical control block (PCBd) and a payload section as shown in Fig. 2.20.

The concepts of G-PON media access control is depicted in Fig. 2.21. The PCBd contains media access control information in the *upstream* (US) *bandwidth* (BW) *map*. As illustrated in the figure, the OLT specifies the start time and end time that each T-CONT can use to transmit upstream data using pointers in the US BW map. The pointers are given in units of bytes. This allows the upstream bandwidth to be controlled with 64-kb/s granularity. However, the standard allows vendors to implement larger granularities.

Figure 2.22 shows the details of downstream frame formats. The PCBd header consists of a fixed part and a variable part. The fixed part contains the Physical Sync field, Ident field, and a PLOAM field. These fields are protected by a 1-byte bit-interleaved parity check. The 4-byte unscrambled physical synchronization pattern indicates the beginning of a downstream frame. The 4-byte Ident field indicates whether FEC is used. In addition, it also implements a 30-bit wrap-around superframe counter, which can be used to provide a low-rate synchronization reference signal. The 13-byte PLOAM field in PCBd is used to communicate the physical layer OAM messages to ONUs. PLOAM functions include ONU registration and deregistration, ranging, power leveling, cryptographic key update, physical layer error reports, etc.

The variable part of the PCBd header contains duplicated payload length downstream (PLend) descriptor, which specifies the length of the upstream bandwidth map and that of the ATM partition in the T-CONT. As mentioned before, each ONU may be configured with multiple T-CONTs. The US BW map specifies the upstream bandwidth allocation with access entries. Each 8-byte access entry in the US BW map contains the Alloc-ID of a T-CONT, the start time and end time to transmit that T-CONT in the upstream direction and a 12-bit flag indicating how the allocation should be used. The rest of the downstream frame is downstream payload. Since each downstream frame has a duration of 125 μs, the length of downstream frames are different for 1.24416-Gb/s and 2.48832-Gb/s speeds, which are 19,440 and 38,880 bytes respectively. The PCBd block is the same for both speeds.

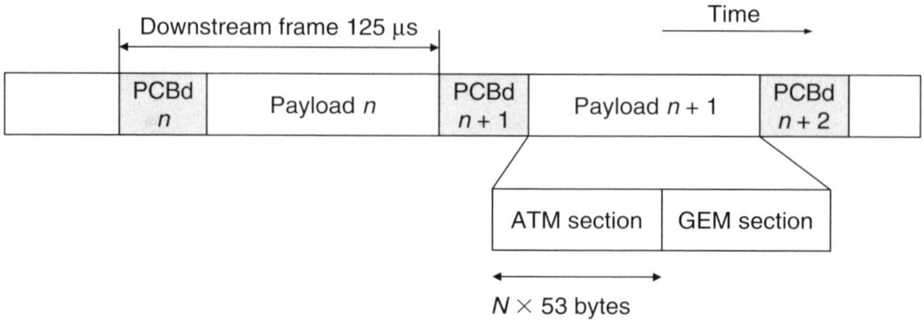

Figure 2.20 GTC downstream signal consists of 125-μs frames with a PCBd header and a payload section.

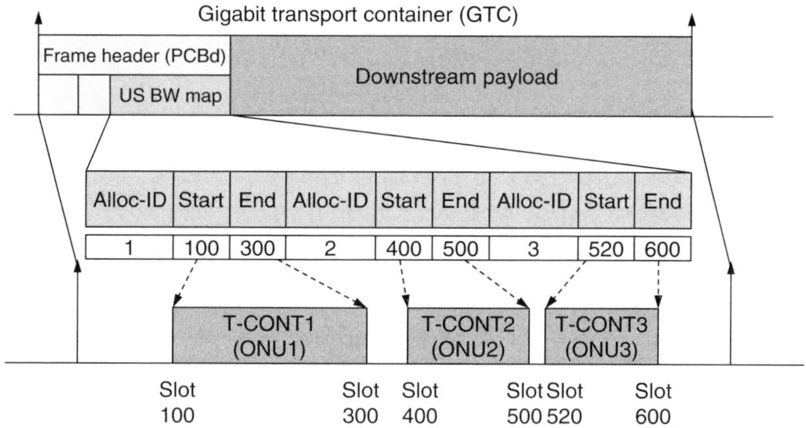

Figure 2.21 GTC downstream frame and media access control concept (from [24], reproduced with kind permission from ITU).

2.6.4.2.1.2 GTC Upstream Framing

The GTC upstream transmission consists of upstream virtual frames of 125-μs duration as shown in Fig. 2.23. Upstream virtual frames are made up of bursts from different ONTs. Each burst starts with a physical layer overhead upstream (PLOu).

The PLOu begins with a preamble to help the burst mode receiver at the OLT to synchronize with the ONT (Fig. 2.24) transmitter. A delimiter following the preamble signifies the beginning of an upstream burst. As in APON, the length and format of preamble and delimiter are specified by the OLT using downstream PLOAM messages. An indication field (Ind) in PLOu provides real-time ONU status reports to the OLT.

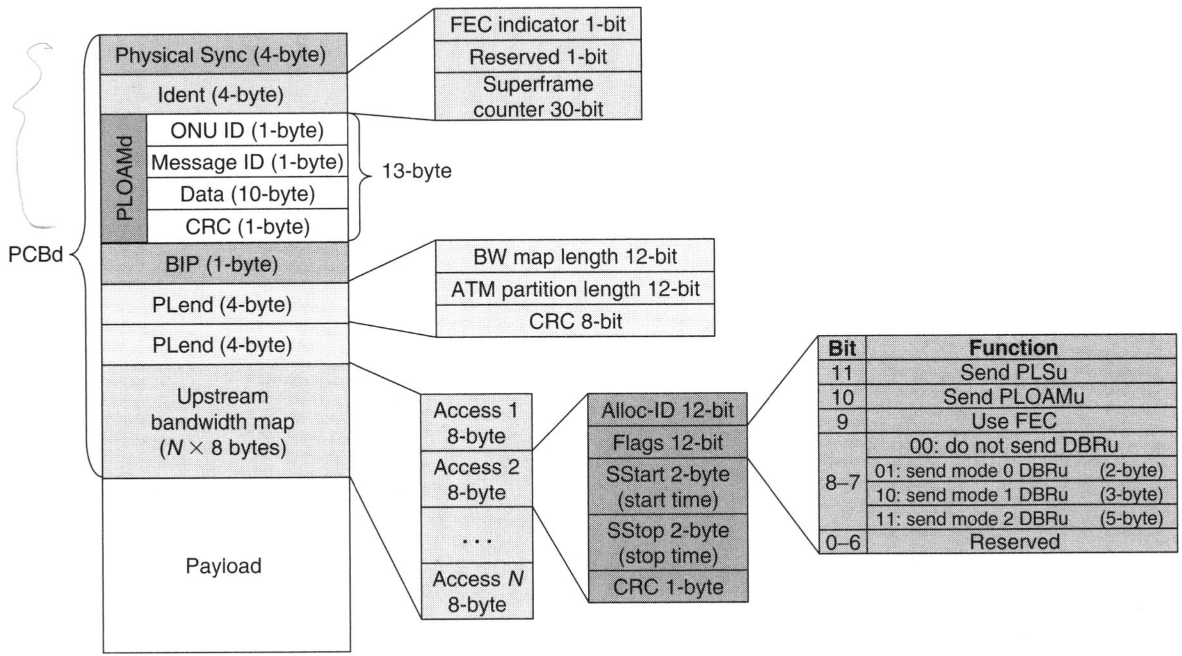

Figure 2.22 GTC downstream frame formats.

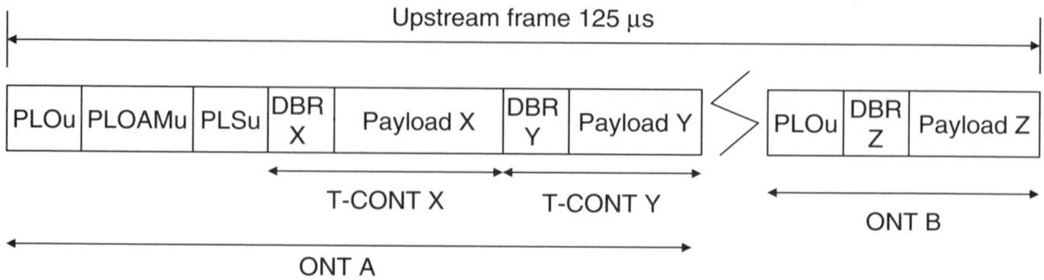

Figure 2.23 GTC upstream framing: each ONT starts the upstream transmission with PLOu. An ONT assigned with two Alloc-ID in two consecutive upstream allocations only needs to transmit PLOu once.

As mentioned earlier, a single ONU may be assigned multiple T-CONTs. If one ONU is allocated contiguous time slots for its multiple T-CONTs with different Alloc-IDs, PLOu only needs to be transmitted once. This is shown in Fig. 2.23 for ONT A.

Following the PLOu field, there are three optional overhead fields in each burst:

1. Physical layer operation, administration, and management upstream (PLOAMu)
2. Power leveling sequence upstream (PLSu)
3. Dynamic bandwidth report upstream (DBRu)

Transmissions of these fields are dictated by the OLT through flags in the US BW map in PCBd. When requested by the OLT, the 120-byte PLSu field is sent by ONU for power measurement purposes.

A DBRu field is tied to each T-CONT for reporting the upstream traffic status of the associated T-CONT. The DBRu field contains the DBA report which is used to indicate the upstream queue length for dynamic bandwidth allocation. G-PON supports status-reporting (SR) and nonstatus-reporting (nSR) DBA mechanisms. In nSR DBA, the OLT monitors the upstream traffic volume and there is no protocol required for ONU to communicate queue status to the OLT. In SR DBA, an ONU informs OLT the status of traffic waiting in the queuing buffer.

There are three ways that ONUs may communicate the status of pending traffic to the OLT:

1. Status indication bits in PLOu (as shown in Fig. 2.24). Status indication gives OLT a quick and simple way of knowing the type of upstream traffic waiting, without much details.

2. Piggyback DBA report in the DBRu. The DBA report have three modes (Mode 0, Mode 1, and Mode 2) corresponding to DBA field lengths of 1, 2, or 4 bytes (i.e. 2-, 3-, or 5-byte DBRu) in Fig. 2.24. Transmission of DBRu and its format are specified by the OLT using bits 8 and 7 of the flags in the US BW map (Fig. 2.22). Piggyback DBA in DBRu allows ONUs to continuously update the traffic of a specific T-CONT. The DBA report contains the number of ATM cells or GEM blocks (which are 48-byte in length) waiting in the upstream buffer.
3. DBA upstream payload. An ONU can send a dedicated whole report of traffic status on any or all its T-CONT in the upstream payload. Details of DBA upstream payload can be found in ITU-T G.984.3.

A G-PON system does not have to support all the reporting modes. Reporting capabilities are negotiated and agreed upon between ONU and OLTs using OMCI handshakes.

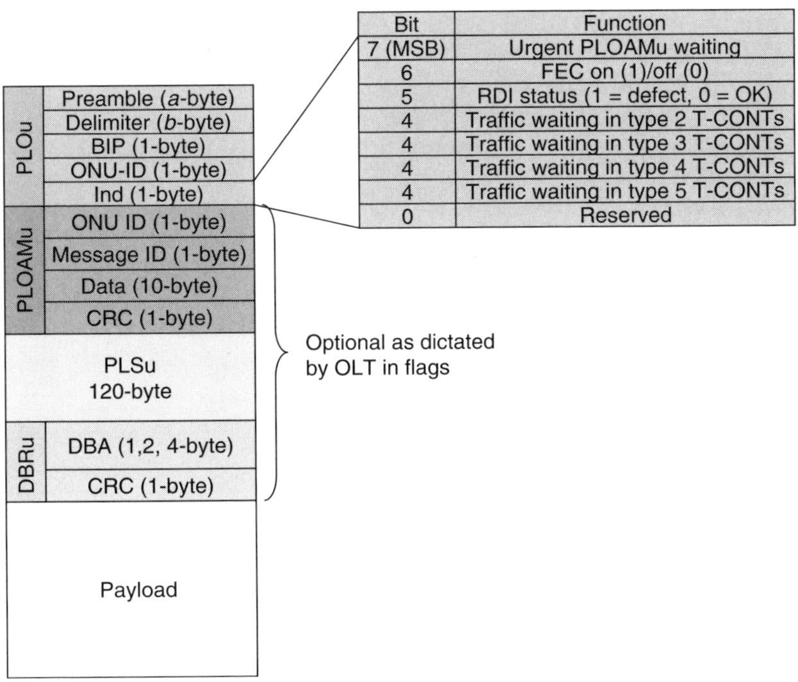

Figure 2.24 GTC upstream frame format.

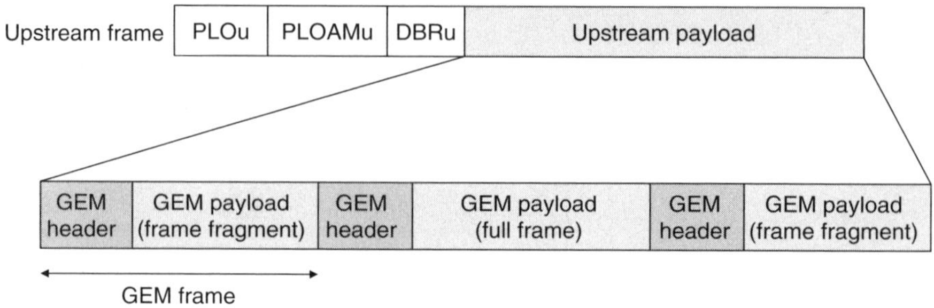

Figure 2.25 GEM frames in upstream payload (from [24], reproduced with kind permission from ITU).

2.6.4.2.1.3 GEM Encapsulation

In the downstream direction, GEM frames are carried in the GEM section of downstream payloads (Fig. 2.20). In the upstream direction, GEM frames are carried in upstream payloads as shown in Fig. 2.25.

GEM frame encapsulation serves two functions: (1) multiplexing of GEM ports and (2) payload data fragmentation. The format of GEM encapsulation is shown in Fig. 2.26. The GEM header contains a 12-bit payload length indicator (PLI) which

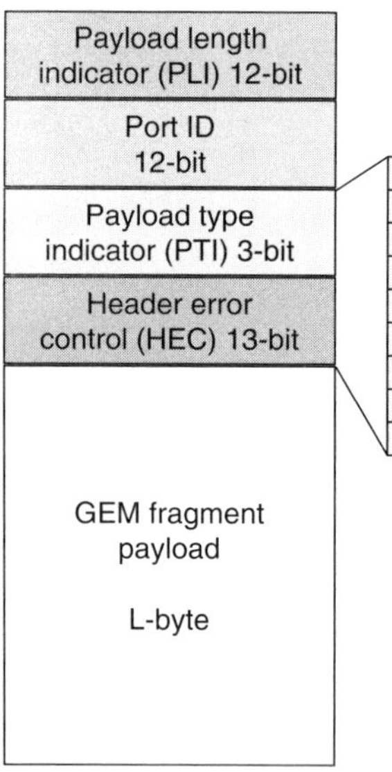

PTI code	Meaning	
000	User data fragment, no congestion, not EOF	
001	User data fragment, no congestion, EOF	
010	User data fragment, congestion, not EOF	FFS
011	User data fragment, congestion, EOF	FFS
100	GEM OAM	
101	Reserved	
110	Reserved	
111	Reserved	

EOF = End of frame
FFS = For future study

Figure 2.26 GEM encapsulation formats.

specifies the GEM payload length in bytes. This allows a maximum payload length to be 4095 bytes. Any user-payload structure longer than 4095 bytes must be fragmented. The 3-bit payload type indicator (PTI) indicates whether the payload contains user data frames or GEM OAM frames. It also indicates whether the user data frame in the payload is the last fragment. The 12-bit port ID allows 4096 port numbers for traffic multiplexing.

The header error control (HEC) field uses a BCH code for header signal integrity protection. This field is also used to maintain GEM frame synchronization.

Figure 2.27 shows the fragmentation process in GEM encapsulation. Fragments of a user data frame must be transmitted contiguously. An advantage of the GEM fragmentation process is the convenience to support high-priority traffic. Urgent frames with high priorities can be inserted at the beginning of each GTC frame (Fig. 2.28) with 125-μs latency corresponding to the frame period. This makes it very easy to support TDM services with deterministic data rate in a G-PON system.

To decouple user data rate, an idle GEM with all zero payload has also been defined. When there is no user data to send to, idle GEM frames are sent to fill up empty slots and maintain receiver synchronization.

2.6.4.3 Forward Error Correction (FEC) in G-PON

To improve optical link budgets in G-PON, ITU-T G.984.3 defines an optional FEC capability using RS(255, 239) code in GTC framing. The same code is also used in the ITU-T G.709 optical transport network (OTN) standard [27]. FEC gives about 3–4-dB extra margin in optical budget but reduces the user data throughput by 6% because the symbol rate and GTC frame length are both

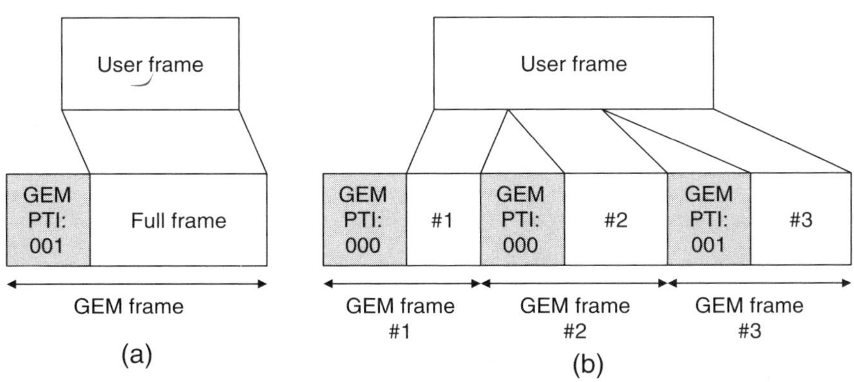

Figure 2.27 Mapping and fragmentation of user data frames into GEM payload (from [24], reprinted with kind permission from ITU).

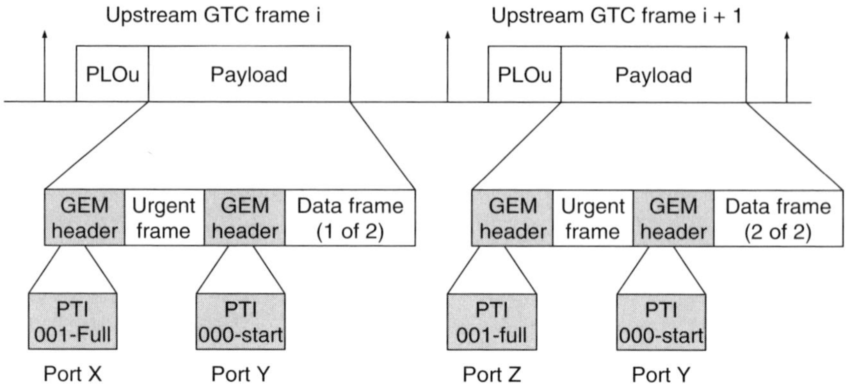

Figure 2.28 Multiplexing of urgent data using GEM fragmentation process (from [24], reprinted with kind permission from ITU).

kept unchanged when FEC is turned on. Therefore, the effective payload data rate has to be reduced in order to accommodate FEC overheads.

2.6.5 APON/G-PON Protection Switching

The reference models for APON/G-PON resiliency and redundancy are specified in the ITU-T G.983.5 standard [28]. ITU-T G.983.6 [29] covers the control and management interface to support APON/G-PON with protection features.

PON protection switching details will be covered in more detail in Chap. 6.

2.7 EPON

EPON is a new addition to the Ethernet family. The work of EPON was started in March 2001 by the IEEE 802.3ah study group and finished in June 2004 [30].

2.7.1 Ethernet Layering Architecture and EPON

Ethernet covers the physical layer and data link layer of the open system interconnect (OSI) reference model. Figure 2.29 shows a comparison of the layering model of the traditional point-to-point (P2P) Ethernet and the point-to-multipoint (P2MP) EPON architecture [9].

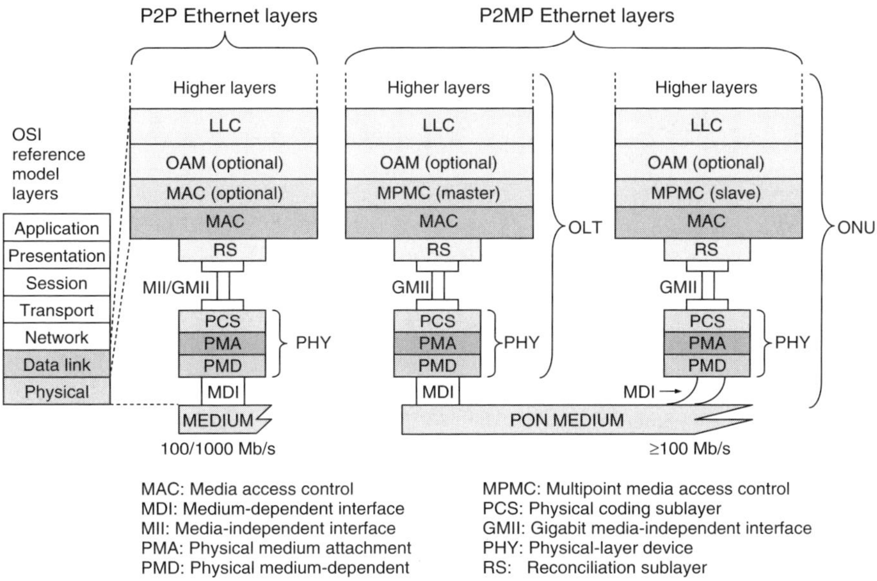

Figure 2.29 Point-to-point (P2P) Ethernet and point-to-multipoint (P2MP) EPON layering architecture.

It can be seen from the figure that EPON layering is very similar to that of traditional P2P Ethernet. The Ethernet standard further divides the OSI physical layer and data link layer into multiple sublayers. The physical layer (PHY) is connected to the data link layer using the media-independent interface (MII) or gigabit media-independent interface (GMII).

The optional MAC sublayer in P2P Ethernet is replaced with a mandatory multipoint media access control (MPMC) layer in EPON [9, Clause 64]. The MPMC layer runs the multipoint control protocol (MPCP) to coordinate the access to the shared PON medium among EPON ONUs. Although the OLT and ONU stacks look nearly identical (Fig. 2.29), the MPCP entity in an OLT functions as the master and the MPCP entity in ONU as the slave.

It will be explained later that the reconciliation sublayer has also been extended in EPON to handle a P2P emulation function so that IEEE 802.1-based bridging protocols [31] will continue to function with EPON.

2.7.2 EPON PMD Layer

The PMD layer specifies the physical characteristics of the optical transceivers. Ethernet has the tradition of adopting mature low-cost designs to promote mass deployments. This philosophy has been the key to the tremendous commercial successes of Ethernet.

As opposed to the 1:32 splitting ratio used in G-PON, the IEEE 802.3ah standard specified a minimum splitting ratio of 1:16. It should be realized that the reconciliation sublayer can support up to 32,768 different logical ONUs through a 15-bit logical link identifier (LLID).

Two different reaches between the OLT and ONU, 10 km and 20 km, have been defined in EPON standards [9, Clause 60]. The 1000BASE-PX10-D PMD and 1000BASE-PX10-U PMD define the OLT and ONU transceiver characteristics for a 10-km reach. The 20-km-reach OLT and ONU PMDs are defined as 1000BASE-PX-20-D and 1000BASE-PX20-U. EPON basically inherited the ITU-G.983.3 wavelength allocation plan. Tables 2.2 and 2.3 present selected properties of 1000BASE-PX10 and 1000BASE-PX20 PMDs from the IEEE802.3ah standards. One can notice that the ONU properties for 10-km and 20-km transmission distances are almost identical. Most of the changes are made at the OLT when the transmission distance is doubled from 10 km to 20 km. This helps to improve the ONU device cost by creating a large economy scale. It also improves the chance that end users can continue to use the same ONU when the transmission distance is increased.

2.7.3 Burst Mode Operation and Loop Timing in EPON

Ethernet protocol is a burst mode protocol. However, modern P2P Ethernet uses dedicated transmitting and receiving channels between a hub and Ethernet workstations. Such a system maintains the clock synchronization between the receiver and transmitter by transmitting idle symbols when there is no data to be sent. Therefore, even though the Ethernet protocol itself is bursty, the physical layer of modern P2P Ethernets is no longer bursty. Although the preamble has been preserved in modern P2P Ethernet, they have no practical significance except for backward compatibility with first-generation Ethernet devices.

Since EPON upstream physical connectivity is bursty, preambles are needed again to help the OLT burst mode receiver to synchronize with the ONU. Moreover, preambles are modified in EPON to carry the logical link ID (LLID) used in P2P emulation [9, Clause 65].

Table 2.2

Selected properties of IEEE802.3ah EPON transmitters

	1000BASE-PX10-D (OLT)	1000BASE-PX10-U (ONU)	1000BASE-PX20-D (OLT)	1000BASE-PX20-U (ONU)	Unit
Transmitter type	Long-wave Laser	Long-wave Laser	Long-wave laser	Long-wave Laser	
Signaling speed (range)	1.25 ± 100 ppm	1.25 ± 100 ppm	1.25 ± 100 ppm	1.25 ± 100 ppm	GBd
Wavelength (range)	1,480–1,500	1,260–1,360	1,480–1,500	1,260–1,360	nm
RMS spectral width (max)	Depends on wavelengths				nm
Average launch power (max)	+2	+4	+7	+4	dBm
Average launch power (min)	−3	−1	+2	−1	dBm
Average launch power of OFF transmitter	−39	−45	−39	−45	dBm
Extinction ratio (min)	6	6	6	6	dB
RIN (max)	−118	−113	−115	−115	dB/Hz
Launch OMA (min)	0.6	0.95	2.8	0.95	mW
Ton (max)	N.A.	512	N.A.	512	ns
Toff (max)	N.A.	512	N.A.	512	ns
Optical return loss tolerance	15	15	15	15	dB
Transmitter dispersion penalty	1.3	2.8	2.3	1.8	dB

Table 2.3
Selected properties of IEEE 802.3ah receiver characteristics

	1000BASE-PX10-D (OLT)	1000BASE-PX10-U (ONU)	1000BASE-PX20-D (OLT)	1000BASE-PX20-U (ONU)	Unit
Signaling speed (range)	1.25 ± 100 ppm	1.25 ± 100 ppm	1.25 ± 100 ppm	1.25 ± 100 ppm	GBd
Wavelength (range)	1,260–1,360	1,480–1,500	1,260–1,360	1,480–1,500	nm
Average receive power (max)	−1	−3	−6	−3	dBm
Receive sensitivity	−24	−24	−27	−24	dBm
Receiver reflectance (min)	−12	−12	−12	−12	dB
Stressed receive sensitivity	−22.3	−21.4	−24.4	−22.1	dBm
Vertical eye closure penalty (min)	1.2	2.2	2.2	1.5	dB
T_receiver-settling (max)	400	N.A	400	N.A	ns

To maintain low cost, traditionally all Ethernet transmitters are running asynchronously on their own local clock domains. There is no global synchronization. A receiver derives the clock signal for gating the received data from its received digital symbols. Mismatches between clock sources are accounted for by adjusting the interframe gap (IFG) between Ethernet frames.

In an EPON system, the downstream physical link maintains continuous signal stream and clock synchronization. In the upstream direction, in order to maintain a common timing reference with the OLT, ONUs use loop timing for the upstream burst mode transmission, i.e. the clock for upstream signal transmission is derived from the downstream received signal.

2.7.4 PCS Layer and Forward Error Correction

The physical coding sublayer (PCS) is the layer that deals with line-coding in Ethernet physical layer devices. EPON defines a symmetric throughput of 1.0 Gbps both in the upstream and downstream directions, and adopted the 8B/10B line coding used in the IEEE802.3z gigabit Ethernet standard [9, Clause 36]. The 8B/10B code adds an overhead of 25% to limit the running disparity[5] to 1. Besides conveying physical layer control sequences, the 8B/10B code produces a DC balanced output and enough transitions for easy clock recovery. Nonetheless, it increases the symbol rate to 1250 Mbaud/s.[6]

The use of FEC is optional in EPON. The IEEE 802.3ah standard defines RS(255, 239) block codes in the EPON PCS layer [9, Clause 65]. This is the same code used in G-PON. Parity bits are appended at the end of each frame. Similar to G-PON, since the clock rate does not change when FEC parities are appended, the data throughput is decreased when FEC is used. The RS(255, 239) block code does not change the information bits. This allows ONUs which do not support FEC to coexist with ONUs supporting FEC coded frames. An ONU with no FEC support will simply ignore the parity bits albeit running at a higher bit error rate (BER).

2.7.5 Ethernet Framing

EPON transmits data as native Ethernet frames in the PON section. Ethernet frames are variable size frames. Standard Ethernet frame format [9] is shown in

[5] The running disparity represents the difference in the number of 0s and 1s in the transmitted symbols.

[6] Many EPON marketing materials fraudulently claim EPON data rate as 1.25 Gb/s.

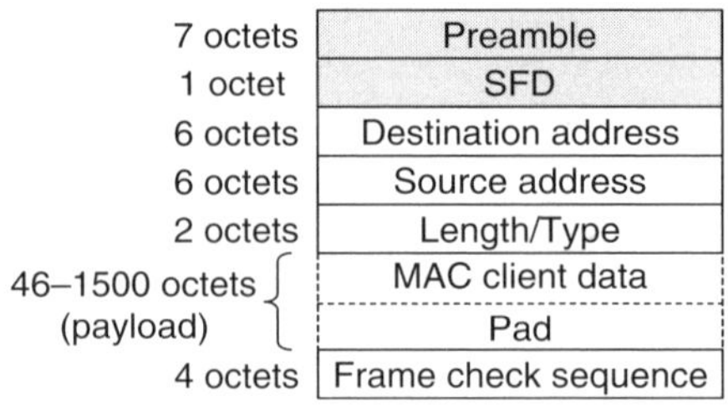

Figure 2.30 Standard Ethernet frame format.

Fig. 2.30. It starts with a preamble and a 1-octet start frame delimiter (SFD) to signify the beginning of a frame.[7] The EPON standard makes use of the preamble field with modifications to carry LLIDs for ONUs [9, Clause 65].

Each frame carries the destination and source MAC addresses, which are both 6-octet fields. A 2-octet length/type field is used to represent the length of the payload when its value is between 0 and 1500 (the maximum payload length). When its value is between 1536 and 65,535, it is used to represent the type of Ethernet frames. The use of this field to represent length and type is therefore mutually exclusive. Ethernet frames have a variable payload size from 46 to 1500 octets. Following the payload is a 4-octet frame check sequence (FCS) using cyclic redundancy check (CRC).

It can be seen from Fig. 2.30 that Ethernet frames carry minimum overhead bytes to convey management and protocol information. In the Ethernet field, management and OAM information are carried using protocol data units (PDUs) and OAM frames, which are standard Ethernet frames identified by special length/type values. Protocol and OAM information is carried in the payload field of PDUs and OAM frames. These frames are multiplexed in-band with other Ethernet frames carrying actual user data payloads.

2.7.6 Multipoint Control Protocol (MPCP)

The MPCP [9, Clause 64] in the MPMC sublayer uses multipoint control protocol data units (MPCPDUs) to perform ONU discovery and ranging functions. It also provides the arbitration mechanism for upstream medium access control among multiple ONUs.

[7] Ethernet frame lengths are counted without including the preamble and SFD fields.

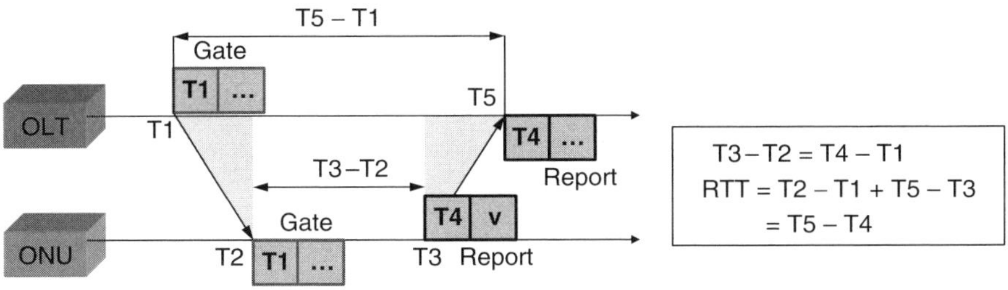

Figure 2.31 EPON ranging process.

2.7.6.1 Ranging in EPON

The ranging process measures the RTT between the OLT and ONUs so that the OLT can appropriately offset the upstream time slots granted to ONUs to avoid upstream collision. It is achieved through *Gate* and *Report* MPCPDUs as shown in Fig. 2.31. Both the OLT and ONU maintain their own local 32-bit counters that increment in 16-ns time quantum. In the ranging process, the OLT sends a *Gate* message with time stamp T1 (Fig. 2.31) which represents the absolute time. The ONU receives the Gate message at T2 after the transmission delay and resets its timer to T1. After some processing delay, the ONU sends a Report message at time T3, with time stamp T4 = T1 + (T3–T2). The OLT receives the Report message with time stamp T4 at absolute time T5 as indicated in Fig. 2.31. It can be seen that the RTT is simply T5–T4.

2.7.6.2 Gate and Report Operation

The Gate operation provides the mechanism for the OLT to specify the time slots that ONUs can transmit. Unlike in G-PON where the OLT specifies the start and stop time in increments of 1-byte, the Gate operation specifies the start time and length of time slots in increments of 16 ns (equivalent to 2-byte in gigabit EPON) time quanta [9, Clause 64].

Figure 2.32 shows the EPON Gate operation. The *Gate* MPCPDU contains time stamp, the start time, and length of upstream transmission. Upon receiving the Gate frame, the ONU updates its time stamp register, slot-start register, and slot-length register.

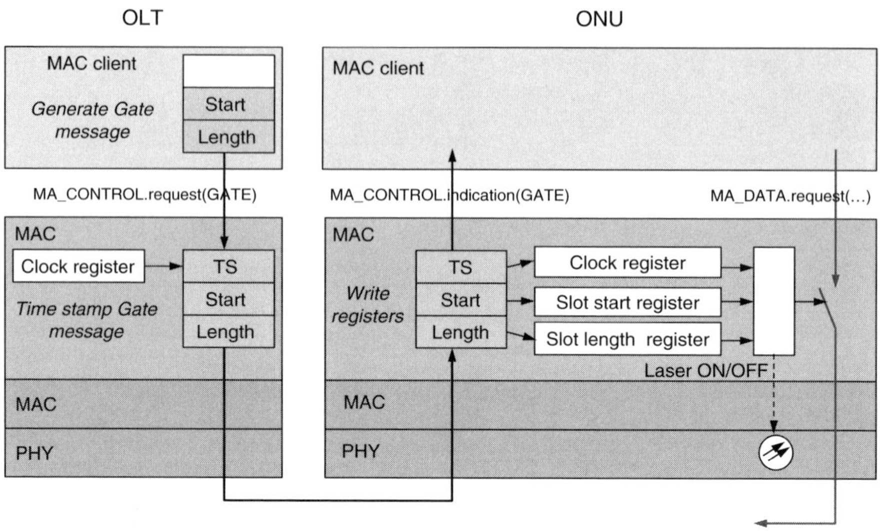

Figure 2.32 EPON Gate operation.

In the report operation, an ONU informs the OLT with its queue lengths and provides the timing information to calculate RTT by sending the Report MPCPDU. Upon receiving the Report MPCPDU, the OLT updates ONU queue length registers and RTT register as shown in Fig. 2.33.

2.7.6.3 Dynamic Bandwidth Allocation

The EPON standard does not specify the implementation details of DBA. However, the Gate and Report operation provides the necessary interface and mechanism for controlling ONU bandwidth. It is up to vendors to design the DBA strategies and algorithms. Chapter 5 discusses DBA mechanisms in more details.

2.7.6.4 Multipoint Control Protocol Data Unit (MPCPDU)

MPCPDUs are 64-byte-long MAC frames with no VLAN tagging. The generic format of MPCPDU is shown in Fig. 2.34. MPCPDUs are recognized by the MAC frame type 0×88–08 in the length/type field. The 2-octet op-code field identifies the type of MPCPDU message. Figure 2.34 also shows the op-codes for all the MPCPDU types defined in EPON. Every MPCPDU message contains a 4-octet time stamp field so that OLT and ONUs can continuously

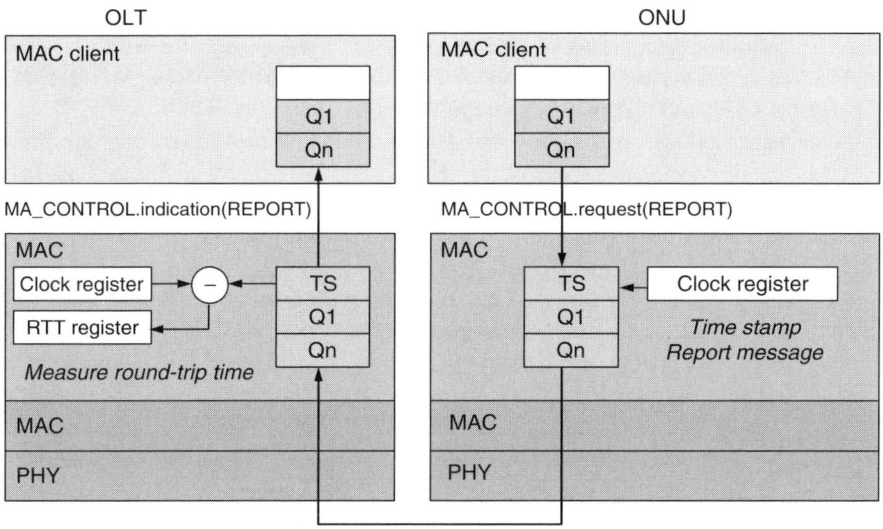

Figure 2.33 EPON Report operation.

update each other to correct for timing drifts (e.g. due to changes in fiber temperatures and stresses).

The data/pad portion of the MPCPDU contains the MAC parameters used in MPCP and the necessary zero padding to maintain the 64-byte frame size. More details of MPCPDU are given in Appendix II. Interested readers should refer to Clause 64 of the IEEE802.3 standard for the exact details of MPCPDU.

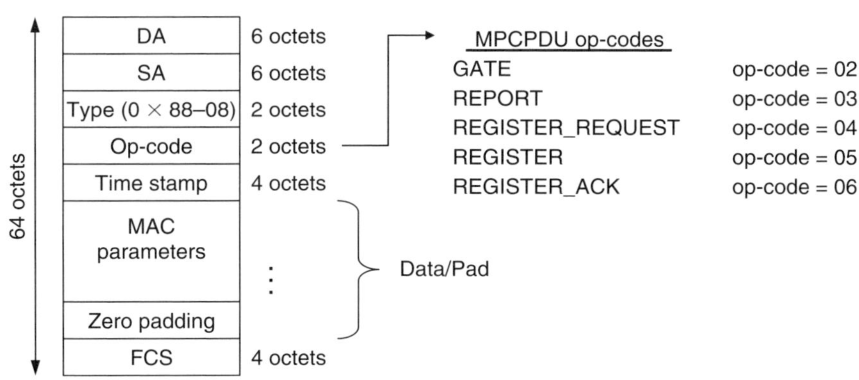

Figure 2.34 Generic format of MPCPDU.

2.7.6.5 Autodiscovery of EPON ONU

The autodiscovery process allows EPON ONUs to register and join the system after powering up. The OLT allocates the virtual MAC and assigns LLIDs to ONUs for point-to-point emulation (P2PE).

During autodiscovery, ONU and OLTs exchange each other's capabilities such as the *synchronization time* of the OLT burst mode receiver. *Synchronization time* is the time required by the OLT, after receiving a burst of data, to lock itself to the ONU transmitter clock and adjust its decision threshold to account for the differences in received power levels from different ONUs.

To perform autodiscovery, the OLT broadcasts discovery Gate frames periodically. As shown in Fig. 2.35, a reserved discovery window is granted by the discovery Gate. The OLT also transmit its burst mode receiver synchronization time to ONUs in the discovery gate so that ONUs know to format the upstream signal with idle symbols during the initial burst synchronization time. An ONU intending to register receives the discovery gate and sends a Register Request after waiting for a random delay. The OLT receives the Register Request and allocates the ONU with the LLID. The OLT then sends another Gate frame to grant the upstream time slot for the ONU with the newly allocated LLID to transmit a Register Acknowledgment frame, which signifies the finish of the registration process.

During the discovery window, multiple ONUs may attempt to register at the same time and cause collisions at the OLT. Contention in the discovery window

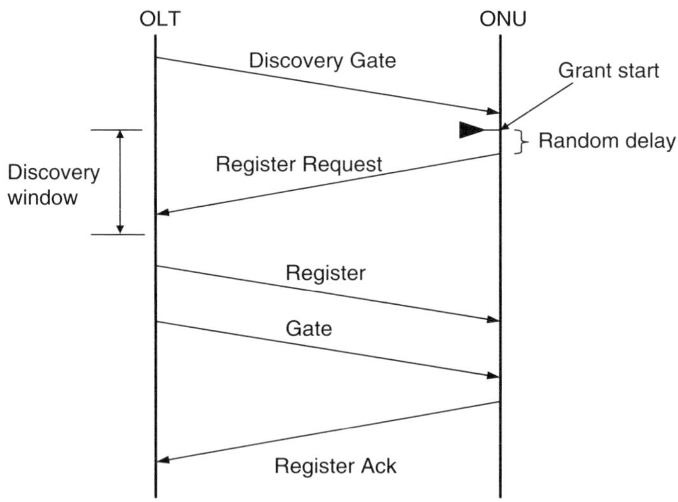

Figure 2.35 Autodiscovery process (from [9], reprinted with permission from IEEE Std. IEEE Std 802.3, 2005, carrier sense multiple access with collision detection (CSMA/CD) access method and physical layer specifications, copyright [2005], by IEEE).

is resolved by adding a random delay before each ONU transmits the Register frame. It is also possible for an OLT to receive multiple valid Register frames in a discovery window.

2.7.7 Point-to-Point Emulation (P2PE) in EPON

Ethernet MAC interfaces are interconnected with Layer-2 (L2) switches or Layer-3 (L3) routers. L2 switches directly interconnect network devices on the Ethernet layer and are therefore more efficient, less costly, and easier to manage in a LAN environment. On the other hand, routing is used in a wide area network to interconnect multiple Layer-2 networks [13]. In an Ethernet environment, L2 connection is achieved using IEEE802.1 based bridges[8] or switches [31, 32]. An Ethernet switch forwards the traffic among its multiple ports. Each port is connected to a different broadcast domain containing one or more MAC devices. A switch assumes that within the same broadcast domain MAC devices can forward traffic directly toward each other without its help. It performs the L2 switching function by examining the SA and DA of each received frame. If both of them belong to the same domain (i.e. connected to the same port), it filters out the packet without forwarding it. This helps to preserve the bandwidth in other parts of the network and improves the network performance.

In an EPON system, the P2P symmetric Ethernet connectivity is replaced by the asymmetric P2MP connectivity. Because of the directional nature of the remote node, ONUs cannot see each other's upstream traffic directly (Fig. 2.36). In a subscriber network, this directional property provides an inherent security advantage. Nevertheless, it also requires the OLT to help forwarding inter-ONU transmissions.

Without any treatment, an IEEE 802.1 switch connected to the OLT would see all the inter-ONU frames with SA and DA belonging to MAC entities connected to the same switch port and thus they would be within the same domain. As a result, the switch would not forward the traffic between different ONUs connected to the same OLT.

To resolve this issue, a P2PE function has been created in the RS layer. The P2PE function maps EPON frames from each ONU to a different MAC in the OLT, which is then connected to a higher-layer entity such as L2 switch (Fig. 2.37).

The P2PE function is achieved by modifying the preamble in front of the MAC frame to include an LLID. The modified preamble with the LLID is used in the PON section. The format of the modified EPON preamble is shown in Fig. 2.38.

[8] Bridge and switch are interchangeable terms used to describe L2 frame forwarding devices. Bridge is the original term. Early bridges were implemented using software. Switch is more commonly used to describe bridges implemented in Silicon chips.

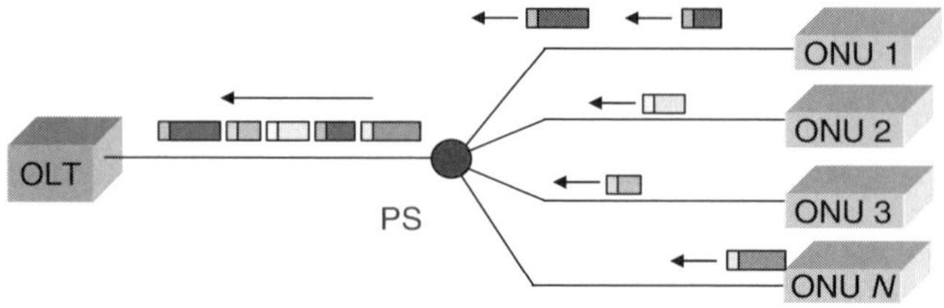

Figure 2.36 Although all the ONU traffic arrives at the same physical port at the OLT, because of the directional power-splitting coupler used at the remote node, ONUs cannot see each other's traffic without the forwarding aid of OLT.

It starts with a start LLID delimiter (SLD) field, followed by a 2-byte offset and a 2-byte LLID. A 1-byte CRC field protects the data from the SLD to the LLID. The first bit of the LLID is a mode bit indicating broadcast or unicast traffic. The rest of the 15 bits are capable of supporting 32,768 different logical ONUs.

The mode bit is set to 0 for P2PE operation. Figure 2.39 shows principle of EPON P2PE. When the mode bit is set to 1, the OLT uses the so-called single-copy

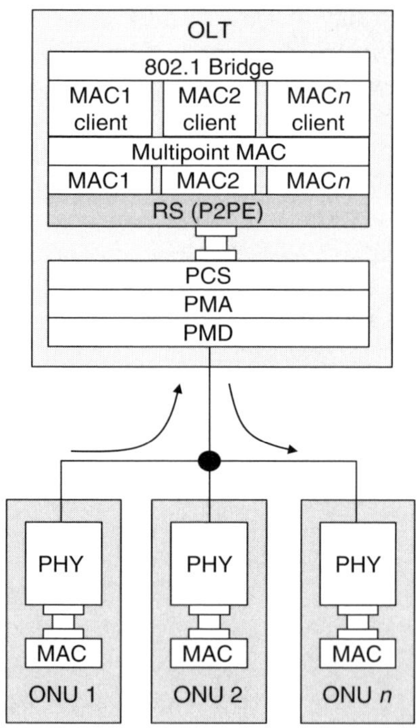

Figure 2.37 Point-to-point emulation in EPON.

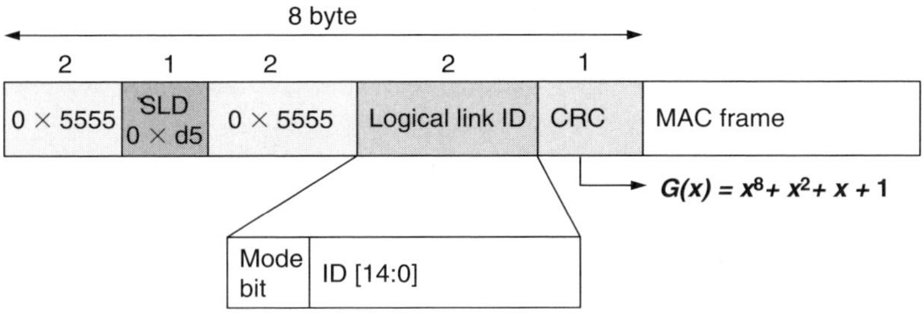

Figure 2.38 Modified preamble with LLID for point-to-point emulation in EPON.

broadcast (SCB) MAC to broadcast traffic to all ONUs. It takes the advantage of native EPON downstream broadcast operation. To prevent broadcast storm in L2 switches, EPON standard recommends avoiding the connection of the SCB port to 802.1 switches, and using it only to connect to L3 routers[9] or servers for disseminating information. Figure 2.40 illustrates the SCB MAC and emulated P2P MACs in an EPON model.

One may have noticed that the concept of LLID in EPON and that of Alloc-ID in G-PON are similar. In a G-PON, single ONU can be allocated with multiple T-CONTs with different Alloc-IDs. As a matter of fact, an EPON ONU may also be allocated with multiple LLIDs. Certain implementations of EPON make use of the multiple LLIDs to implement different quality of service classes [36].

2.7.8 EPON Encryption and Protection

The IEEE802.3ah standard does not specify encryption and protection mechanisms for EPON. Encryption is important to ensure privacy when ONUs are directly connected to users as in FTTH applications. Protection is important when ONUs are shared among groups of users as in FTTB/FTTC applications.

Many implementations of EPON chips include vendor-specific encryption mechanisms which can be enabled by service providers if necessary.

[9] Broadcast storm is caused by loops in a network, allowing packets to replicate and circulate which eventually use up the network bandwidth. L2 switches uses spanning tree protocol (STP) to ensure no multiple paths exist between two nodes and thus avoid the possibility of forming loops. When both the emulated P2P link and SBC link exist between the OLT and ONU, STP will get confused. Unlike L2 switches, L3 routing protocols can make use of multiple signal paths for load balancing. They can also use the time-to-live (TTL) field to avoid loops.

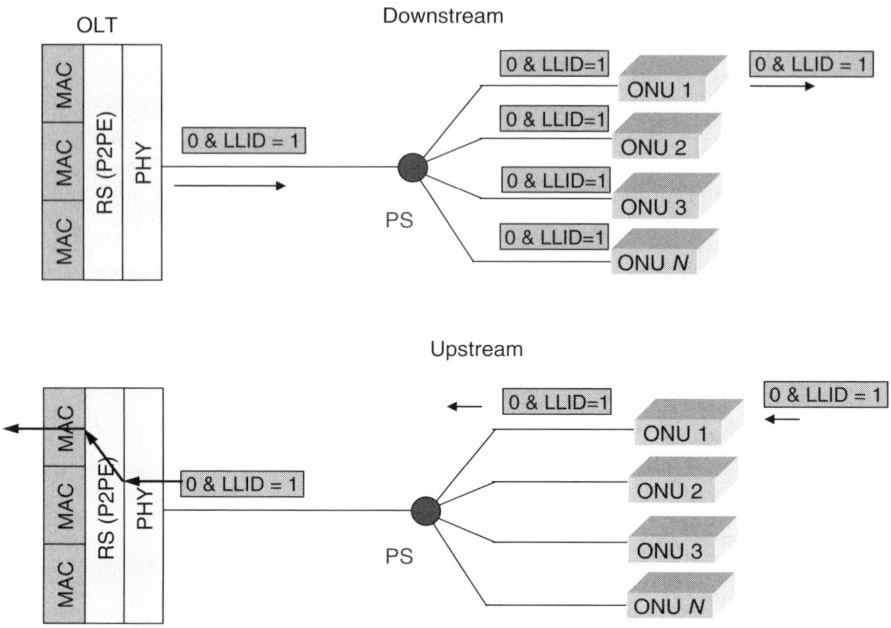

Figure 2.39 EPON point-to-point emulation operation.

2.7.9 Ethernet OAM Sublayer

Besides standardizing the EPON technology, another charter of the IEEE 802.3ah task force was to develop the OAM sublayer for Ethernet. The OAM

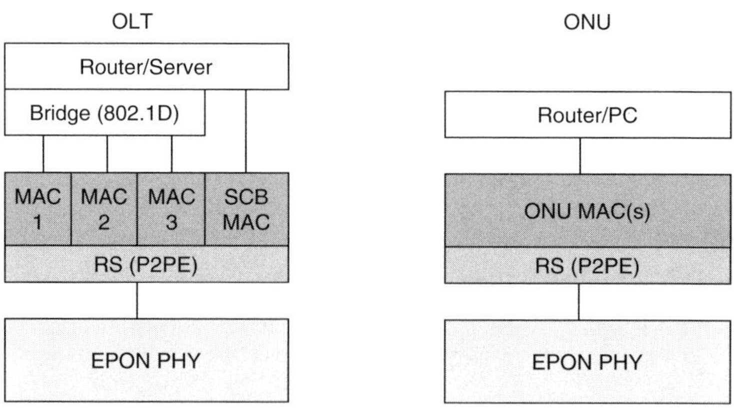

Figure 2.40 Point-to-point and single-copy broadcast (SCB) MACs in an EPON model.

sublayer [9, Clause 57] is an optional Ethernet MAC sublayer (Fig. 2.29) which not only can be built in EPON but also in other Ethernet devices. In fact, EPON only consists of a subset of the objects manageable by the OAM sublayer. Most of the EPON ASICs are built with the OAM sublayer which can be used for managing OLT and ONU.

2.8 G-PON AND EPON COMPARISON

A high-level comparison of EPON and G-PON is given in Table 2.4. From a technical viewpoint, one major difference between G-PON and EPON is the support for TDM circuits. G-PON divides upstream and downstream signals into 125-μs frames. Data frames are encapsulated using the GEM encapsulation with segmentation capability, which allows TDM circuits with guaranteed bandwidth and granularities of 64 kb/s to be created between OLT and ONUs.

On the other hand, in an EPON system, variable length native Ethernet frames are used in the transport layer. Circuit emulation [33–35] is needed to implement fixed-bandwidth TDM circuits.

While EPON has been optimized for native Ethernet packet transport, the GEM encapsulation allows easier adaptation of other format signals. In an EPON system, bandwidth report and grant functions are implemented using MPCPDUs. Each logical ONU port with a different LLID requires its own separate Gate and Report frames, which are full Ethernet frames. The more logical IDs are allocated, the higher the overhead of the MPCP [9, Clause 64]. On the contrary, G-PON bandwidth report and grant functions can be piggybacked into the PCB overhead of the GTC frame [24]. Each PCBd overhead contains the bandwidth allocation information for all the allocated T-CONTs. This makes G-PON overhead very efficient compared to EPON. A very good analysis of the APON, G-PON, and EPON overhead efficiency can be found in [36].

Table 2.4

G-PON and EPON comparison

	EPON	G-PON
Downstream data rate (Mbps)	1,000	1,244 or 2,488
Upstream data rate (Mbps)	1,000	155, 622, 1,244, or 2,488
Payload encapsulation	Native ethernet	GEM
TDM support	Circuit emulation	Yes
Laser on/off	512 ns	≈13 ns
AGC	≤400 ns	44 ns
CDR	≤400 ns	

Assuming 32-way[10] split for both EPON and G-PON, the EPON system offers an average bandwidth of 31.25 Mbps per ONU in both downstream and upstream directions. A G-PON with 2488 Mbps symmetric bandwidth will offer 77.75 Mbps for each ONU. The RBOCs in the United States do not think the EPON capacity meets the requirements in a modern data-centric society (Chap. 1). This is one of the reasons that G-PON has been selected as the technology of choice for the joint request for proposal (RFP) issued by Verizon, SBC, and Bell South.[11] To address the growing demand, at the time of writing, IEEE has started a new task force 802.3av to charter the development of 10 Gb/s EPON [38].

2.9 SUPER PON

Super PON was proposed to achieve better economy by increasing the reach of PON systems beyond 20 km and supporting higher splitting ratios of 1:64 or even 1:128. Most of the super PON studies were performed within the European Community [39–41].

However, super PON faces the following challenges:

1. The bit rate in PON systems will keep increasing as demand for bandwidth grows. A minimum energy per bit is required to keep the BER. This means that by increasing the bit rate, the power required, which equals the product of bit rate and energy per bit, will increase proportionally. To support a bigger share group size with the same average user bandwidth requires increasing the physical layer bit rate. Increasing the splitting ratio and transmission distance will increase the loss between the OLT and ONU. As a result, a significant increase in power budget between the OLT and ONU is necessary for super PONs. In fact, super PON requires amplifiers in the field to provide the needed power budget.
2. Economical fiber amplifiers are not available at the 1.3/1.49-μm upstream and downstream wavelengths. The most mature optical amplification technology, erbium-doped fiber amplifier (EDFA) works in the C-band, which is also the spectral range with minimum fiber loss. The fiber loss at 1.3 μm almost doubles that at 1.55 μm. Although other amplification technologies such as Semiconductor Optical Amplifier (SOA) [42] or Raman Amplifiers [43] can be used, these technologies are still very expensive. Raman amplification requires very

[10] The IEEE 802.3ah standard only specified 1:16 splitting ratio. Most commercial EPON PHY do support the 1:32 splitting ratio.

[11] SBC, AT&T, and Bell South have merged into one single company, which became the new "at&t."

high pump power and poses safety issues, which is tricky in an access network environment. SOAs generally have very high noise figure.

3. Downstream optical amplifications can be easily shared amongst multiple ONUs. However, to share an upstream amplifier could be challenging because ONUs are located at different distance with different losses to the optical amplifier. In addition, upstream data are sent in burst mode, requiring optical amplifier to have fast transient control capabilities.
4. Dispersion effect around 1.5-μm wavelength is nonnegligible. Dispersion effect increases as the square of bit rate [3]. So without using dispersion-tolerant transceivers, when both bit rate and distance increases, dispersion penalty will become significant. In a cost-sensitive access system, this may not make economic sense, at least for the foreseeable future.
5. RTT increases between the OLT and ONU. For a 20-km transmission distance, the RTT is 200 μs. For 60 km, this time increases to 600 μs. At 1-Gb/s transmission speed, this is equivalent to 750,000 bytes of data. Our previous discussion showed that the size of the discovery window reserved for new ONU registration and ranging needs to cover the maximum differential distances between ONUs. Increasing the PON size means that more bandwidth needs to be reserved for autodiscovery of ONUs, unless the frequency of autodiscovery windows is decreased. Also, a bigger share group will increase the chances of collision due to ONU registration and increase the time required for ONU registration. The increased RTT will also increase the latency from bandwidth request to bandwidth grant, which might be problematic for real-time applications.
6. In an EPON system, separate Gate and Report frames are required for each LLID. Increasing the share group size will increase the bandwidth overhead required for the MPCP used in a super EPON system [37].
7. Better protection mechanisms are required as the share group size increases because a failure will have a more significant business impact than that with a small share group size.

These challenges of super PON in both the physical and MAC layers may demand advanced technologies which eventually outweigh the economic advantages achieved through longer transmission distances and bigger share groups.

2.10 WDM-PON

An alternative to expand PON capabilities besides super PON is WDM-PON. A WDM-PON is characterized by a WDM coupler, which replaces the power splitter at the remote terminal [44].

2.10.1 Advantages and Challenges of WDM-PON

WDM-PON offers the following advantages:

1. The optical distribution plant is still passive and therefore has the same low-maintenance and high-reliability properties of PS-PON.
2. Each user receives its own wavelength. Therefore WDM-PON offers excellent privacy.
3. P2P connections between OLT and ONUs are realized in wavelength domain. There is no P2MP media access control required. This greatly simplifies the MAC layer. There will not be distance limitation imposed by ranging and DBA protocols.
4. Easy pay-as-you-grow upgrade. In a PS-PON, if the OLT speed is increased, all ONUs need to be upgraded at the same time. Such a problem does not exist with WDM-PONs. Each wavelength in a WDM-PON can run at a different speed as well as with a different protocol. Individual user pays for his or her own upgrade.

The challenges of WDM-PON include:

1. High costs of WDM components. However, the costs of WDM components have dropped tremendously in recent years which made WDM-PONs economically more viable. For instance, Korean Telecom has already begun its WDM-PON trial. Interests in WDM-PON standardization has also been generated in ITU-T SG15 recently.
2. Temperature control. WDM components' wavelengths tend to drift with environmental temperatures. Temperature control consumes power and requires active electronic parts in the optical distribution network. To remove the need for temperature control, tremendous component and architectural progresses have been made in producing athermal WDM components and systems which are temperature-agnostic.
3. Colorless ONU operation. In a WDM-PON, each ONU needs a different wavelength for upstream connection. This presents a rather serious operational and economical issue. Wavelength specific ONU introduces significant challenges in managing production lines, inventory stocks, sparing, and maintenance. A lot of the solutions have been invented to realize colorless ONUs in the last 20 years.

Technological progresses to materialize economical WDM-PON system will be reviewed in detail in Chap. 3.

2.10.2 Arrayed Waveguide Grating (AWG) Router

The AWG router is a key element in many WDM-PON architectures. A conventional N-wavelength WDM coupler is a $1 \times N$ device as shown in Fig. 2.41 (a).

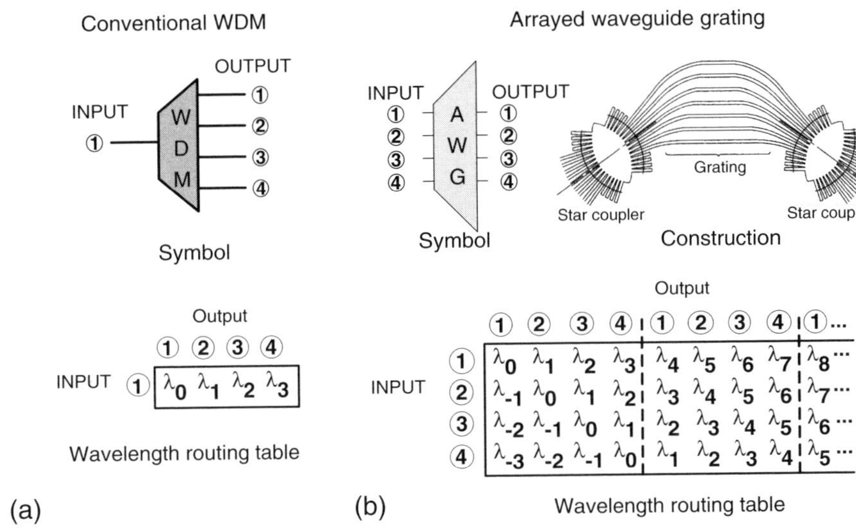

Figure 2.41 Conventional WDM coupler vs arrayed waveguide grating.

Components of an input optical signal are directed to specific output ports according to their wavelengths as indicated in Fig. 2.41 (a). A general AWG router consists of two star couplers joined together with arms of waveguides of unequal lengths as shown in Fig. 2.41 (b) [45]. Each arm is related to the adjacent arm by a constant length difference. These waveguides function as an optical grating to disperse signals of different wavelengths.

Another name of AWG in literature is wavelength grating router (WGR). A very important and useful characteristic of the AWG is its cyclical wavelength routing property illustrated by the table in Fig. 2.41 (b) [2]. With a normal WDM multiplexer in Fig. 2.41 (a), if an "out-of-range" wavelength, (e.g. λ_{-1} or λ_4,λ_5) is sent to the input port, that wavelength is simply lost or "blocked" from reaching any output port. An AWG device can be designed so that its wavelength demultiplexing property repeats over periods of optical spectral ranges called free spectral ranges (FSR). Moreover, if the multi-wavelength input is shifted to the next input port, the demultiplexed output wavelengths also shift to the next output ports accordingly. Cyclical AWGs are also referred to as colorless AWGs, although the author does not like this term personally.

The wavelength-cyclic property of AWG can be exploited to enable many clever architectural innovations. For example, by using two adjacent ports for upstream and downstream connections, the same wavelength can be "reused" for transmission and reception at an ONU. Chapter 6 illustrates many novel protection schemes enabled by cyclical AWGs.

2.10.3 Broadcast Emulation and Point-to-Point Operation

Broadcast is a very efficient way to distribute a large amount of information to multiple users. Figure 2.42 illustrates a method to emulate a broadcast service in WDM-PON with a broadband optical source such as LED [46, 47].

In Fig. 2.42, a broadband amplified spontaneous emission (ASE) light source such as an LED is first modulated with the broadcast data. The output of the LED is filtered (by filter F) to one FSR of the AWG before being delivered to the end users. After going through the AWG, each user receives a slice of the modulated broadband source with the same data modulation, albeit at different wavelengths. The same figure also shows P2P services being delivered in a different FSR using laser diodes which are line sources. The broadcast and P2P services are multiplexed at the CO using a coarse WDM (CWDM) device.

2.10.4 2-PONs-In-1

A problem with spectral slicing is the limitation due to spontaneous–spontaneous emission beat noise in ASE sources, which is proportional to the ratio of the receiver's electrical bandwidth to the received optical spectral bandwidth [48].

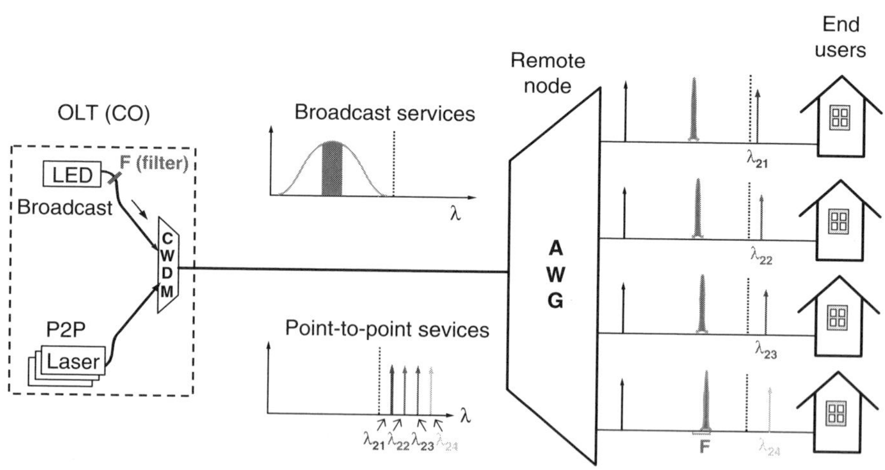

Figure 2.42 Emulation of broadcast services on a WDM-PON with a broadband source.

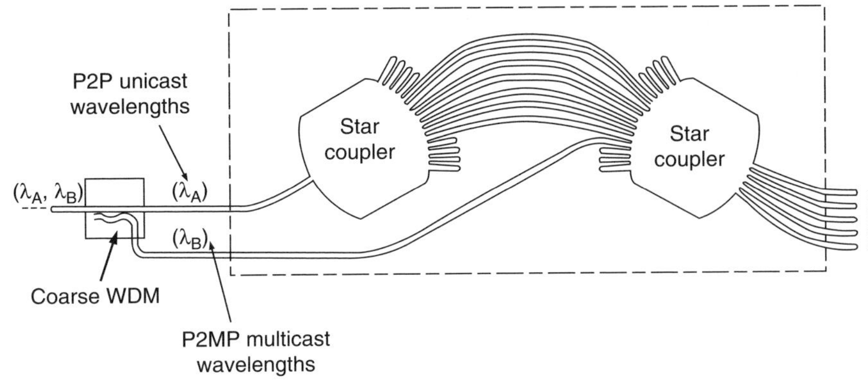

Figure 2.43 A modified AWG for 2-PONs-in-1.

As the bit rate increases and channel spacing decreases, the BER is eventually limited by the spontaneous–spontaneous emission beat noise.

An alternative to support broadcast operation on a WDM-PON is to use the 2-PONs-in-1 architecture. A way to realize this is to use the device shown in Fig. 2.43 [49]. In this arrangement, the broadcast wavelength (for PS-PON) is separated from the routed wavelengths using a coarse WDM multiplexer. The broadcast wavelength only goes into the second start coupler and is broadcast to all the output ports while the P2P WDM wavelengths are routed to their respective output port.

2.10.5 WDM-on-WDM

By using multiple FSRs of an AWG, one can deliver multiple wavelength services to the same ONU. A coarse wavelength division multiplexer (CWDM) is used at the OLT to combine services occupying different FSRs (Fig. 2.44). A similar CWDM at an ONU separates the individual services. In effect, on one physical fiber distribution infrastructure, one has spawned multiple WDM-PONs segregated by different spectral domains defined by the AWG FSRs. This WDM-on-WDM architecture offers the path to achieve unparalleled flexibility and growth ability [50, 51].

The concept of WDM-on-WDM has been demonstrated on the split-net test-bed by AT&T Labs in 1997 [50]. Figure 2.45 describes the test-bed setup. An 8×8 AWG with two-fiber configuration was used in this pioneer demonstration. The power budget has been designed to simulate a 16-way split.

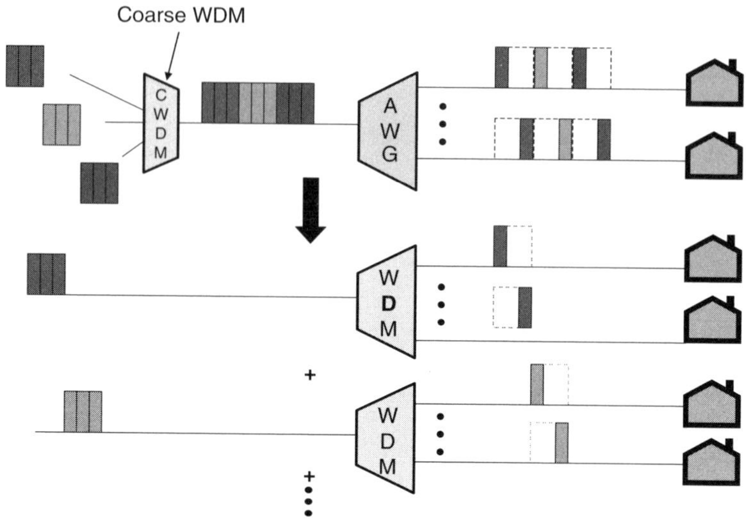

Figure 2.44 By using CWDM devices to combine and separate optical signals in multiple FSRs of an AWG device, a highly flexible WDM-on-WDM system can be achieved.

Three services were "offered" in this early-day setup. (1) A 50 Mbps low-cost baseband data, voice and video service carried in one FSR near the 1.3-μm wavelength. Uncooled LEDs and spectral slicing were used to carry this service. (2) Two sets of 80-channel digital broadcast TV service carried on two FSRs centered at 1545-nm and 1565-nm wavelengths. These services used uncooled LEDs which were directly modulated with QPSK RF subcarriers. The LED outputs were amplified with enough power to serve over 1000 people, demonstrating the sharing of optical backbone equipment. (3) 2.5-Gbps OC-48 high-speed "future" services carried by DFB lasers with wavelengths in the 1.5 μm region.

2.10.6 Hybrid WDM/TDM-PON

Another way to increase PON-system scalability besides the brute-force super-PON approach is to use a hybrid WDM/TDM-PON architecture. A demonstration of such a system is given in [52] and shown in Fig. 2.46. Using a 16-wavelength AWG with 1 × 8 power splitter, operation of 128 ONUs was demonstrated on the same feeder plant. Colorless ONU and OLT transmitters are realized by Fabry–Perot lasers whose output wavelengths are injection-locked to spectrally sliced ASE sources and therefore automatically

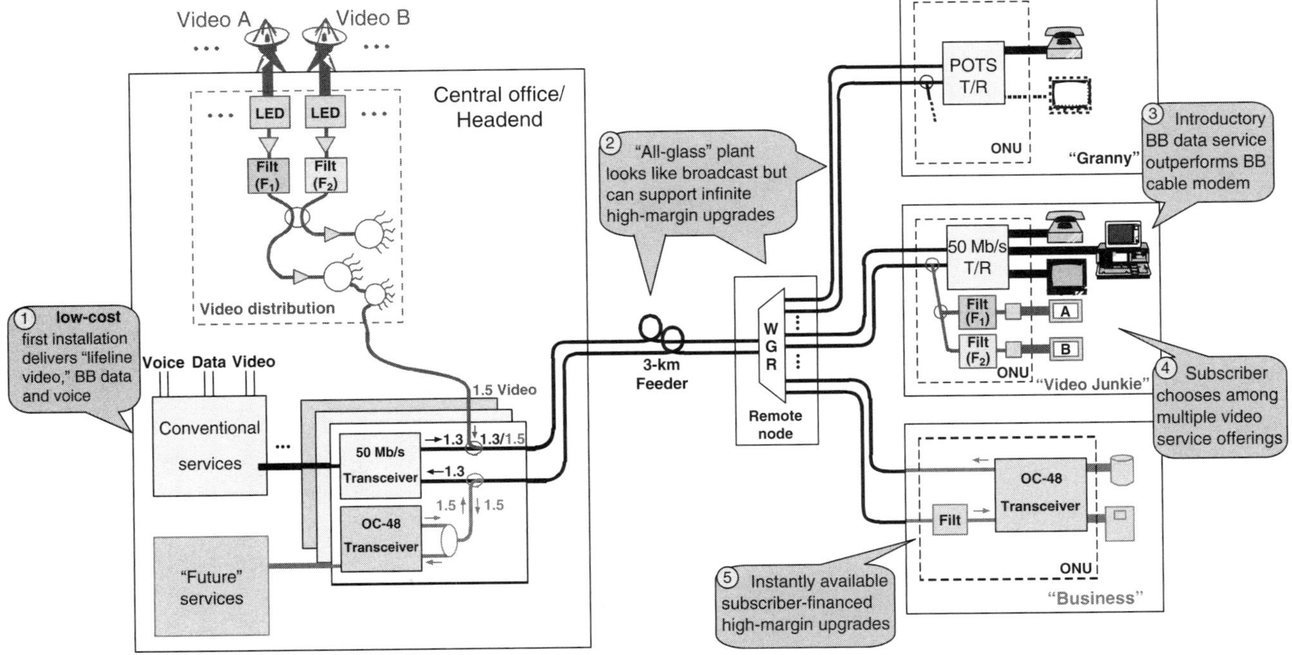

Figure 2.45 Split-net test-bed demonstrating the WDM-on-WDM concept (from [50], copyright [1997] by IEEE, reprinted with permission).

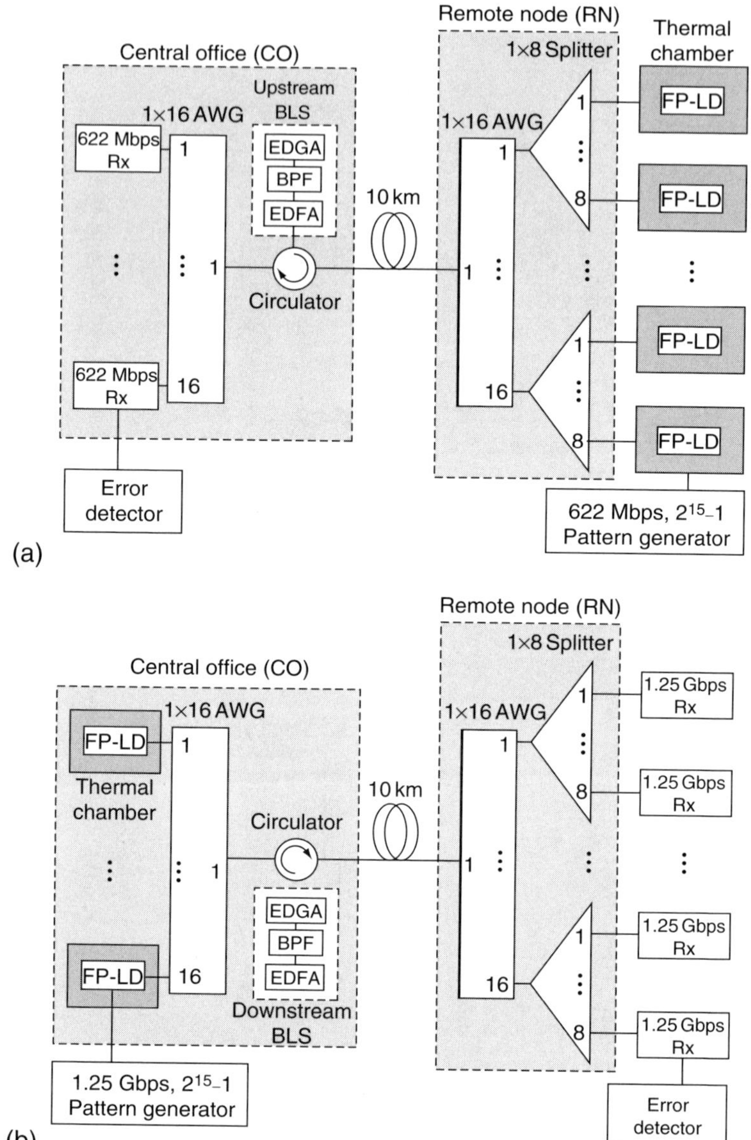

Figure 2.46 Demonstration of a hybrid WDM/TDM-PON with wavelength-selection-free transmitters: (a) downstream link and (b) upstream link (from [52], copyright [2005] by IEEE, reprinted with permission).

aligned with the AWG wavelength grid. Burst mode operations over a temperature range of 0–60 °C were demonstrated in this setup.

The WDM Ethernet-PON (WE-PON) project jointly developed by Electronics and Telecommunications Research Institute, Korea (ETRI) and Korea Telecom [53] combines EPON and WDM-PON technologies together. Using 32 wavelengths and 1 × 32 splitters, 1000 users per WE-PON is possible.

In the WE-PON test-bed, two types of upstream transmitters were used. The optical layer block diagrams of these two designs are shown in Fig. 2.47. The first one uses a gain-saturated Reflective SOA (RSOA) (Fig. 2.48). The partially modulated downstream optical signal is reverse-modulated by a feed-forward current signal in the RSOA to remove the downstream modulation (Fig. 2.49). This same-wavelength signal is then modulated with upstream data and transmitted back to the OLT. Two fibers are used at the OLT to separate the downstream and upstream signal as shown in Fig. 2.47 (a).

The second type of upstream stream transmitter uses an integrated PLC-ECL planner lightwave circuit external cavity tunable laser (Fig. 2.50). The downstream and upstream signals are separated into the L-band and C-band respectively, making use of the multiple FSRs of a cyclical AWG. A single-feeder fiber is used to connect the OLT to the RT (Fig. 2.47 (b)).

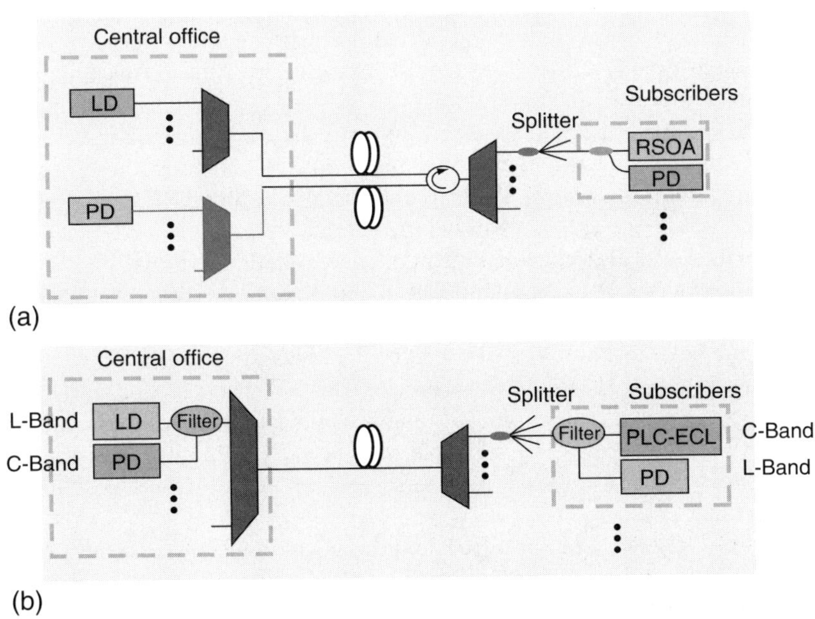

Figure 2.47 WE-PON optical layer block diagram. (Courtesy of ETRI and Korea Telecom.)

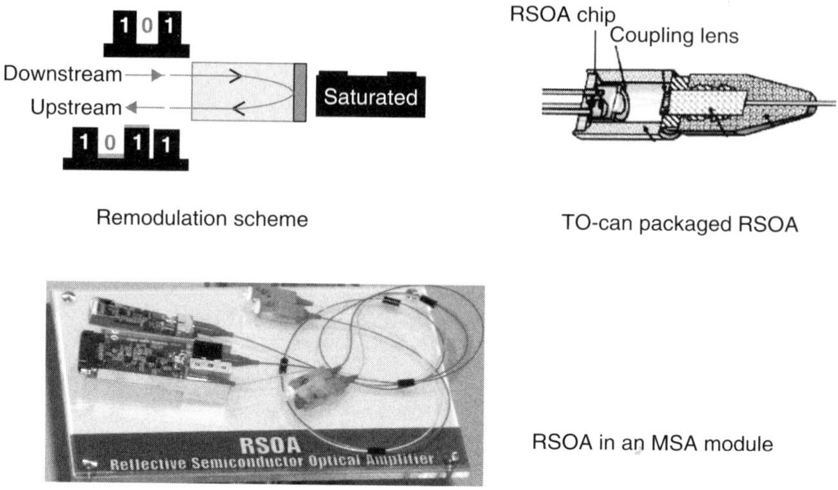

Figure 2.48 RSOA module used in ETRI WE-PON prototype. (Courtesy of ETRI and Korea Telecom.)

2.10.7 Multiply Fiber Plant Utility by WDM

Fiber plant comprises the most important infrastructure of a fiber-optic access network. Constructing a fiber plant requires extensive investments in capital, time, labor, license approval, and complicated right-of-way negotiations. Techniques to improve fiber plant utilization are therefore very valuable to infrastructure owners.

WDM provides a path to multiply the utility of a fiber plant by spectral segregation [54]. For example, one can imagine a broadband WDM wavelength interleaver which we call a "bandwidth sorter" here. The bandwidth sorter segregates the optical spectrum into interleaved bands as shown in Fig. 2.51. Such band sorters may be economically realized using optical thin-film filters, fiber Bragg gratings, or planner light waveguide technologies.

Different PON technologies can then be deployed on the same physical fiber plant using interleaved bands. For example, the "−" band in Fig. 2.51 can be occupied by an EPON while the "+" band, by an APON or G-PON, which is shown in Fig. 2.52. Thus, by a simple modification of the transceiver wavelengths and using low-cost band sorters, one can realize two PONs for one infrastructure.

The ITU-T G.694.2 standard [55] defines optical bands on a 20-nm CWDM (coarse WDM) grid, which are wide enough to allow uncooled DFB laser operation in each band. These CWDM bands could form the basis for band-sorter designs.

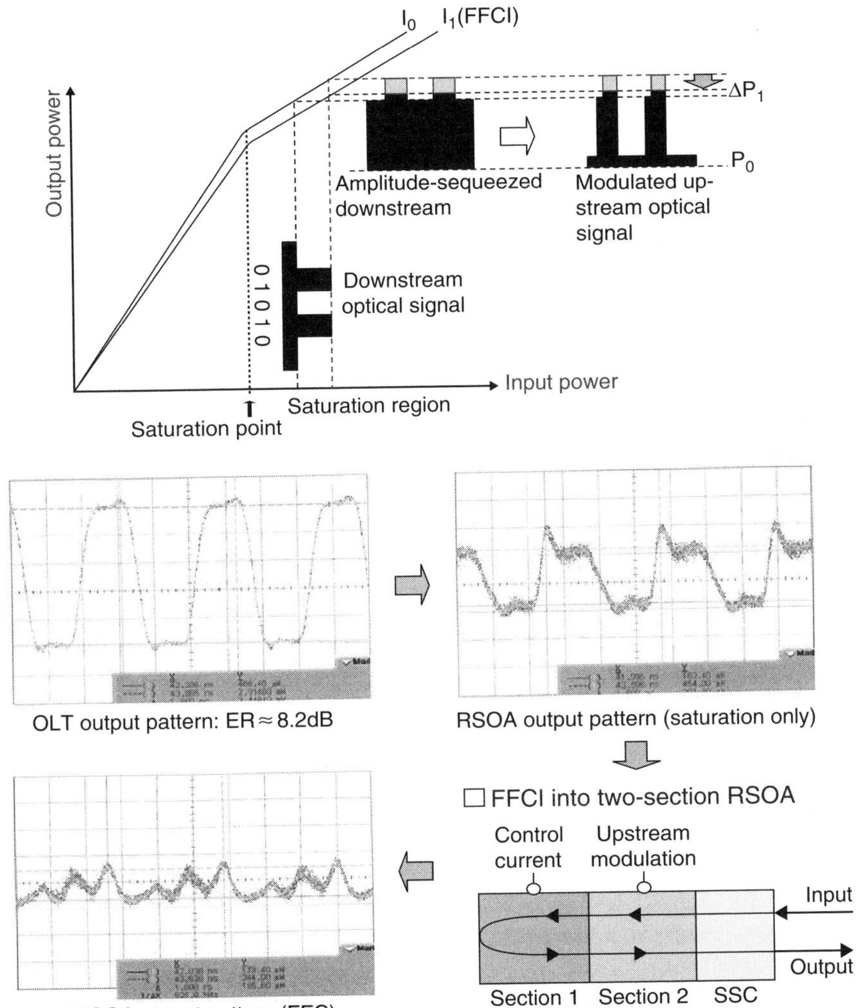

Figure 2.49 Feed-forward current injection to RSOA in WE-PON. (Courtesy of ETRI and Korea Telecom.)

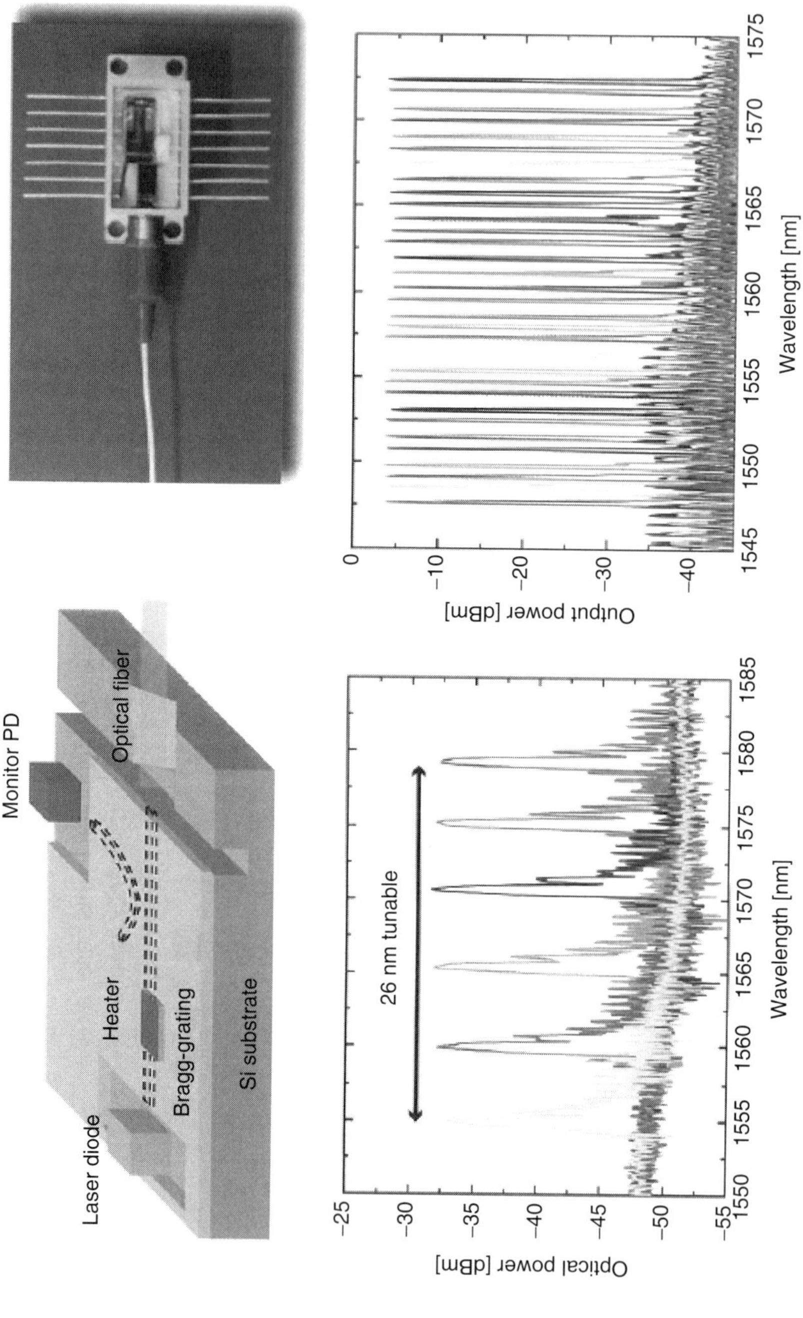

Figure 2.50 PLC-ECL used in WE-PON prototype. (Courtesy of ETRI and Korea Telecom.)

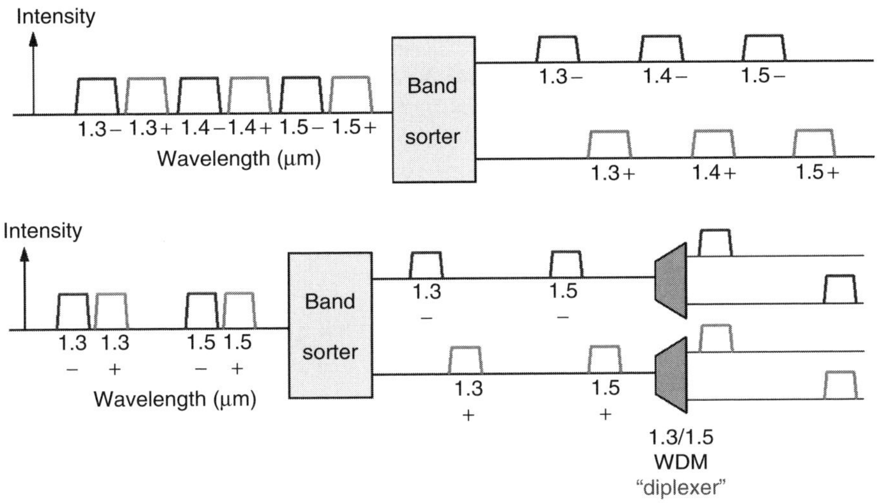

Figure 2.51 A band sorter interleaves the optical spectrum into bands (top) which can be further separated with diplexers (bottom) into bands for upstream and downstream PON connections.

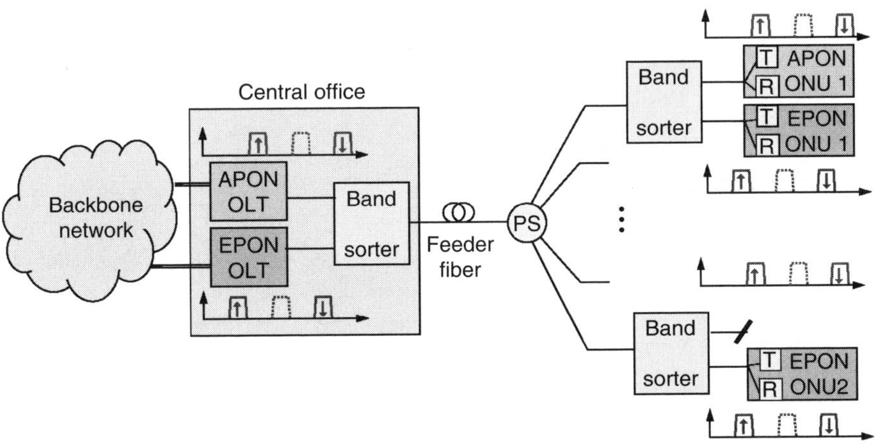

Figure 2.52 Overlaying APON and EPON on the same fiber plant using "band sorters."

2.11 SUMMARY

In summary, we have reviewed various PON architectures in this chapter. PON has become an import access technology to offer next generation broadband services. It offers long-reach and high-bandwidth passive local loops using single-mode optical fiber. There are two major ways to realize a passive distribution plant: power splitting (TDM-PON) and wavelength multiplexing (WDM-PON).

Time division multiplexing is used for multiplexing data from different ONUs on a power-splitting PON. A ranging process is required to set up ONUs at different transmission distances from the OLT with a common logical timing reference, which is used in scheduling ONU transmissions. In a power-splitting PON, the PON bandwidth is shared among ONUs. Dynamic bandwidth allocation can be used to improve the bandwidth efficiency and user experience.

Power budget eventually limits the speed, distance, and ONU count in a power-splitting PON. A WDM-PON provides an upgrade path to overcome the limitations of a power-splitting PON. In a WDM-PON, a WDM multiplexer is used at the RT to combine the signals from different ONUs. Since each user is allocated with its own wavelengths, WDM-PON offers better capacity, security, privacy, as well as easier upgrade because each user can be individually upgraded.

The WGR is an important optical element for WDM-PON remote node. It enables many scalable and flexible WDM-PON designs. By using CWDM technology on a PON infrastructure, one can reuse the fiber plant infrastructure through different FSRs in a WDM-PON or overlaying multiple PON technologies.

REFERENCES

[1] N.J. Frigo, P.P. Iannone, and K.C. Reichmann, "A view of fiber to the home economics," *IEEE Opt. Commun.*, vol.2, no.3, ppS16–S23, Aug., 2004.

[2] N.J. Frigo, "A Survey of Fiber Optics in Local Access Architectures," in *Optical Fiber Telecommunications, IIIA*, edited by I.P. Kaminow and T.L. Koch, Academic Press, pp461–522, 1997.

[3] G.P. Agrawal, *Fiber-Optic Communication Systems*, 2/e, John Wiley & Sons, 1997.

[4] E. Desurvire, *Erbium-Doped Fiber Amplifiers*, John Wiley & Sons, 1994.

[5] W.I. Way, *Broadband Hybrid Fiber/Coax Access System Technologies*, Academic Press, 1999.

[6] Corning SMF-28e Fiber Data Sheet, available from http://doclib.corning.com/OpticalFiber/pdf/pi1463.pdf

[7] P. Bohn, S. Das, "Return loss requirements for optical duplex transmission," *J. Lightwave Tech.*, vol.5, pp234–254, 1987.

[8] ITU-T Recommendation G.983.1, Broadband Optical Access Systems Based on Passive Optical Networks (PON), 2005.

[9] IEEE Standard 802.3, 2005 Edition, Carrier Sense Multiple Access with Collision Detection (CSMA/CD) access method and physical layer specifications.
[10] L.A. Coldren and S.W. Corzine, *Diode Lasers and Photonic Integrated Circuits*, John Wiley & Sons, 1995.
[11] J. Wellen, "High-speed FTTH technologies in an open access platform—the European MUSE project," in *Broadband Optical Access Networks and Fiber-to-the-Home*, edited by C. Lin, pp139–166, John Wiley & Sons, 2006.
[12] IEC 60825–1, Safety of laser products—Part 1: Equipment classification, requirements and user's guide, 2001.
[13] S. Tanenbaum, *Computer Networks*, 3/e, Prentice Hall, 1996.
[14] ITU-T G.984.1, Gigabit-capable Passive Optical Networks (GPON): General Characteristics, 2003.
[15] ITU-T G.984.2, Gigabit-capable Passive Optical Networks (GPON): Physical Media Dependent (PMD) layer specification, 2003.
[16] ITU-T G.983.3, A broadband optical access system with increased service capability by wavelength allocation, 2001.
[17] F. Coppinger, L. Chen, and D. Piehler, "Nonlinear Raman cross-talk in a video overlay passive optical network," paper TuR5, *OFC 2003 Tech. Digest*, vol.1, pp285–286, 2003.
[18] ITU-T G.983.2, ONT management and control interface specification for B-PON, 2005.
[19] FSAN website: http://www.fsanweb.org/
[20] ITU-T SG15 website: http://www.itu.int/ITU-T/studygroups/com15/index.asp
[21] ITU-T G.982, Optical access networks to support services up to the ISDN primary rate or equivalent bit rates, 1996.
[22] ITU-T G.983.4, A broadband optical access system with increased service capability using dynamic bandwidth assignment, 2001.
[23] W.J. Goralski, *Introduction to ATM Networking*, McGraw-Hill, 1995.
[24] ITU-T G.984.3, Gigabit-capable Passive Optical Networks (G-PON): Transmission convergence layer specification, 2004.
[25] S.M. Sze, *Physics of Semiconductor Devices*, 2/e, John Wiley & Sons, 1981.
[26] ITU-T Recommendation G.7041/Y.1303 (2003), *Generic framing procedure (GFP)*.
[27] ITU-T G.709/Y.1331, *Interfaces for the Optical Transport Network (OTN)*, 2003.
[28] ITU-T G.983.5, A broadband optical access system with enhanced survivability, 2002.
[29] ITU-T G.983.6, ONT management and control interface specifications for B-PON system with protection features, 2002.
[30] Ethernet for the first mile (EFM) website: http://www.ieee802.org/3/efm/
[31] IEEE Standard 802.1d, *Media Access Control Bridge*, 1993.
[32] J.J. Roese, *Switched LANs*, McGraw-Hill, 1999.
[33] Metro Ethernet Forum White Paper, "Introduction to circuit emulations services over Ethernet," available from http://www.metroethernetforum.org/PDFs/WhitePapers/Introduction-to-CESoE.pdf
[34] IETF draft "Emulation of structure agnostic (unstructured) PDH services (T1/E1/T3/E3)," available from http://www.ietf/org/internet-drafts/draft-ietf-pwe3-satop-01.txt
[35] IETF PW3 working group, available from http://www.ietf.org/html.charters/pwe3-charter.html
[36] M. Hajduczenia, H.J. da Silva, and P.P. Monteiro, "Flexible logical-link-identifier assignment policy for Ethernet passive optical networks based on extended multipoint-control-protocol DU flow control," *J. Optical Networking*, vol.5, no.9, pp681–698, Sept., 2006.

[37] M. Hajduczenia, H.J. da Silva, and P.P. Monteiro, "EPON versus APON and G-PON: a detailed performance comparison," *J. Optical Networking*, vol.5, no.4, pp298–319, Apr., 2006.
[38] IEEE 802.3av 10GEPON task force website: http://www.ieee802.org/3/av/index.html
[39] Van de Voorde, C. Martin, J. Vandewege, and X.Z. Qiu,"The SuperPON demonstrator: an exploration of possible evolution paths for optical access networks," *IEEE Commun. Mag.*, vol.38, no.2, pp74–82, Feb., 2000.
[40] C. Bouchat, C. Martin, E. Ringoot, M. Tassent, I. Van de Voorde, B. Stubbe, P. Vaes, X.Z. Qiu, and J. Vandewege, "Evaluation of superPON demonstrator," 2000 IEEE/LEOS Summer Topical Meetings, Broadband Access Technologies, pp25–26, July 24–28, Aventura, Florida, 2000.
[41] J.M.P. Delavaux, G.C. Wilson, C. Hullin, B. Beyret, and C. Bethea, "QAM-PON and Super PON for Access Distribution Networks," paper WN2, *OFC 2001 Technical Digest*.
[42] L. Spiekman, "Applications of Semiconductor Optical Amplifiers," Tutorial We 1.3, Proceedings of ECOC 2003, vol.5, pp268–297, Remini, Italy, 2003.
[43] C. Headley and G.P. Agrawal, *Raman Amplification in Fiber Optical Communication Systems*, Academic Press, 2005.
[44] A. Banerjee, Y. Park, F. Clark, H. Song, S. Yang, G. Kramer, K. Kim, and B. Mukerjee, "Wavelength-division-multiplexed passive optical network (WDM-PON) technologies for broadband access: a review," *J. Optical Networking*, vol.4, no.11, pp737–756, Nov., 2005.
[45] Dragone, "A NxN optical multiplexer using a planar arrangement of two star couplers," *IEEE Photon. Technol. Lett.*, vol.3, pp812–815, 1991.
[46] P.P. Iannone, K.C. Reichmann, and N.J. Frigo, "Broadcast Digital Video Delivered over WDM Passive Optical Networks," *IEEE Phot. Technol. Lett.*, vol.8, pp930–932, 1996.
[47] C.F. Lam, K.C. Reichmann, and P.P. Iannone, "Cascadable Modular Transmitter and Receiver for Delivering Multiple Broadcast Services on WDM Passive Optical Networks," *Proceedings of 26th European Conference on Optical Communication*, vol.1, pp109–110, Munich, Germany, Sept. 3–7, 2000.
[48] C.F. Lam, M.D. Feuer, and N.J. Frigo, "Performance of PIN and APD receivers in high-speed WDM data transmission systems employing spectrally sliced spontaneous emission sources," *Elect. Lett.*, vol.36, no.8, pp1572–1574, Aug. 31, 2000.
[49] Y. Li, US Patent 5,926,298, Optical multiplexer/demultiplexer having a broadcast port, July, 1999.
[50] N.J. Frigo, K.C. Reichmann, P.P. Iannone, J.L. Zyskind, J.W. Sulhoff, and C. Wolf, "A WDM PON architecture delivering point-to-point and multiple broadcast services using periodic properties of a WDM router," post-deadline paper PD24, *OFC 97*.
[51] M.Birk, P.P. Iannone, K.C. Riechmann, N.J. Frigo, and C.F. Lam, US Patent 7,085,495, "System for flexible multiple broadcast service delivery over a WDM passive optical network based on RF block-conversion of RF service bands within wavelengths bands," Aug., 2006.
[52] D.J. Shin, D.K. Jung, H.S. Shin, J.W. Kwon, S. Hwang, Y. Oh, and C. Shim, "Hybrid WDM/TDM PON with wavelength-selection-free transmitters," *IEEE J. Lightwave Tech.*, vol.23, no.1, Jan., 2005.
[53] J. Yoo, H. Yun, T. Kim, K. Lee, M. Park, B. Kim, and B. Kim, "A WDM-Etherent hybrid passive optical network architecture," *8th International Conference on Advanced Communication Technology* (*ICACT 2006*), Feb., 2006.
[54] N. Frigo and C.F. Lam, "WDM Overlay of APON with EPON—a carrier's perspective," http://grouper.ieee.org/groups/802/3/efm/public/sep01/lam_1_0901.pdf
[55] ITU-T G.694.2, Spectral grids for WDM applications: CWDM wavelength grid, 2002.

Chapter 3

Optical Technologies in Passive Optical Access Networks

Elaine Wong

The University of Melbourne

3.1 INTRODUCTION

The optical technologies in the access network environment have been developing and advancing at a rapid rate with goals of achieving ease of installation, manageability, upgradeability, and customer friendliness, in addition to performance and high reliability. Each of these goals bears a direct contribution to the overall capital and operational expenditures for the network. For widespread uptake and deployment, it is imperative that the network be cost-competitive to current access technologies. In this chapter, optical technologies that have been developed to achieve these goals are reviewed.

The optical power splitter is a central component in a power-splitting passive optical network (PON) where its primary function is to split the optical power at the common port equally among all its output ports. Significant improvements in optical power splitters have been achieved by the introduction of planar lightwave circuit (PLC) technology where high reliability, low cost per port, low insertion loss, and high splitting ratio uniformity have been achieved. PLC technology not only allows for chip-size devices but also enables the integration of multiple functions on a single chip, e.g. the dual function power splitter and mux/demux devices. Nonetheless, there are still areas of improvement that are currently being focused on in power-splitter designs and some of these targeted areas will be reviewed.

Currently, each subscriber in fiber-to-the-home (FFTH) access network is projected to demand an average bandwidth of 50–60 Mbps so that simultaneous broadband services such as high definition television (HDTV), HD video streaming, and interactive online gaming over the broadband Internet can be supported. Though such speeds may be accommodated in already-deployed power-splitting PONs, the trend of personal websites (e.g. YouTube, Myspace, Bit Torrent) will necessitate speeds exceeding 1 Gbps and more importantly in the upstream direction in the near future. Many leading service providers and component manufacturers agree a wavelength division multiplexing (WDM) network will be the preferred upgrade solution in order to satisfy such high and increasing bandwidth demands. In a WDM-PON, each user is assigned their own downstream/upstream wavelength channels, enabling dedicated and potentially symmetric downstream/ upstream bandwidths. An arrayed waveguide grating (AWG) serves as the passive WDM mux/demux or a passive WDM routing component at the remote node, replacing the optical power splitter in a power-splitting PON. Together with its unique cyclic property in which downstream and upstream wavelengths assigned to a particular optical network unit (ONU) can enter and exit through the same AWG port, the AWG has facilitated many additional features of the WDM-PON including colorless ONU, wavelength reuse, and protection schemes.

Nonetheless, the central wavelength of a conventional silica-based AWG shifts by as much as 0.0125 nm/°C, and for outside passive plant environment, this necessitates temperature stabilizing elements which in turn require continual power supply. For passive operation in outside plants, the AWG must exhibit good temperature stability, hence athermal operation. Athermal operation in an AWG can be largely achieved using two techniques which will be reviewed in this chapter. They incorporate: (a) the usage of guiding materials that exhibit negative thermo-optic coefficient with opposite temperature dependence to silica; and (b) alternatively, incorporating a mechanically movable compensation plate in the AWG structure. Both techniques compensate for the thermal refractive index change and thermal expansion of the optical path length in silica.

One of the dominant cost factors in FTTH is the installation of the ONU site where highly skilled workers are trained to ensure that the splicing between indoor and outdoor fibers, the wiring of indoor fibers, and the installation of the optical termination box, is within optical coupling loss and bending loss constraints. This is because conventional silica-based single mode fiber (SMF) has a very small fiber core (5–10 μm) that requires precise fiber-to-fiber and fiber-to-source connections. A slight displacement in the connection can yield very large coupling loss. Compounding the issue is that conventional SMF suffers from attenuation if bent with a small radius (< 30 mm). To remove the need for fusion as well as mechanical splicing that necessitate splicing machines and fixtures for fiber alignment, novel connectors based on a plug and socket configuration, have been developed. Further, such a connector allows for easy and quick connection and disconnection between the outdoor and indoor fibers, thereby improving fiber fault detection and isolation between the outdoor and indoor fibers. Angled physical contact (APC) connectors which are normally reserved for analog hybrid fiber coax (HFC) networks are also being considered for access networks especially those with analog video overlay. With APC, the return losses, especially at the end of an unused and unterminated distribution fiber, can be reduced to as low as 65 dB. Further, indoor connectors that can bend from 0 degrees to 90 degrees at the wall socket were also developed to increase customer handling, optical wiring workability, safety, and a good appearance.

To further increase the uptake on FTTH, research activities have also been focused on developing high performance, highly functional, yet cost-effective indoor optical fibers that are more susceptible to customer handling with lower bending loss constraints. Bend insensitive silica-based single-mode fibers that have a bending radius of only 15 mm have been developed for indoor fiber installations which allow for cleaner installation with bent sections that are less prone to accidental snagging by customers. An optical curl cord which exploits the properties of hole-assisted fibers to achieve very small bending radius and reduced bending losses as compared to the SMF, has also been introduced in the FTTH market. Such a curl cord can be stretched several times in length and retracted like the one used with a telephone, thus providing flexible and reliable connection between communication devices at home or in the office. Another promising optical fiber candidate to achieving low-cost and user-friendly optical connections within a home or an office is the graded index polymer optical fiber (GI-POF), which has a large core (0.5–1.0 mm) that can tolerate a higher displacement in fiber connections.

Another cost-prohibitive factor in the FTTH environment lies in the inventory of the wavelength-specific transmitter sources, especially at the ONU site. The high initial component and installation costs along with operational costs involving administration and management of the distinct ONU wavelengths, must be overcome in order to increase the uptake of WDM-PONs. Though still at the research stage with a few exceptions of deployments in field trials, cost-effective ONU configurations are currently being actively researched worldwide. Cost-effective ONU configurations can largely be categorized into three groups. The first category of ONUs is known as wavelength specific ONUs such as distributed feedback (DFB) and distributed Bragg reflector (DBR) lasers. These single-mode lasers are suitable candidates as ONU transmitters due to their high-speed direct modulation property. Nonetheless, due to the temperature drift of these lasers (with a temperature coefficient of 0.1 nm/°C), active temperature control and wavelength feedback and monitoring circuitry are required to ensure that emission wavelengths are maintained at their designated wavelengths, thus contributing to high power consumption, bulk size, system complexity, and bottom-line cost. A cost-effective alternative to DFB and DBR lasers is the vertical-cavity surface-emitting laser (VCSEL) that permits on-wafer testing. Due to its short cavity length of typically 2–5 μm, VCSELs exhibit excellent single-mode behavior and its lasing light beam is circular rather than elliptical, allowing efficient coupling into a single-mode fiber. As the volume of the VCSEL is small, it also exhibits low-threshold currents in the sub-mA range, thus consuming very low power. Nonetheless, the temperature coefficient of VCSELs is identical to that of DFB and DBRs, necessitating wavelength stabilization circuitry. In that regard, injection-locking schemes for VCSELs have been proposed to align their emission wavelengths to the WDM grid to avoid active temperature control or wavelength monitoring.

The second category of ONUs and perhaps the category with the most potential of widespread deployment is known as colorless ONUs. Here, the term color refers to the color of the wavelength. The emission wavelength of a colorless ONUs is nonspecific and can be determined by external factors such as the filtering properties of the AWG in the remote node or the wavelength of an injection/seeding light into the ONU. As such, identical ONUs can be mass-produced and deployed across the network. Colorless ONUs can be further divided into three categories. In one category, each ONU incorporates a broadband optical source such as LED, which is modulated with upstream data. The broadband spectrum is then spectrally sliced at the AWG in the remote node. The modulation speeds in colorless ONU configurations based on broadband sources are limited due to several inherent noise sources, whose effect increases as a function of narrower optical channel bandwidth and increasing data rate. Alternatively, using an external spectrally sliced broadband optical light injected into a multimode laser such as a Fabry–Perot laser diode (FP-LD) can result in the excitation of only one stable mode for high-speed modulation. Likewise, injecting an external seeding light into a reflective semiconductor optical amplifier (RSOA) enables the generation of a frequency-stable, low-intensity, low-phase noise yet high-power optical source. The emission wavelength of an injection-locked FP-LD or a wavelength-seeded RSOA follows that of the seeding light. Recently, schemes that remove the need for centralized broadband sources at the central office (CO) was proposed based on self-injection-locking FP-LDs and self-seeding RSOAs. Typically, these schemes employ an optical feedback element in the remote node to reflect part of the upstream spectrum from either an FP-LD or RSOA back to itself. Once self-injection-locking or self-seeding has been established, the FP-LD or RSOA can be directly modulated with upstream data.

The final category of ONUs is known as source-free ONUs. In this particular category, the ONUs do not contain optical sources, with the optical light required for upstream transmission originating from the CO. The upstream data can be modulated onto the downstream carrier and sent back upstream to the CO. Carrier-reuse schemes often adopt different modulation formats for downstream and upstream transmissions to prevent residual downstream modulation on the upstream data. Recently, source-free ONU schemes have been proposed based on downstream data that is modulated on a subcarrier that is offset from a base-band carrier. The base-band carrier is used at the ONU for modulation of the upstream data. This category of ONUs, along with others, will be reviewed in detail in this chapter. Collectively, the advancements in the above-mentioned areas have contributed to the rapid uptake and deployment of power-splitting PONs, with future-proof WDM-PONs following suit in the near future.

3.2 OPTICAL TECHNOLOGIES IN PASSIVE OUTSIDE PLANT

3.2.1 Planar Lightwave Circuit (PLC)-Based Optical Power Splitter

In a power-splitting PON, an optical power splitter is the passive device in the outside plant that physically connects to the CO with a feeder fiber. It also connects to a number of ONUs via a series of distribution fibers. In the past few years, significant improvements in reliability, cost per port, insertion loss, and splitting-ratio nonuniformity, have been demonstrated with planar lightwave circuit (PLC)-based splitters. Central to the splitter is a PLC chip comprising of optical waveguides fabricated on a planar substrate, typically made of silicon or quartz, to form a cascade of Y-branches. For a $1 \times N$ splitter, one side of the PLC chip is aligned to a fiber whereas the opposite side is aligned to an array of N fibers, as shown in Fig. 3.1 [1]. The number of power-splitting fanouts in a PON is typically $N = 16$ and $N = 32$, but with an increasing demand of up to $N = 64$, thereby making the alignment of the fiber array to the PLC chip more challenging [2]. Compared to fused biconical-taper-based splitters, PLC

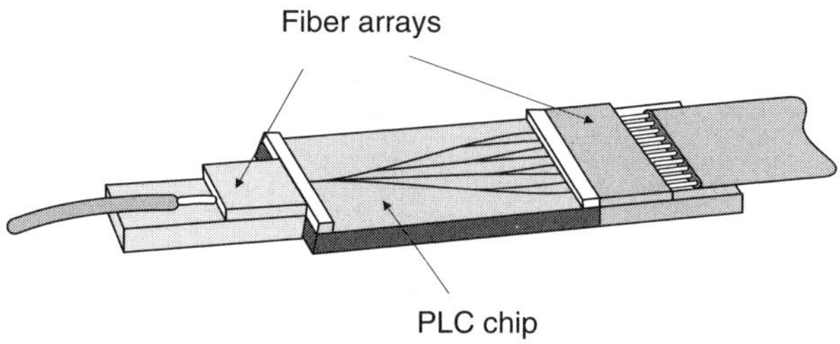

Figure 3.1 A PLC splitter is manufactured using two fiber arrays and one PLC chip all aligned within one package. (From Ref [1])

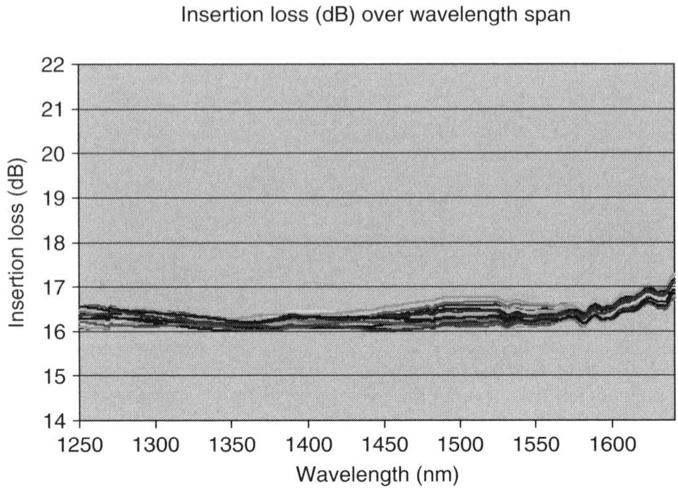

Figure 3.2 Measured insertion loss of a 1×32 PLC splitter as a function of wavelength, showing uniform insertion loss (≤1.25-dB variation) and low excess insertion less (≤1.25 dB) above the theoretical value. (From Ref [3])

technology allows for chip-size devices with the potential of integrating multiple functions, e.g. WDM coupler, onto a single chip. It also enables a more uniform loss over a wide operating range of wavelengths from 1250 nm to 1625 nm, and operation of a wide range of temperatures from −40 °C to + 80 °C [1–6]. Figure 3.2 illustrates the measured insertion losses from samples of 1×32 PLC-based splitters approved by AT&T Labs for use in the Project Lightspeed FTTH trial, showing uniform loss over a wide wavelength range [3].

Aside from uniform loss, the insertion loss of PLC-based splitters is another important parameter in network implementations that will influence system performance and the overall cost per drop. Lower insertion loss PLC-based splitters will extend the reach and number of customers that can be accommodated within the same PON, yielding higher revenue per PON for service providers. Aside of the theoretical splitting loss attributed to the division of optical power at the input port equally into N output ports, and given by the formula:

$$\text{Theoretical splitting loss (dB)} = 10 \times \log_{10}(1/N)$$

A PLC-based splitter suffers from excess insertion loss from fiber array alignment to the PLC chip, fiber array uniformity caused by pitch and depth inaccuracies in the v-grooves of fiber array block that holds the fiber array, splitting ratio uniformity caused by imperfections in the PLC chip due to manufacturing, inherent chip material loss, and connector loss. The targeted areas for improvement of insertion loss in PLC-based splitters have been in reducing connector losses, and improving fiber array and splitting ratio nonuniformity [1]. The connector loss can be improved from 0.5 dB to 0.15 dB through using high-quality ferrules and an excellent polishing method. With advances in manufacturing processes of the fiber array block and PLC chip, insertion losses from fiber array nonuniformity and splitting-ratio nonuniformity can be reduced from 0.7 dB to 0.4 dB, and 1.8 dB to 1.0 dB, respectively [2, 4]. Collectively, the excess insertion losses of PLC-based splitters are currently 1–1.5 dB above the ideal theoretical splitting loss with a nonuniformity within 2 dB over the specified range of operating wavelengths from 1250 nm to 1625 nm.

3.2.2 Arrayed Waveguide Gratings

3.2.2.1 Introduction

As briefly discussed in Sect. 3.1, the WDM technology has been considered as one of the most graceful upgrade paths beyond power-splitting PONs to support more users at higher bandwidths. For an upgrade in the outside plant, the power splitter in the remote node of a power-splitting PON is replaced with an AWG. In a $1 \times N$ configuration, an AWG serves as a wavelength router or a demultiplexer because a composite WDM signal launched into the input port is separated into individual channels by the device [7–10]. Due to reciprocity, the AWG can be used as a multiplexer in the reverse direction by upstream signals in a WDM-PON. AWGs based on the silica-on-silicon technology are most commonly used for their low propagation loss (<0.05 dB/cm) and high fiber coupling efficiency (losses in the order of 0.1 dB). AWGs based on the InP semiconductor technology are selected when small footprints and integration with other functions on a single chip are required, but at the expense of relatively high adjacent wavelength crosstalk (e.g. 4-channel AWG with a footprint of $230 \times 330\,\mu m^2$ and -12 dB crosstalk [11]).

The schematic layout of a $1 \times N$ AWG is shown in Fig. 3.3 [7]. The free-propagating regions or slabs, act as a lens whilst the waveguide array acts as a grating. The difference in length between two neighboring waveguide arms is constant. The beam of an incoming WDM signal, consisting of multiple channels of different wavelengths at a constant channel spacing, will become divergent in the first free-propagating region and will be directed into each waveguide. In each of the waveguides, the WDM signal experiences a different phase shift because of the different lengths of waveguides. The net result is that each wavelength channel will be focused on a specific output port after propagating through the second free-propagating region, thus forming a demultiplexer.

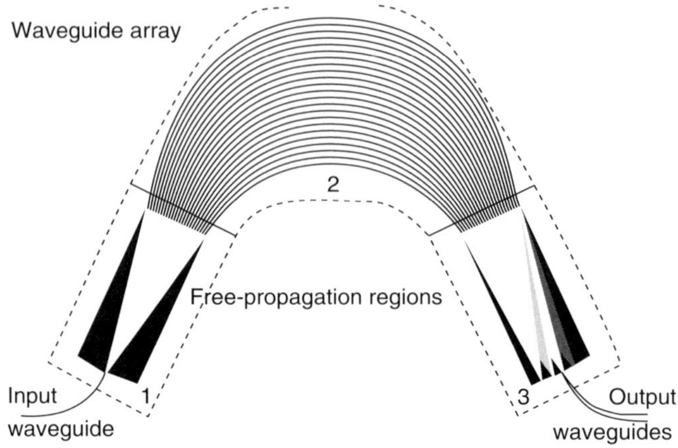

Figure 3.3 Schematic illustration of arrayed waveguide grating comprising two free-propagating regions and a waveguide array. (From Ref [7])

However, the requirements of an AWG used in a PON differ from that of a conventional WDM system. Aside from very low crosstalk level between adjacent channels to provide high channel isolation, an AWG that is located in the remote note of a PON must be completely passive for cost reduction, exhibit low-loss for extended reach, and offer reliable operation with stable characteristics over a wide range of temperatures.

An important parameter in the AWG is the free spectral range (FSR) which defines the wavelength periodicity of fixed width [12]. In turn, a cyclic AWG is one in which wavelengths that emerge at a particular output port is spaced by an integer of the FSR [12]. This cyclic property enables the downstream and upstream wavelengths assigned to a particular ONU to enter and exit through the same AWG port. The cyclic property also plays an important role in an $N \times N$ AWG which is used in WDM-PON schemes that facilitate colorless ONUs and wavelength reuse (Sect. 3.4 and Chap. 2, Sect. 2.10), in addition to WDM-PON protection schemes (refer to Chap. 6). An $N \times N$ AWG has N inputs and N outputs, and differs from the $1 \times N$ configuration by incorporating an additional $N-1$ waveguides at the input (Fig. 3.4). The way in which a signal propagates through an $N \times N$ AWG at each input waveguide is identical to that described in the preceding paragraph for the $1 \times N$ configuration. Figure 3.5 illustrates the cyclic property of an $N \times N$ AWG with $N = 5$ [12]. In the example, signals on five different wavelengths are incident on each of the five input ports. The wavelengths, $\lambda_1, \lambda_2, \ldots \lambda_5$, incident on input Port 1 are distributed amongst output Ports b, a, e, d, and c, respectively. The wavelengths, $\lambda_1, \lambda_2, \ldots \lambda_5$, if incident on input Port 2 will however be cyclically rotated and distributed amongst output Ports a, e, d, c,

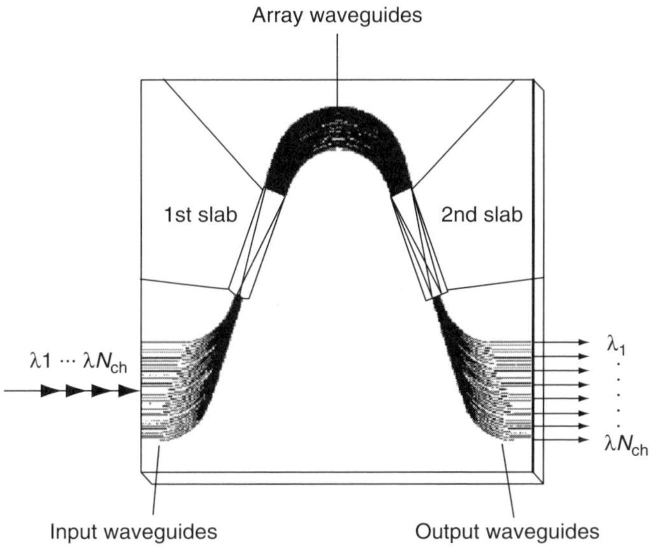

Figure 3.4 Schematic layout of $N \times N$ arrayed waveguide grating router. (From Ref [12])

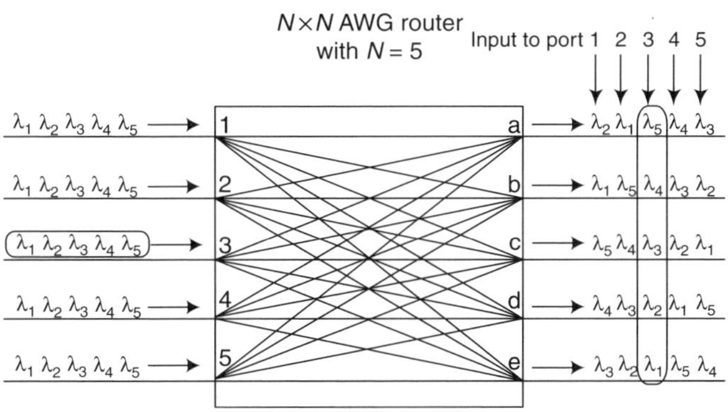

Figure 3.5 Schematic illustration of the cyclic property of an arrayed waveguide grating router. (From Ref [12])

and b, respectively [12]. The symmetric and cyclic property enables simultaneous transmission of signals on *N* wavelengths into *N* different input ports of the AWG, with each output port receiving signals on *N* different wavelengths, one from each input port. This property also enables the simultaneous multiplexing/demultiplexing of wavelengths in a nonblocking manner, finding applications in optical add drop multiplexers and optical cross-connects.

Dual-function AWG configurations have been proposed to achieve both optical power splitting and wavelength routing [13, 14]. Such a function will be essential to WDM and power-splitting PONs that are overlaid on the same physical fiber plant. The logical point-to-point topology on the WDM network is facilitated through the wavelength routing function, whereas the logical point-to-multipoint topology on the power-splitting network that is optimized for broadcast video services, is facilitated through the power-splitting function at the AWG. In [13], a PLC-based 14-channel optical splitter/router in a reflection-type AWG configuration as shown in Fig. 3.6 was proposed. For the WDM network with wavelengths $\lambda_1, \lambda_2, \ldots \lambda_N$, a signal incident on the AWG will first be diffracted into the slab, then divided equally in optical power into each waveguide. Signals in each of the waveguides experience a different phase shift because of different waveguide lengths. At the end face of the AWG, each signal is reflected by a multilayered dielectric filter back into the slab through the waveguides. As the signal wavefront is tilted due to the path difference between the arrayed waveguides, the signal is focused at a different output waveguide, depending on its wavelength. In contrast, for the overlaid power-splitting network, a signal on a broadcast wavelength λ_0 is incident on one of the AWG waveguide from the side of the multilayered dielectric filter (which is transparent to λ_0). The signal travels into the slab through the input waveguide, which divides it into each output.

Figure 3.7 shows the experimental measurements from a dual-function AWG, combining a 1×14 power-splitting function and a 14-channel wavelength router

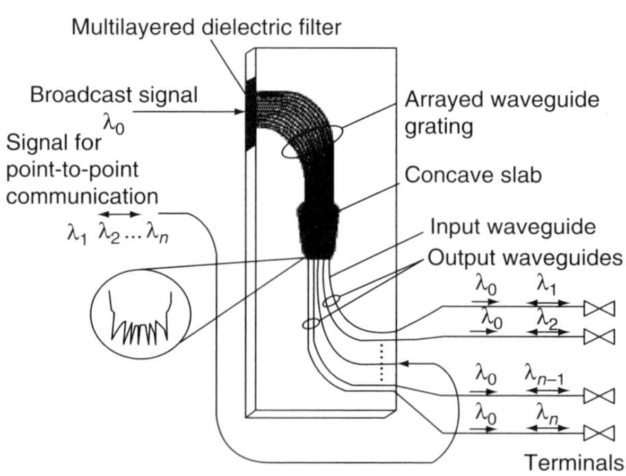

Figure 3.6 Schematic illustration of dual-function reflective arrayed waveguide grating. (From Ref [13])

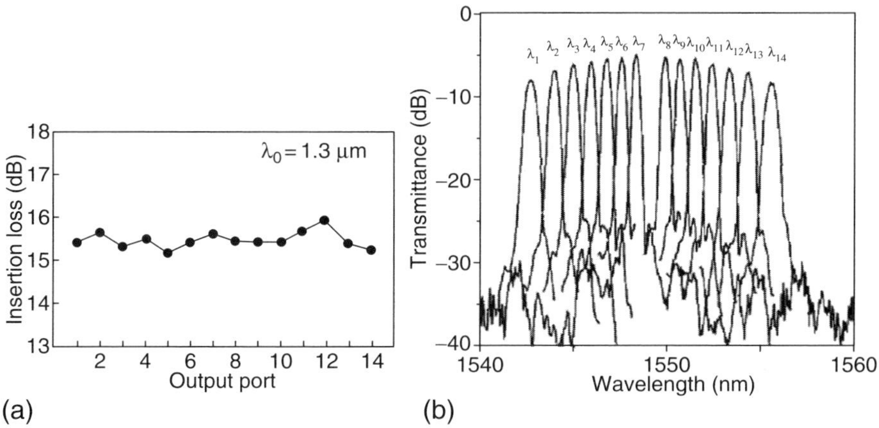

Figure 3.7 (a) Fiber-to-fiber insertion loss of 1×14 power splitter function; (b) transmission spectra of 14-channel wavelength-router function. (From Ref [13])

function. The broadcast wavelength λ_0 was allocated to 1.3 μm whilst the WDM wavelengths $\lambda_1, \lambda_2, \ldots \lambda_N$ were allocated to the 1.55 μm band. Figure 3.7 (a) plots the insertion losses of the splitter showing a deviation of 0.2 dB across the output ports, while Fig. 3.7 (b) shows the transmission spectra of the 14-channel wavelength router with insertion losses ranging from 5.2 dB to 8.5 dB and channel spacing between 0.8 nm and 1.3 nm. Funnels with nonuniform widths as shown in the inset of Fig. 3.6 were intentionally added to the interface between the slab and the waveguides to guarantee uniform insertion losses across the outputs for the power-splitting function, but at the expense of nonuniform channel spacing. Nonetheless, improvements to the fabrication of the end face for the multilayered dielectric filter are expected to reduce insertion losses arisen from both splitting and wavelength routing functions.

Another dual function AWG was proposed in [14] combining an AWG operating around the 1.55-μm band and an optical power splitter operating around 1.31-μm band. The dual-function device known as a two-PONs-in-one (2P1) device is formed on a single monolithic chip. Figure 3.8 shows the schematic diagram of a 1×8 2P1 device where both WDM and power-splitting signals are input through the same input fiber. An input coarse WDM coupler (CWDM-1) is used to separate the two types of signals. The coarse WDM coupler has two outputs ports with pass bands centered at 1.55 μm and 1.31 μm, and the output signals are directed to a 1×8 AWG and a 1×8 power splitter, respectively. A second coupler, CWDM-R, is used to provide additional isolation of the 1.31-μm signals from the 1.55-μm signals. At the output of the 2P1 device, a final coarse WDM coupler, CWDM-2, combines each demultiplexed

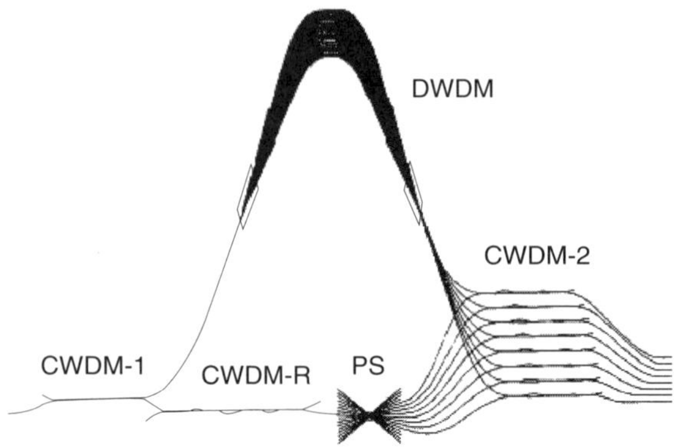

Figure 3.8 Schematic illustration of 1×8 two-PONs-in-one (2P1) device. For clarity, the vertical scale has been expanded by four times. (From Ref [14])

signal from the AWG with an output of the power splitter. Figure 3.9 shows the spectra from the output ports of the 1×8 2P1 device measured around the 1.3-μm waveband (Fig. 3.9 (a)) and 1.55-μm waveband (Fig. 3.9 (b)). For wavelengths in the range from 1.26 μm to 1.36 μm, the average insertion loss is 12 dB with a 2-dB variation around the value measured at 1.30 μm. The transmission spectra measured in the 1.55-m window show two free spectral ranges

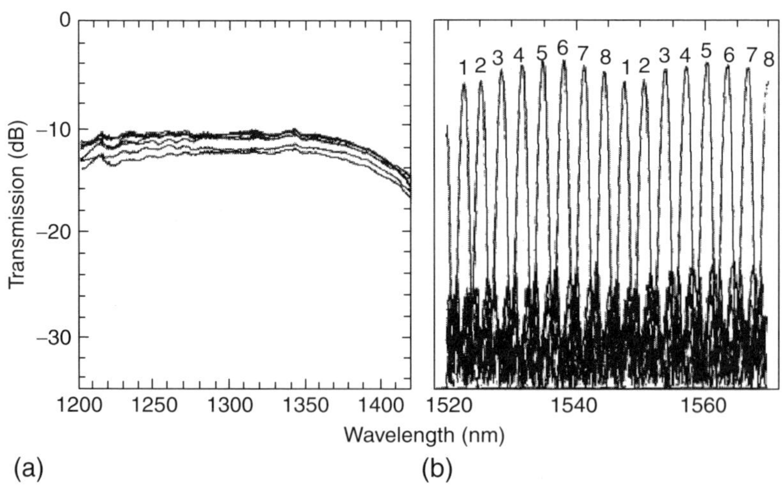

Figure 3.9 Spectra from the output ports of a 1×8 2P1 device measured around the (a) 1.3-μm waveband and (b) 1.55-μm waveband. (From Ref [14])

with 3.23-nm (400 GHz) channel spacing, and the insertion losses from 4 dB to 6 dB. Excess insertion losses in the optical power splitter and AWG are expected to be reduced with further design optimizations [14].

3.2.2.2 Athermal Arrayed Waveguide Grating

Existing temperature-controlled AWGs require continual power supply to the temperature-stabilizing elements, thus making them unsuitable for use in outside passive fiber plant environments. Traditionally, temperature-stabilizing elements comprising thermoelectric cooler and control circuitry are incorporated in AWG packaging as silica-based waveguides are extremely sensitive to ambient temperatures. The central wavelengths of conventional silica-based AWGs shift by as much as 0.0125 nm/°C. This wavelength shift is attributed to the thermal refractive index change and thermal expansion of the optical path length in silica. For passive operation in outside plants, the AWG must exhibit good temperature stability without requiring the need for temperature monitoring and control.

Temperature-insensitive or athermal silica-based AWGs have been extensively studied. The earliest investigations to achieve athermal operation in an AWG were based on employing guiding materials that exhibit negative thermo-optic coefficients with opposite temperature dependence to silica [15–17]. In [16], a triangular groove formed in the waveguide array and filled with a silicone resin with negative thermo-optic coefficient was implemented to compensate for the temperature-dependent optical path difference in the silica-based arrayed waveguides. The proposed AWG structure is shown in Fig. 3.10. The polyamide half-waveplate inserted in the arrayed waveguide was incorporated to eliminate polarization dependence [18]. Figure 3.11 compares the temperature-dependent transmission spectrum of the central wavelength, Channel 4, over the temperature range of 0–85 °C between a conventional silica-based 1×8 AWG with 200-GHz channel spacing (Fig. 3.11 (a)) and an athermal 1×8 AWG with 200-GHz channel spacing (Fig. 3.11 (b)). The latter shows minimal temperature dependence. The crosstalk level was measured to be −29 dB with insertion losses from 3.9 dB to 4.2 dB, depending on the output port. Figure 3.12 shows the temperature-dependent wavelength shift of Channel 4 in the temperature range of 0–85 °C, with the results confirming a reduction in wavelength shift from 0.95 nm using the conventional AWG structure to 0.05 nm using the proposed athermal AWG structure.

An improvement over the structure in [16] was proposed in [17] to reduce the excess loss (~2 dB) caused by the triangular groove in the arrayed waveguide. It was shown that the excess loss increases quadratically with the width of the groove. By replacing the triangular groove with multiple grooves of smaller widths (refer to Fig. 3.13), a reduction of excess loss from 2 dB to 0.4 dB was

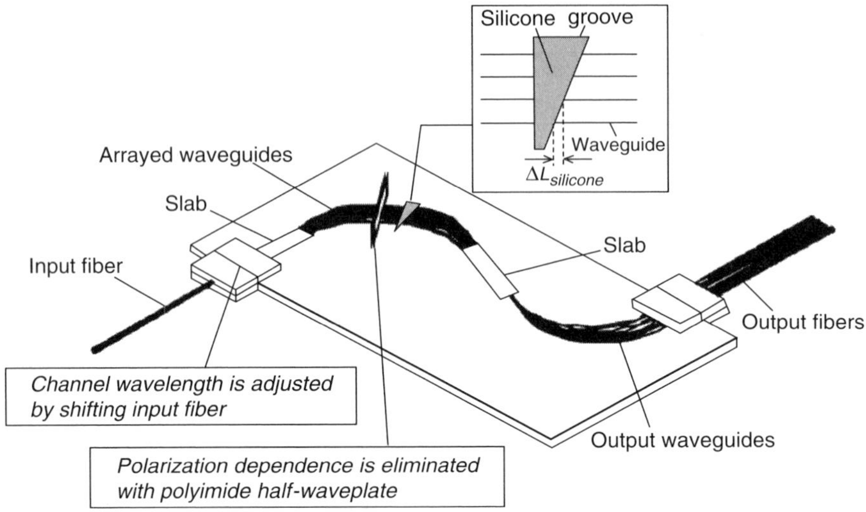

Figure 3.10 Schematic diagram of athermal AWG with silicone resin-filled triangular groove. (Adapted from Ref [15])

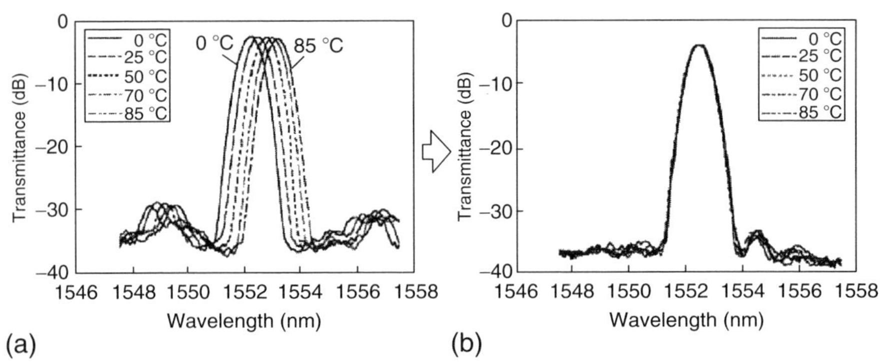

Figure 3.11 Temperature-dependent transmission spectrum of Channel 4 over the temperature range of 0–85 °C for (a) conventional silica-based AWG and (b) athermal AWG with silicon resin-filled triangular groove. (From Ref [16])

achieved whilst retaining low loss (~3.5 dB), low crosstalk (<−30 dB), and higher temperature stability (~0.02 nm from 0 °C to 85 °C) [17].

To lower costs in addition to those associated with the removal of the temperature control and power source, the compactness and size of the AWG chip must be taken into account. Given a fixed-size wafer, a smaller footprint will reduce the overall manufacturing, testing, and packaging costs per chip [19].

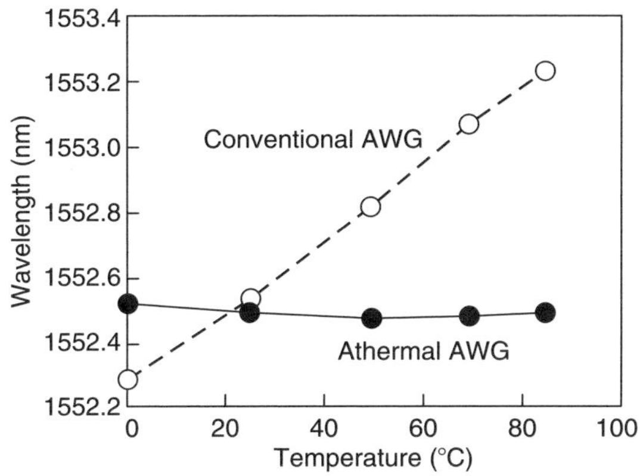

Figure 3.12 Temperature-dependent wavelength shift of Channel 4 in the temperature range of 0–85 °C. (From Ref [16])

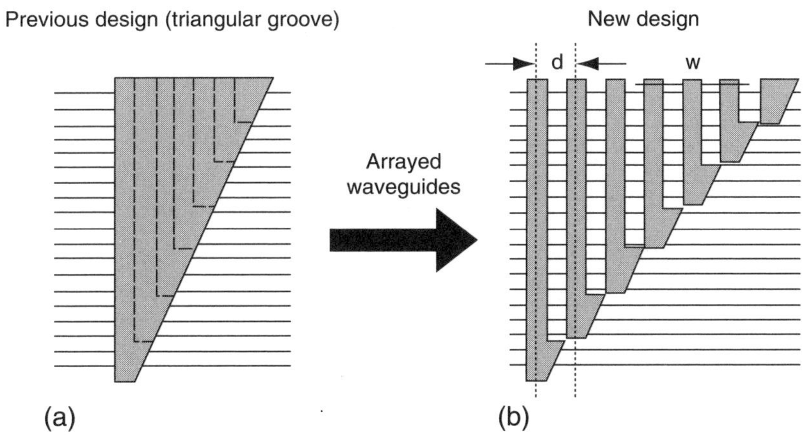

Figure 3.13 Reduction of excess loss by replacing triangular groove of wider width with several grooves of smaller widths. (From Ref [17])

One approach to reducing chip size is to increase the relative index difference Δ between the core and cladding. For temperature-controlled AWGs, a one-tenth reduction in chip size was achieved using 2.5%-Δ waveguides, compared to the conventional 0.8%-Δ [20]. However, high-Δ waveguides in athermal AWGs will result in increased diffraction losses at the trenches due to the reduced spot size. In 1.5%-Δ athermal AWG waveguides, a modified low-loss groove and array structure was proposed to reduce the lateral diffraction loss from 1.9 dB to

0.4 dB [21]. The combination of a wider-arrayed waveguide, unequally divided groove, and waveguide gaps adjacent to the grooves, shown schematically in Fig. 3.14, results in an increase in lateral spot size that in turn has reduced power leakage at the grooves due to smaller diffraction angle. In another 1.5%-Δ athermal AWG configuration shown schematically in Fig. 3.15, silicon resin-filled trenches were inserted in the slab region to reduce the lateral diffraction loss (~0.7 dB) [20].

The reduction in diffraction loss in the vertical direction is also important, especially for higher-Δ athermal AWGs. A configuration composed of spot size converters based on a segmented core in conjunction with silicone resin-filled

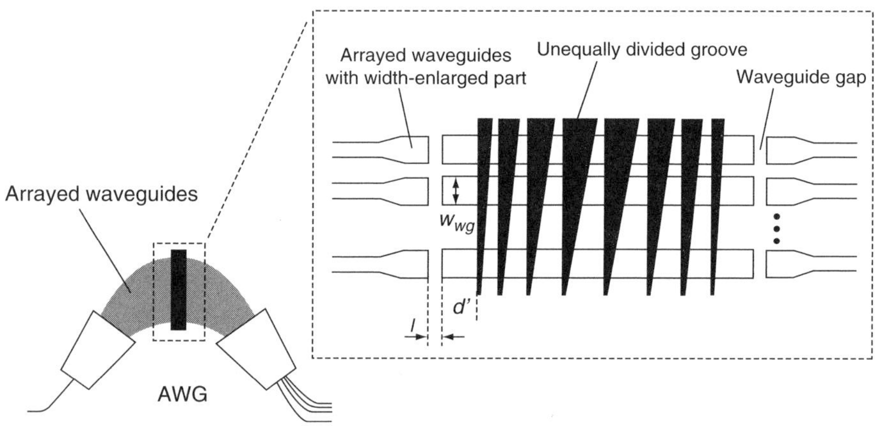

Figure 3.14 Schematic diagram of modified low-loss groove and array structure for a 1.5%-Δ athermal AWG. (From Ref [21])

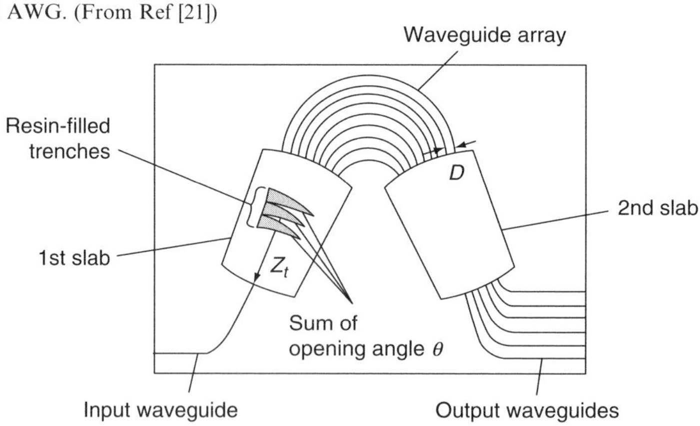

Figure 3.15 Schematic diagram of silicon resin-filled trenches in the slab region of 1.5%-Δ athermal AWG. (From Ref [20])

trenches in the slab region was proposed in [19] (refer to Fig. 3.16) to reduce the diffraction loss in both the lateral and vertical directions in a 2.5%-Δ athermal AWG. A diffraction loss of 0.9 dB was achieved as opposed to the 2.4 dB using a regular 2.5%-Δ athermal AWG structure.

A slight drawback to the techniques mentioned above for athermal AWG operation is the further need to insert a polyamide half-waveplate in the waveguide array to eliminate polarization dependence [18]. One approach to achieving simultaneous temperature and polarization insensitivity during the waveguide forming process is to form a polymer overcladding with negative thermo-optic coefficient on a silicon core and undercladding [22]. The waveguide structure is shown in Fig. 3.17 (a). In this design, the size of the rectangular core is $6 \times 6\ \mu m$ and commercially available fluoracrylate type-polymer was used to form the polymer overcladding of a 1×16 AWG with a channel spacing of 100 GHz. The measured polarization-dependent wavelength shift for Channel 9 (central wavelength) is shown in Fig. 3.17 (b). A comparison with an AWG with a silica overcladding shows that the polymer overcladding minimized the wavelength shift to 0.05 nm. The wavelength shift can be further minimized with the optimization of the cladding thickness. In the temperature range of 26~100 °C, the central wavelength shift of 0.98 nm in conventional silica overcladding AWG was minimized to 0.21 nm by the polymer overcladding, agreeing with the simulated results shown in Fig. 3.17 (c). However, the measured insertion losses were 6~7 dB depending on the channel output and the crosstalk level was −25 dB, attributed to the incompleteness of the fabrication process [22].

An alternative to achieving simultaneous temperature and polarization insensitivity during the waveguide forming process is to use polymer materials in both the waveguide structure as well as the substrate [23]. In [23], the polymer waveguide materials used were fluoracrylate-type polymers where its large negative thermo-optic coefficient was matched to the large positive coefficient of

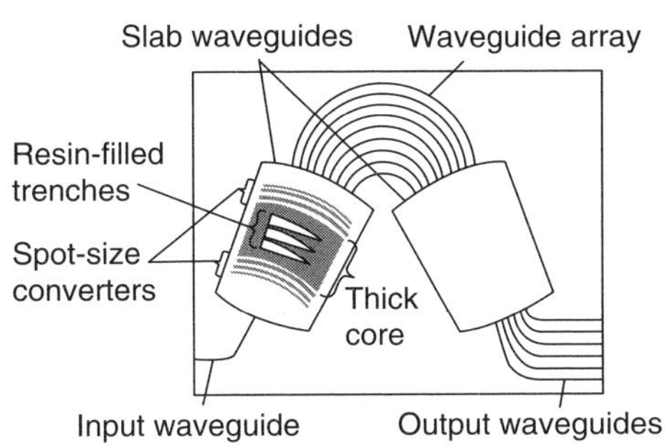

Figure 3.16 Schematic diagram of silicon resin-filled trenches with spot size converters based on segmented core in the slab region of 2.5%-Δ athermal AWG. (From Ref [19])

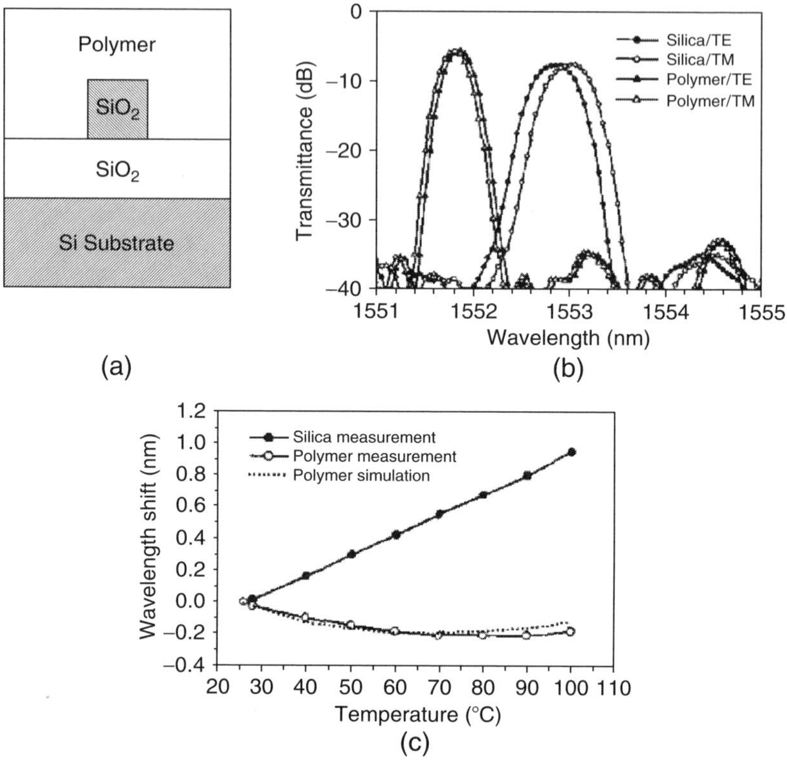

Figure 3.17 (a) Cross-sectional view of silica/polymer hybrid waveguide; (b) comparison of measured TE/TM polarization shift of the central wavelength, Channel 9, with polymer and silica overcladding; (c) measured temperature-dependent wavelength shift of Channel 9 with polymer and silica cladding. (Adapted from Ref [22])

thermal expansion of the polymer substrate. Figure 3.18 (a–c) show the measured spectra, the TE/TM polarization shift, and the temperature-dependent transmission spectrum of Channel 4 respectively, of an all-polymer 8×8 AWG with 200 GHz channel spacing. The measured crosstalk level was below −30 dB but with insertion losses from 5.8 dB to 7.5 dB depending on the channel port, which is predominantly due to the high optical loss 0.8 dB/cm of the polymer waveguide. The polarization shift was measured to be less than 0.02 nm whilst the temperature-dependent wavelength shift was ± 0.05 nm over the entire temperature range of 25–65 °C.

Aside from using materials with negative thermal coefficients that are opposite to that of the silica waveguide arrays, a completely different approach is to incorporate a mechanically movable compensation plate in the AWG structure (Fig. 3.19). In [24], a circuit of the AWG is cut at one of the slab waveguides into

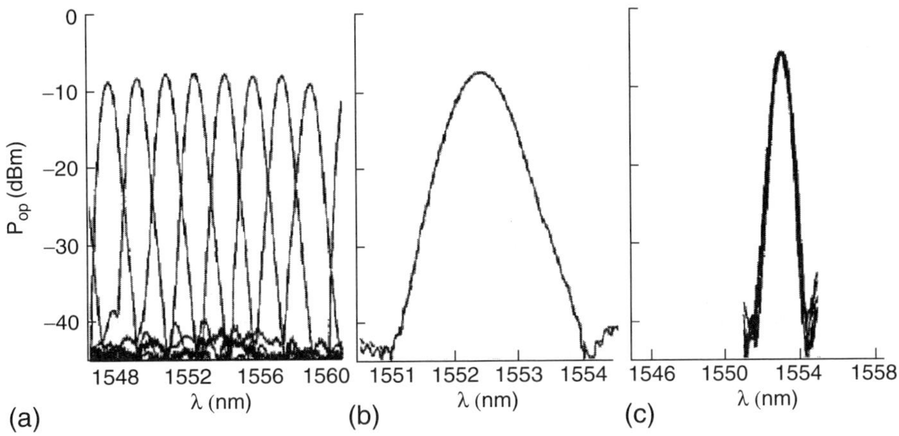

Figure 3.18 All-polymer 8×8 AWG with 200 GHz channel spacing: (a) measured spectra; (b) measured TE/TM polarization shift of Channel 4; (c) measured temperature-dependent transmission spectrum of Channel 4 for the temperature range of 25–65 °C. (Adapted from Ref [23])

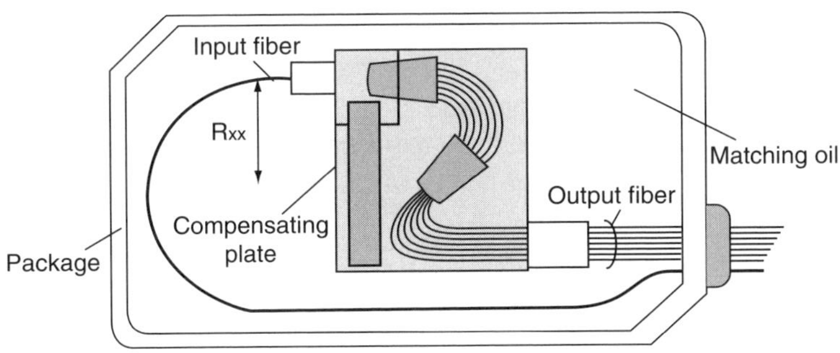

Figure 3.19 Schematic diagram of an athermal AWG with temperature compensating plate. (From Ref [24])

two parts which are then connected by a compensating plate. The compensating plate is made out of a metal which thermally contracts and expands with temperature, thus offsetting the center wavelength drift of the silica-based AWG by moving part of the slab waveguide. The AWG package is filled with index matching oil to minimize loss at the separated slab waveguide. To minimize the size of the chip which is dependent on the length of the compensating plate, a metal such as aluminum with a large thermal expansion coefficient was chosen. Figure 3.20 shows the optical spectra of a 200 GHz channel spaced 1×16 athermal AWG based on the compensating plate. At 1.2~1.4 dB, the measured

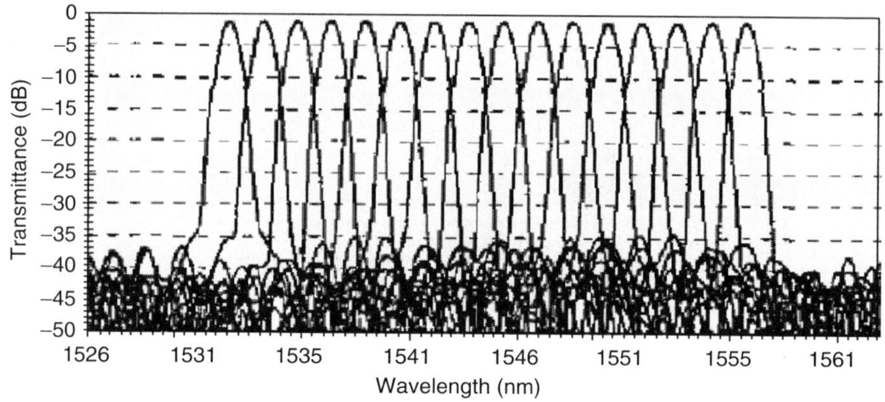

Figure 3.20 Optical spectra of the athermal AWG with temperature compensating plate shown in Fig. 3.19. (From Ref [24])

insertion loss was considerably less than athermal AWGs based on negative thermal coefficient materials. Furthermore, the measured crosstalk was less than −30 dB and the wavelength drift was measured to be less than ± 0.015 nm for all channels from −5 °C to 70 °C.

In a similar structure shown in Fig. 3.21, a metal rod, which is attached to the input fiber, contracts and expands with temperature to stabilize wavelengths [25]. A 32-channel 100-GHz athermal AWG based on this structure was demonstrated to have low loss (5∼5.3 dB) and relatively low crosstalk levels (−26.8 dB to −28 dB)

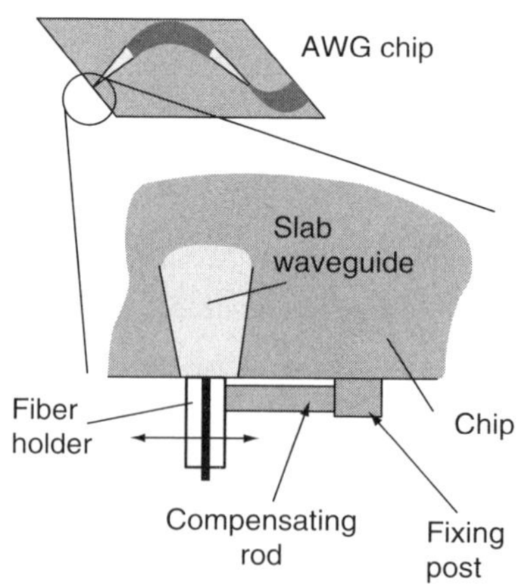

Figure 3.21 Schematic diagram of an athermal AWG chip with temperature-compensating rod. (From Ref [25])

over the L-, C-, S-, and E-bands. Figure 3.22 (a) shows the optical spectra in the C- and L-bands of the athermal AWG with the temperature compensating rod. Figure 3.22 (b) shows the temperature-dependent wavelength shift of four channels in the C-band over the range of temperatures from −30 °C to +70 °C. These results show a maximum wavelength shift of less than 0.034 nm with a spread of less than 0.008 nm across individual channels.

In another temperature-compensating approach using mechanical movable parts, a bimetal plate is attached to a conventional AWG to induce temperature-dependent stress to the waveguide [26]. Figure 3.23 shows the schematic illustration where an x-direction strain ε_x is applied by the bimetal plate to the waveguide layer at a specific bending radius R of the waveguide layer to the bimetal plate. The resultant temperature-dependent stress on the waveguide layer causes a change in the grating pitch, thereby compensating for the wavelength shift due to thermal refractive index change. By carefully choosing the thickness, width and the material of the bimetal plate, the temperature dependence

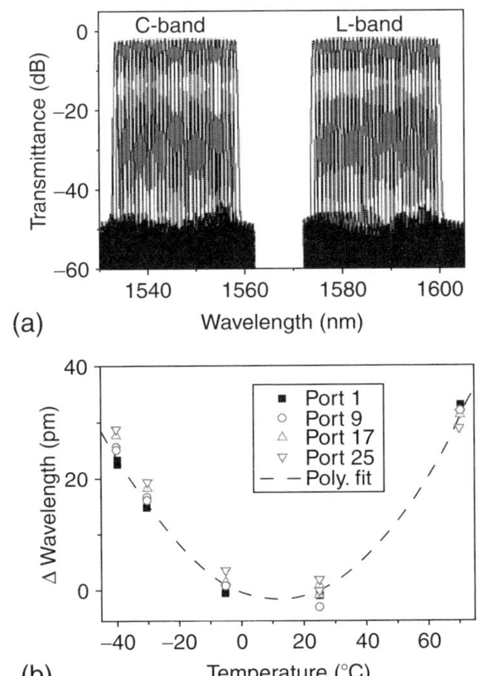

Figure 3.22 (a) Optical spectra in C- and L-bands of athermal AWG using temperature-compensating rod; (b) temperature-dependent wavelength shift of four channels in the C-band from −30 °C to +70 °C. (Adapted from Ref [25])

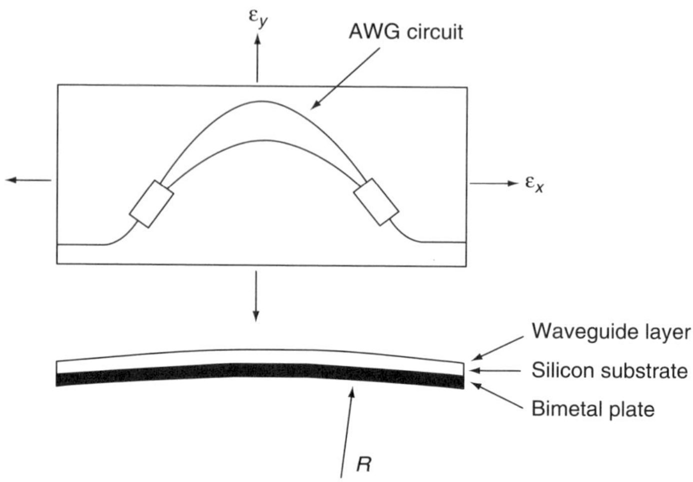

Figure 3.23 Schematic diagram of AWG with bimetal plate. (From Ref [26])

of the central wavelength can be made zero. In [26], the temperature dependence of the TM and TE mode of the central wavelength of a polarization dependent AWG with an attached bimetal plate was measured and is shown in Fig. 3.24. With the bimetal plate, the TM and TE modes had an average temperature dependence of less than 0.001 nm/°C and 0.004 nm/°C, compared to 0.013 nm/°C for the conventional AWG. A side effect due to the attachment of the bimetal plate to the silicon substrate of the AWG is the shift in central wavelength due to the initial stress, which needs to be accounted for in an AWG layout.

Athermal AWGs of up to 40 channels [27–30] and 100/200 GHz channel spacings are now commercially available [27–31] with good crosstalk performance (<30 dB), low polarization-dependent loss (<0.4 dB [27]), low insertion loss (<2.0 dB for Gaussian type passband, <4.5 dB for semi-flat type passband [28]), and good temperature stability over a wide operating range (<0.02 nm from 0 °C to 60 °C [27]).

3.3 PON TECHNOLOGIES FOR INDOOR INSTALLATION

3.3.1 Field Assembly and Indoor Connectors

In an FTTH environment, the fiber connection between the outdoor drop fiber and indoor fiber is usually performed via fusion or mechanical splices. For mechanical splicing, splicing machines and fixtures are used to align and connect

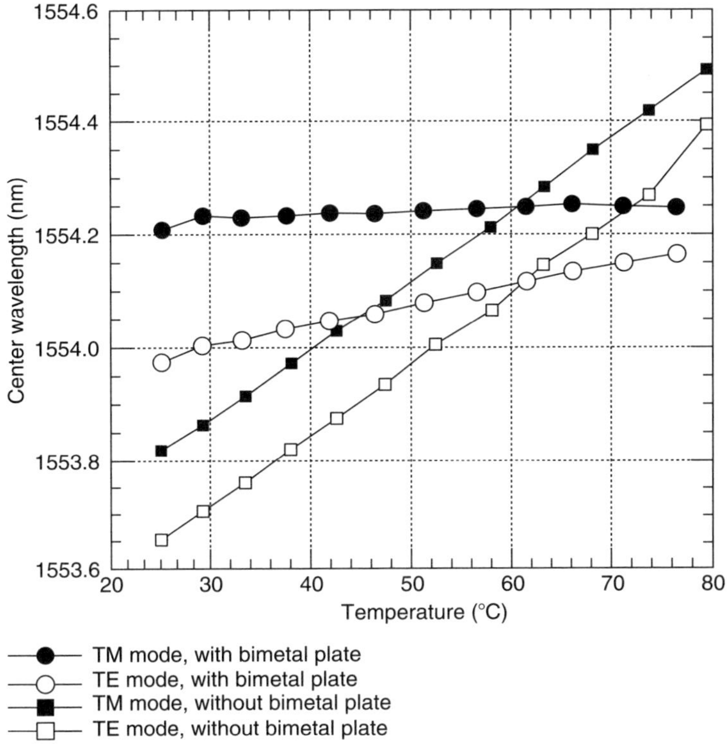

Figure 3.24 Center wavelength of AWG as a function of temperature. (From Ref [26])

optical fibers, and materials such as transparent adhesives (epoxy resin) and index-matching gels, provide index matching between the connected fibers. Nonetheless, mechanical splicing does not facilitate easy disconnection and reconnection, thereby making fiber fault detection and isolation between the outdoor and indoor fibers difficult. An outer clasp connector comprising of a field assembly (FA) connector plug and a FA connector socket was proposed in [32] by NTT Corporation, whereby connection and disconnection can be easily achieved. Figure 3.25 (a) shows the structure of a FA connector plug and socket, consisting of connector ferrules, fiber locks, and clasp parts. Only 25 mm of the leading ends of the outdoor and indoor fibers need to be inserted into both the plug and the socket to be locked in place. The FA connector socket is also compatible with SC connectors. The placement of the connectors in the optical cabinet at the customer premises is shown in Fig. 3.25 (b), where connections between the outdoor and indoor fibers no longer require the use of specialized mechanical or fusion splicing equipment.

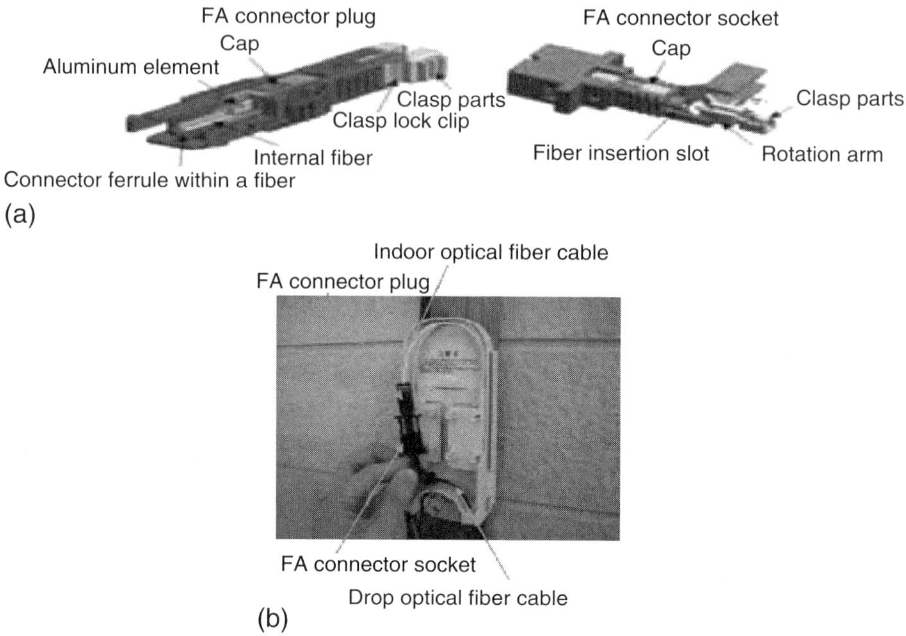

Figure 3.25 Field assembly (FA) connector: (a) structure of plug and socket; (b) in optical cabinet at customer premises. (From Refs [32] and [33])

Results from connection loss and return loss trials, which charts the distribution of the loss in each case are shown in Fig. 3.26 (a) and (b) respectively. The aim is to have as low a connection loss as possible, and as high a return loss as possible to minimize the reflected optical power at the connector. In connection loss trials repeated 500 times using the FA connector plug and socket, an average loss of 0.26 dB was measured. The maximum loss was 0.69 dB, satisfying the specification of 0.7 dB or lower for 1.3 μm transmission. In return loss trials taken over 250 measurements, all return losses were measured to be more than the specified 40 dB, with the average being 48.8 dB. A novel bendable SC connector shown in Fig. 3.27 was also developed by NTT to be used indoors, with good results in splice loss, repeated bending, and heat cycle tests [33]. Its construction enables the connector to be movable between 0 degrees and 90 degrees to the wall socket, allowing increased tolerance to customer handling, optical wiring workability, safety, and a good appearance.

In terms of reducing return losses which are essential in networks transporting analog video services, angled physical contact (APC) connectors which are normally reserved for long haul telecommunication networks are now being

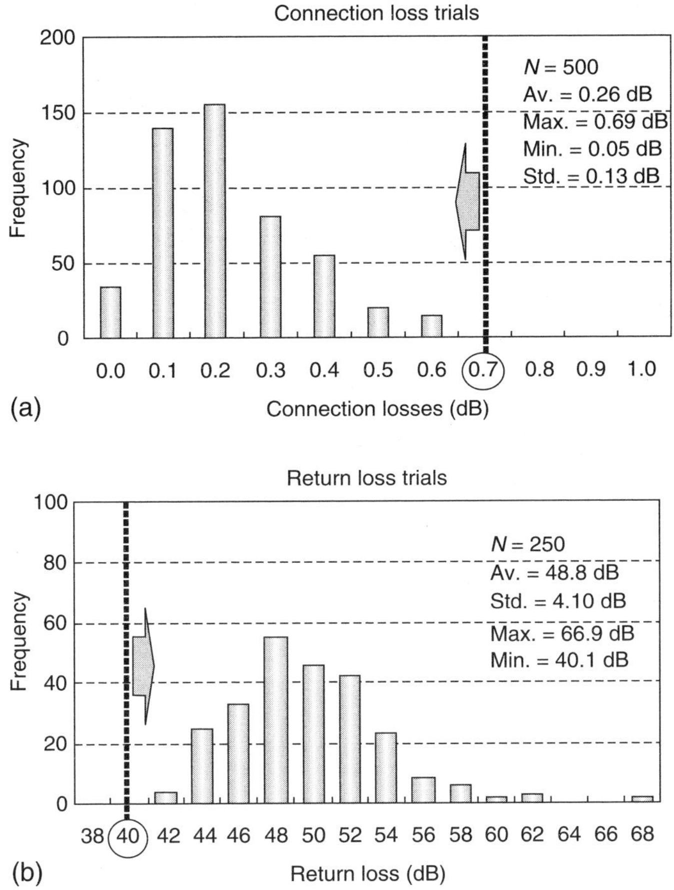

Figure 3.26 (a) Connection loss trials and (b) return loss trials. (From Ref [32])

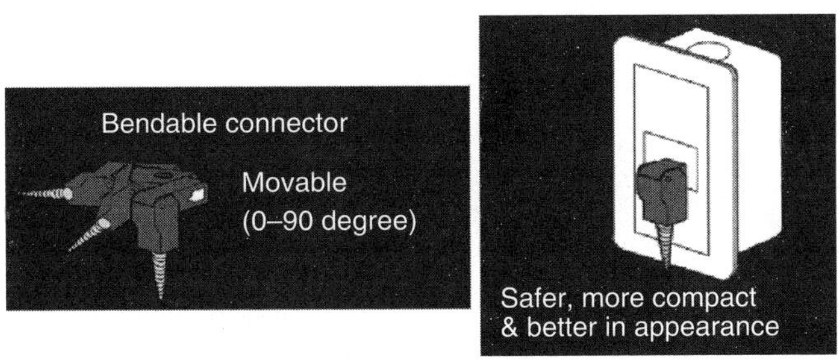

Figure 3.27 Bendable SC connector. (From Ref [33])

considered for the access network with video overlay (refer to Chap. 2) [34, 35]. The return loss of APCs can approach 65 dB. While general physical contact (PC) connectors have a flat fiber end face, an APC is designed to have a fiber end face at a standardized angle of 8 degrees as shown in Fig. 3.28, so that any reflected light is directed into the fiber cladding. The usage of APC connectors with very low return loss is an advantage in newly deployed PONs, where not all of the distribution fibers are connected to ONUs. An otherwise unterminated PC connector would result in Fresnel reflections as high as −14 dB, necessitating a proper terminator for each unused connector [34].

3.3.2 Fiber for Indoor Installations

3.3.2.1 Bend-Insensitive Single Mode Optical Fiber

Optical fiber cables with bending losses lower than conventional optical fibers are highly suited for indoor fiber installation due to increased bending loss tolerance. Conventional standard single mode optical fiber requires a bending radius of at least 30 mm during installation to meet the standard specification of less than 0.25 dB/10 turns. In [32], a bend-insensitive silica-based fiber was developed where studies showed that early adoption and deployment of the bent-resistant fiber can achieve cost levels similar to existing fiber cables.

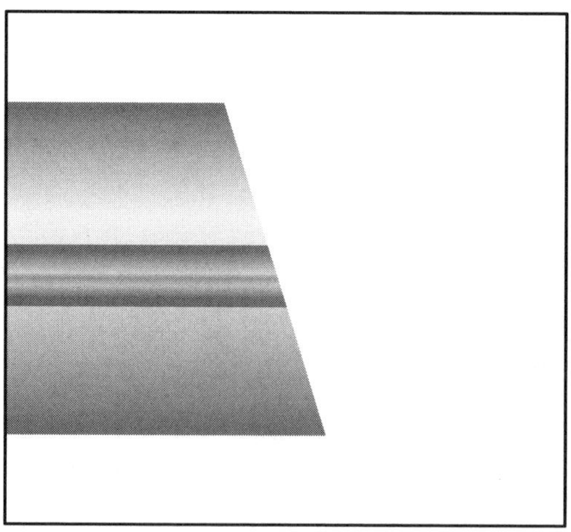

Figure 3.28 Cross-sectional view of angled physical contact (APC) connector. The 8-degree fiber end face directs reflected light into the cladding. (From Ref [34])

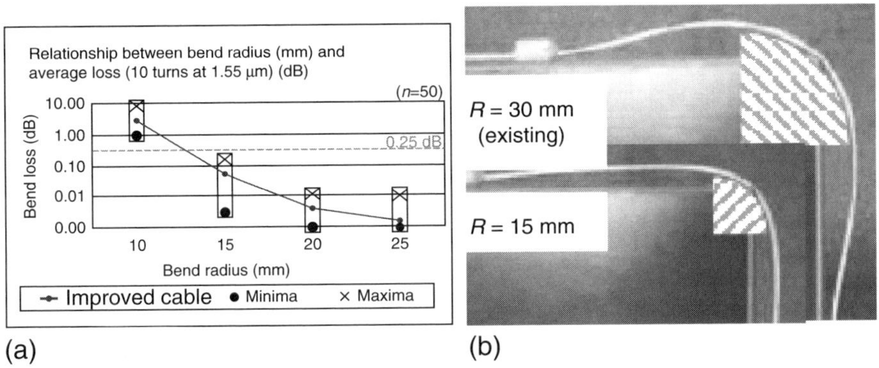

Figure 3.29 (a) Bend loss characteristics of bend resistant fiber; (b) comparison of conventional optical fiber with minimum bending radius of 30 mm and bend-resistant with minimum bending radius of 15 mm. (From Ref [32])

The fiber was also backward-compatible with existing single mode fibers (SMFs). Figure 3.29 (a) charts the average bending loss in dB measured for 10 turns as a function of bend radius. As expected, the smaller the bending radius the higher the bending loss. As discussed previously, a bending radius of 30 mm or more is required to achieve the standard bend loss of 0.25 dB/10 turns for conventional standard cable; but the bend-insensitive fiber is shown to exhibit improved bend characteristics that meet the specifications with a bending radius of only 15 mm. Figure 3.29 (b) compares the standard fiber and the bend-insensitive fiber with the latter showing a cleaner installation with bent sections that are less prone to accidental snagging by customers [32].

Bend-insensitive optical fibers are now commercially available specifically for indoor FTTH installation. For example, the bend-insensitive fiber manufactured by StockerYale, Inc. is an SMF with a moderately higher numerical aperture than conventional SMF to achieve an allowable bending radius of 10 mm at the specific wavelengths of 1550 nm, 1310 nm, and 780 nm [36]. The bend-insensitive SMF, PureAccess-Ultra, manufactured by Sumimoto Electric has an allowable bending radius of 7 mm. An additional feature of this fiber is that it is also a low-water-peak fiber that offers excellent attenuation stability across 1310–1626 nm. Using hydrogen aging, the water-peak absorption at around 1383 nm is substantially reduced, as shown in Fig. 3.30 (a), which plots the attenuation properties of the fiber across the range of wavelengths from 1225 nm to 1675 nm [37]. Figure 3.30 (b) compares the wavelength dependence of bending losses at 7.5 mm radius between the conventional SMF, PureAccess-Ultra, and its predecessor PureAccess (allowable bending loss of 15 mm) fibers, with the PureAccess-Ultra showing the least dependency [37].

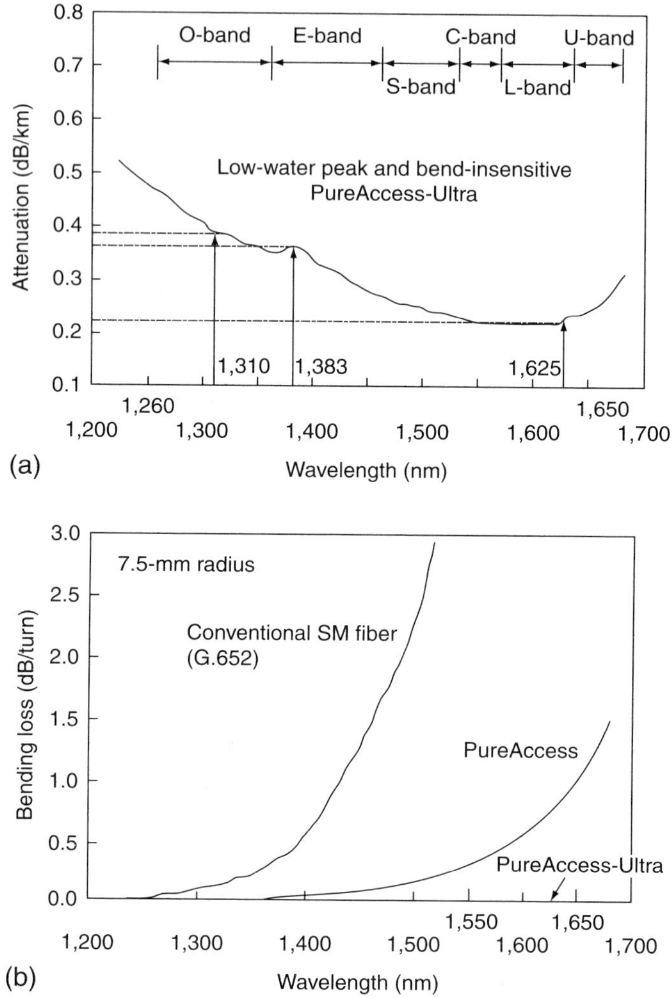

Figure 3.30 (a) Typical spectral attenuation of PureAccess-Ultra; (b) bending loss properties as a function of wavelength. (From Ref [37])

3.3.2.2 Hole-Assisted Fiber-Based Optical Curl Cord

Another type of indoor fiber that has been developed specifically for FTTH usage is the optical curl cord [33, 37–40], which is shown in Fig. 3.31. The cord can be stretched several times in length and retracted like a telephone cord without affecting the optical properties [39]. This enables flexible and reliable

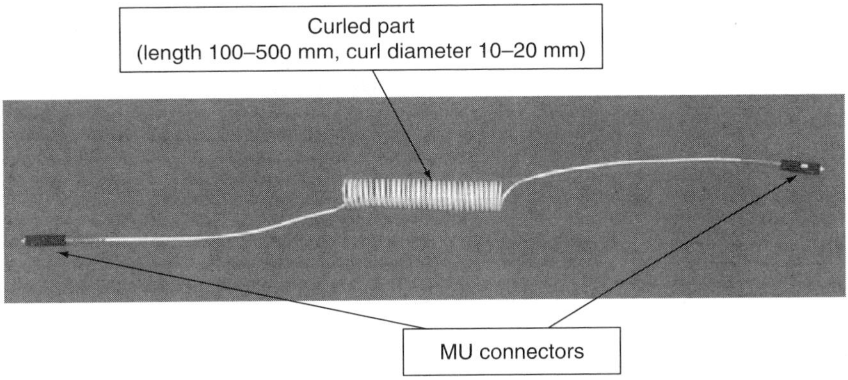

Figure 3.31 Schematic diagram of optical curl cord. (From Ref [39])

connection between communication devices and the optical local area network at offices and homes. Figure 3.32 shows the connection of a laptop terminal to an optical outlet in the wall using the optical curl cord [39]. The optical curl cord exploits the properties of hole-assisted fibers to achieve very small bending radius and reduced bending losses, as compared to conventional SMF.

Hole-assisted fiber has an array of air holes which is arranged around a silica glass core. The air holes in the cladding, which run down the entire length of the fiber, confine the field distribution to the core, thus greatly suppressing the bending loss. An example of the sectional structure of a hole-assisted fiber is shown in Fig. 3.33 (a), where the parameter c is the distance between the center

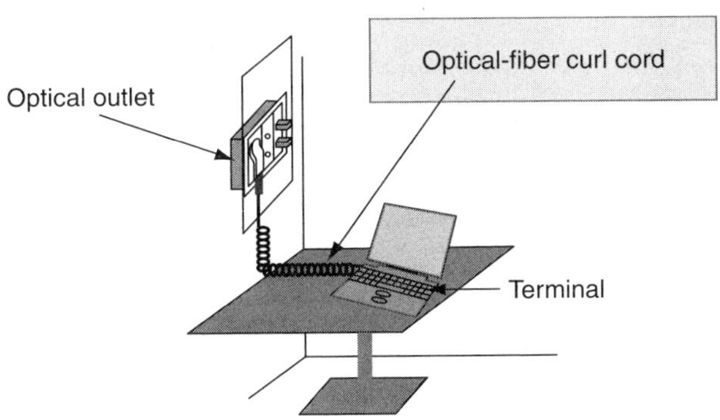

Figure 3.32 Connection of laptop terminal to optical outlet in the wall using optical curl cord. (From Ref [39])

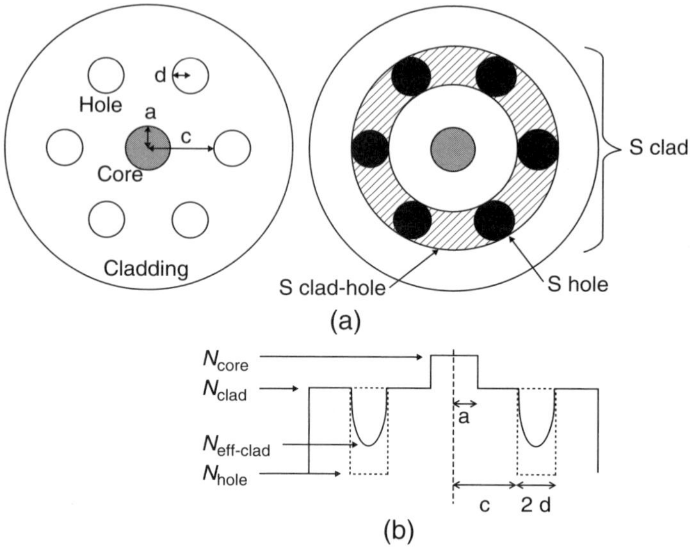

Figure 3.33 Cross section and refractive index profile of hole-assisted fiber. (From Ref [38])

of the core and hole circumference, a is the core radius, and d is the hole radius [38]. The cladding region with several holes in which the refractive index is not circularly symmetrical, is equivalent to a clad ring with circularly symmetrical refractive index distribution, as shown in Fig. 3.33 (b). The resulting refractive index profile of such a hole-assisted fiber has a W-shaped structure.

The bending loss of a hole-assisted fiber at a fixed bending radius decreases with the hole arrangement (c/a) and hole refractive index, N_{hole}, as summarized by the results in Fig. 3.34 (a) [38]. The measurements were obtained using a hole-assisted fiber that was fabricated by drawing a conventional perform of 1.3-μm zero dispersion SMF (39% core and cladding refractive index difference) with precise holes drilled in the cladding. The hole-assisted fiber was subjected to a bending radius of 10 mm at a wavelength of 1550 nm. The bending loss is highest when $N_{hole} = N_{clad}$, which also represents the case of a conventional SMF. Therefore the hole-assisted fiber has lower bending loss as compared to SMF. Minimum bend loss was measured when the holes are vacant (i.e. air, $N_{hole} = 1$). Figure 3.34 (b) compares the bending losses of the hole-assisted fiber and conventional SMF as a function of wavelength. At a bending radius of 10 mm, no significant increase in bending loss was observed with the hole-assisted fiber.

It was also found that the hole arrangement (c/a) affects the splicing loss that is predominately attributed to the mode field diameter (MFD) mismatch

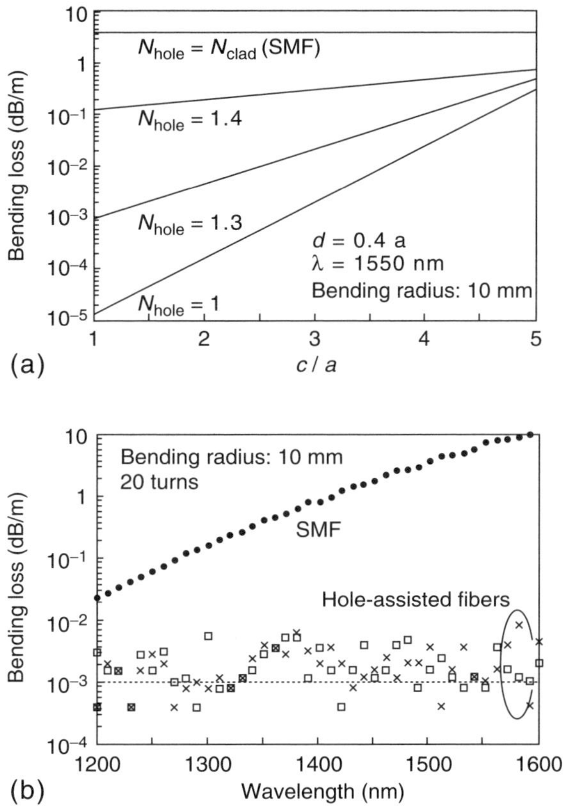

Figure 3.34 Bending loss characteristics of hole-assisted fibers: (a) calculated bending loss performance as a function of normalized hole distance; (b) wavelength dependence of measured bending loss. (From Ref [38])

between the input and output fibers. In Fig. 3.35, the theoretically calculated MFD mismatch (solid lines) as a function of hole arrangement is compared between SMF ($N_{hole} = N_{clad}$) and hole-assisted fiber ($N_{hole} = 1$). Experimental measurements from two hole-assisted fibers (○ symbols) are superimposed on the plot. Results show that MFD mismatch decreases with increasing c/a, especially with $c/a > 2$. The splice loss induced by MFD mismatch is plotted in Fig. 3.35 (b) as a function of refractive index difference between N_{hole} and N_{clad} for two hole-assisted fibers with different hole arrangements. Relatively large loss is observed at $N_{hole} = 1$ when $c/a = 1.2$. Hence, the placement of the hole with regard to the core is important to achieve low splicing loss and bending losses. At a practical bending radius of 10 mm and a splice loss of less than 0.5 dB, the hole arrangement (c/a) must be at least 2.

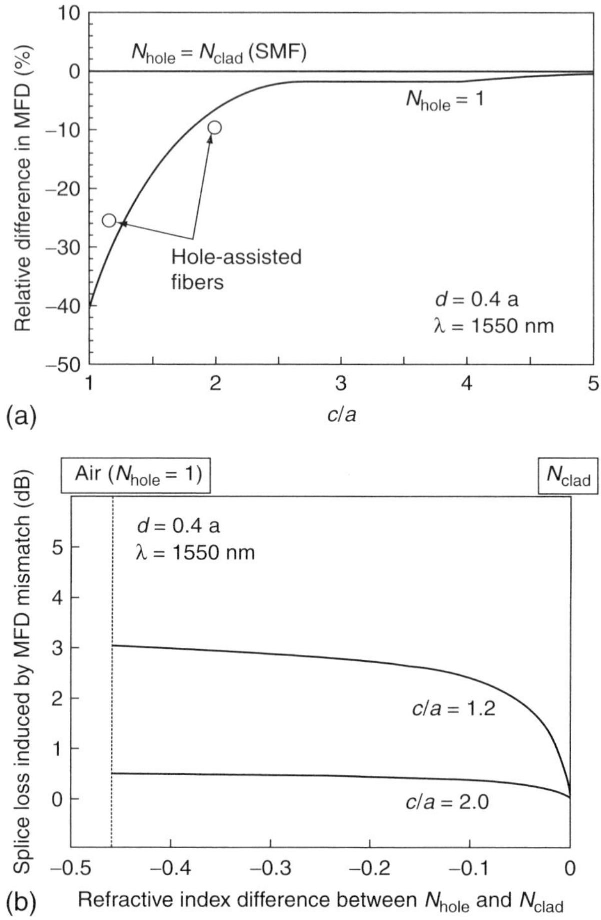

Figure 3.35 (a) Calculated and experimental mode field diameter mismatch as a function of hole arrangement; (b) splice loss induced by MFD mismatch as a function of refractive index difference between N_{hole} and N_{clad}. (From Ref [38])

Optical curl cords (e.g. from NTT corporation [38–40] and Hitachi [40]) are now commercially available with robust features against ambient temperature change (from –10 °C to 40 °C), high humidity (up to 90%), pressure and instantaneous pulling (i.e. snagging). In addition, the curl optical cord's resistance to mechanical stress and temperature can be enhanced by enveloping the core with a longitudinal high-tensile strength aramid fiber and thermoplastic elastomer jacket, as proposed in [39].

3.3.2.3 Graded-Index Polymer Optical Fiber for Home Networking

As discussed in Sect. 3.1, conventional silica-based single-mode fiber requires precise handling and installation by highly skilled personnel due to its small 5–10-μm core diameter. A promising optical fiber candidate to achieving low-cost, user-friendly, yet high-speed short-range network in a home environment is the graded index polymer optical fiber (GI-POF) [41–46]. The GI-POF has a large core, typically 0.5–1.0 mm, that facilitates simple fiber-to-fiber and fiber-to-source connections (due to easier light injection at the source), and thus low-cost transmitters and interconnections. The bandwidth of a GI-POF, typically around 2 GHz.km [43], is dependent on the refractive index profile in the core that is formed in the preform by an interfacial-gel polymerization process [44, 45]. By controlling the core-cladding (index/radius) ratio in the preform through the interfacial-gel polymerization process or by controlling the fiber diameter during the heat-drawing process, the diameter of the core can be changed. Furthermore, by changing chemical parameters during the polymerization process, the refractive index profile and the numerical aperture (NA) of the GI-POF can be changed.

The bending loss of a GI-POF may be reduced by appropriately tailoring the graded index profile, NA, and core diameter [47]. Figures 3.36 (a) and (b) plot the index profiles of three GI-POFs with different core diameters and NAs, and their corresponding bending losses as a function of bending radius [47]. Results indicate that in order to reduce bending loss, the core diameter must be made smaller but the NA should be made higher. In more detail, Fig. 3.37 plots the core diameter and NA dependences of the bending loss at a bending radius of 5 mm. The bending loss increases exponentially when the core diameter increases beyond 150–200 μm or when the NA lowers below 0.25. With the GI-POF set to a core diameter of 200 μm and an NA of 0.24, the bending loss is effectively 0 dB even under 5 mm bending radius, as shown in Fig. 3.38 (a).

The bending loss was experimentally confirmed in [47] to be caused by mode coupling from bending. In turn, the strength of mode coupling prior to bending affects the amount of bending loss, and that the stronger the mode coupling the higher the induced bending loss. The mode coupling strength can be evaluated by the propagation constant difference ($\Delta\beta$) between adjacent modes, which is a function of the core diameter and NA. Based on the calculation of $\Delta\beta$, a set of core diameter and NA with the same $\Delta\beta$ as that of the fiber in Fig. 3.38 (a) are plotted in Fig. 3.38 (b). The plot provides the maximum core diameter and minimum NA at various bending ratios to maintain the bending loss as low as effectively 0 dB. The calculation of the $\Delta\beta$ value therefore enables an appropriate design of GI-POF waveguide parameters to achieve 0 dB bending loss [47]. Though the bending loss of GI-POF can be easily reduced to 0 dB, ongoing work is focused on reducing the attenuation of the GI-POF, especially in

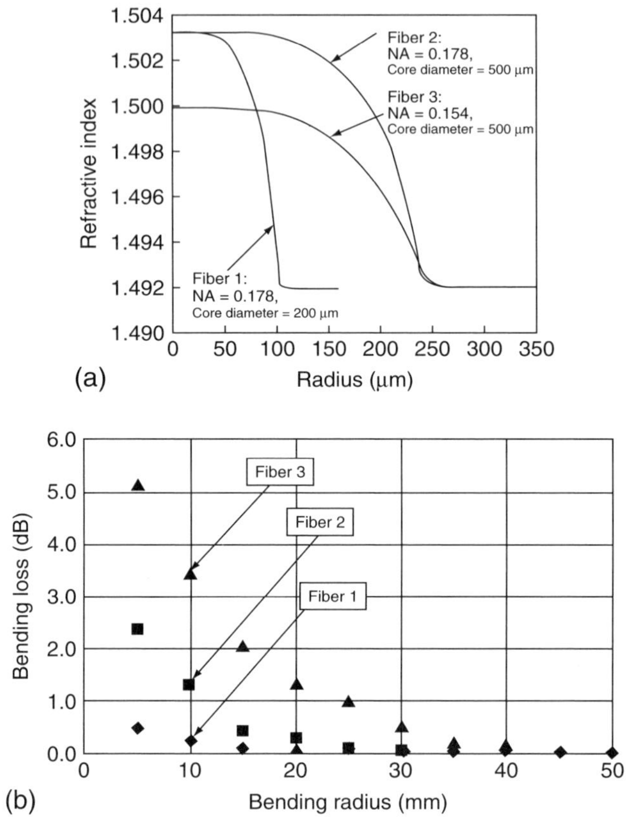

Figure 3.36 Comparison of three GI-POFs with differing core diameter and numerical aperture (NA): (a) refractive index profile; and (b) bending loss as a function of radius. (From Ref [47])

GI-POFs based on perfluorinated polymers. The best reported attenuation is still in excess of 25 dB/km at 850-nm and 1300-nm wavelengths [48].

3.4 TRANSMITTER SOURCES AT SUBSCRIBER PREMISES

3.4.1 Introduction

While wavelength division multiplexed passive optical network (WDM-PON) is actively being developed to meet the ever-increasing bandwidth demands of end users in a cost-effective and future-proof manner, the inventory of

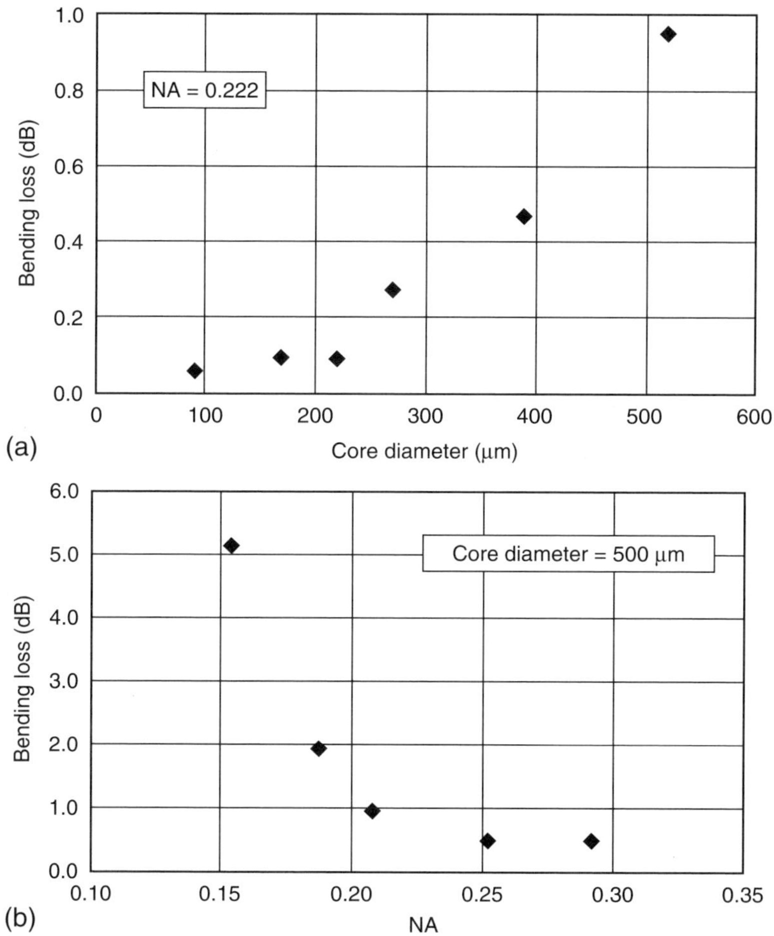

Figure 3.37 (a) Core diameter and (b) NA dependence loss at bending radius of 5 mm. (From Ref [47])

wavelength-specific transmitter sources and the costs associated with the operation, administration, and management of these distinct wavelengths, remains a major issue that needs to be overcome before widespread commercial deployment can be achieved. This is especially critical at subscriber premises (i.e. ONU) where cost is most sensitive. In that respect, cost-effective ONU configurations are actively being researched with most proposals still at the research stage. Cost-effective ONU configurations can largely be categorized into three groups:

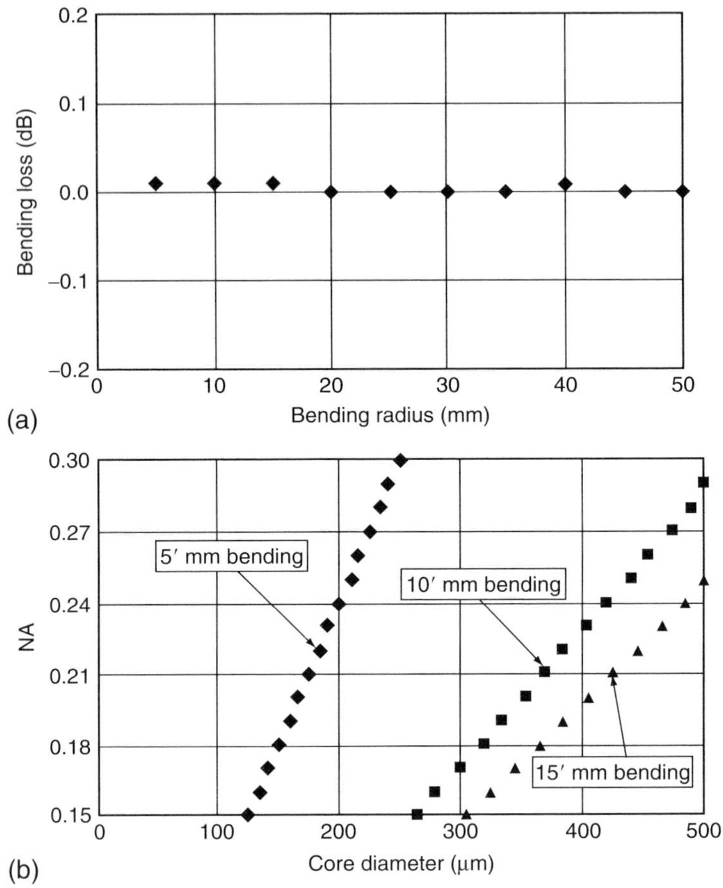

Figure 3.38 (a) Bending loss of GI-POF with a core diameter = 200 μm and NA = 0.24; (b) optimum waveguide parameters of GI-POF to suppress the bending loss. (From Ref [47])

(a) wavelength-specific ONUs, (b) colorless ONUs, and (c) source-free ONUs, which will be described in the following subsections.

3.4.2 Wavelength-Specific ONUs

3.4.2.1 Distributed Feedback (DFB) and Distributed Bragg Reflector (DBR) Lasers

One of the key requirements of transmitter sources in a WDM system is that the emission wavelengths remain stable and fixed to the WDM grid. This key requirement is to minimize crosstalks with other wavelengths and the transmission loss at the WDM remote node, e.g. arrayed waveguide gratings (AWGs). Multimode lasers such as FP-LDs suffer from group velocity dispersion in which each mode propagates in the fiber at a different speed, limiting the bit-rate-distance product to 10 (Gb/s)-km especially in the long wavelength region [49]. The schematic diagram of an index-guided FP-LD with a ridge waveguide structure is shown in Fig. 3.39. A thin active region is sandwiched between *p*- and *n*-type cladding layers of a higher energy bandgap, and an external current injection is required to pump the laser above threshold for stimulated emission. Furthermore, the active region (i.e. gain medium) is placed inside a Fabry–Perot cavity formed by two of its cleaved laser facets that act as mirrors [49]. In Fig. 3.39, parts of the *p*-type semiconductor have been etched away, leaving a strip or ridge which confines the elliptical spot-sized emission light to that region. As the gain spectrum of semiconductor lasers is typically tens of Terahertz wide, more than one longitudinal mode of the FP-LD will experience gain, resulting in an emission light that has several longitudinal modes of the cavity, as shown in the inset of Fig. 3.39. The main mode is the one which coincides with or closest to the gain peak, while the mode spacing, Δf, is a function of the group index, n, and the length of the cavity, L.

By comparison, single-mode lasers such as distributed feedback (DFB) lasers and distributed Bragg reflector (DBR) lasers are suitable candidates as ONU transmitters due to their high-speed direct modulation property [50]. These lasers are designed such that different cavity modes suffer different cavity losses.

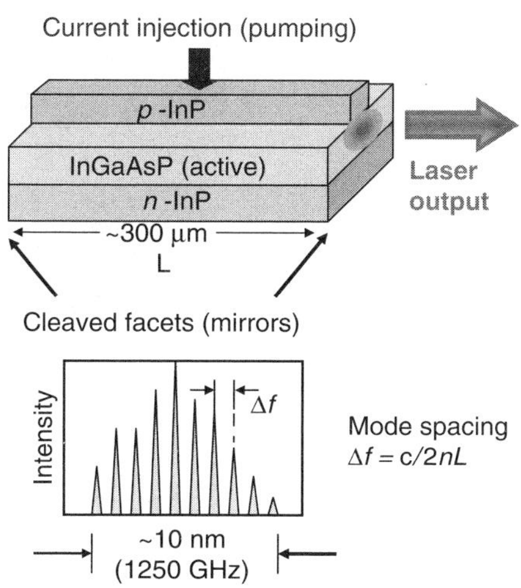

Figure 3.39 Schematic diagram of an index-guided Fabry–Perot laser diode. Optical spectra showing multiple longitudinal lasing modes.

The main mode is the one suffering the least loss that reaches and exceeds threshold condition. Figure 3.40 shows the schematic diagram of a DFB laser. It is similar to the FP-LD with a thin active region sandwiched between *p*- and *n*-type cladding layers of a higher bandgap. However, unlike the FP-LD, the feedback mechanism in a DFB is distributed throughout the cavity length by means of an internal built-in Bragg grating. The grating induces a periodic variation of the mode index which in turn leads to the coupling of waves in forward and backward propagation directions. For a DBR laser, the feedback by means of Bragg diffraction is not continuous throughout the length of the cavity but at both ends of the cavity. In both types of lasers, the selection of a particular mode (or equivalently the specific wavelength of the emission light) depends on the grating period, average mode index, and the first-order Bragg diffraction [49]. The inset of Fig. 3.40 shows a typical optical spectrum of the lasing mode of a DFB laser with excellent single-mode behavior (~1 MHz spectral width). These lasers have temperature coefficients around 0.1 nm/°C, and therefore require active temperature control and wavelength feedback monitoring, e.g. using Fabry–Perot etalons and Bragg filters, to ensure that emission wavelengths are maintained at their designated wavelengths. This contributes to high power consumption, bulky mechanics, system complexity, and bottom-line cost. Thermally tunable DFB modules can also be used as ONU transmitters albeit at an even higher cost, with each tunable DFB module fix-tuned to emit at a different wavelength. With proper thermal design and

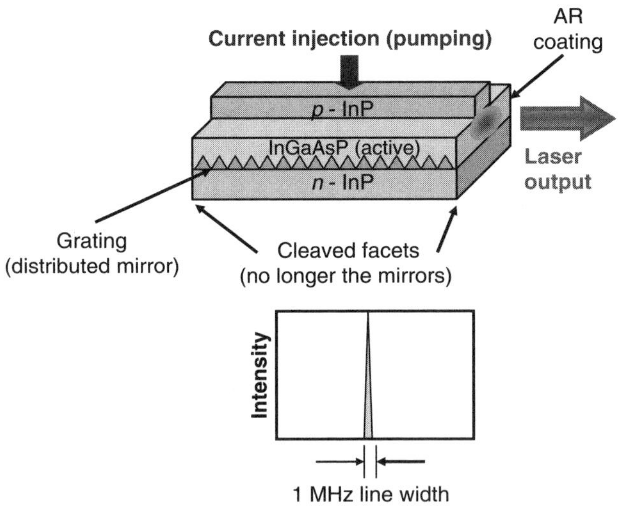

Figure 3.40 Schematic diagram of a distributed feedback laser. Optical spectrum showing excellent single-mode behavior.

controllability, tunable DFB modules have been demonstrated to tune across the entire C-band region whilst maintaining wavelength stability [51].

3.4.2.2 Vertical-Cavity Surface-Emitting Laser (VCSEL)

A promising wavelength-specific alternative to DFB and DBR lasers is the vertical-cavity surface-emitting laser (VCSEL). Due to its short cavity length of typically 2–5 μm, VCSELs exhibit excellent single-mode behavior which favors dispersion mitigation [52]. Figure 3.41 shows the schematic of a VCSEL in which multiple thin layers of semiconductor are grown epitaxially on a substrate. The active region of the VCSEL comprises several quantum wells that are sandwiched between high reflectivity (>99.5%) DBR mirrors that are in turn made up of 10–50 pairs of semiconductor layers with different refractive indices [53]. The lasing light from the VCSEL is emitted in the direction normal to the active layer plane and has a circular beam that facilitates efficient coupling into a single-mode fiber. VCSELs have fundamental cost advantages over traditional edge-emitting lasers (e.g. DFB and DBR) because of their low-cost manufacturing, packaging, aligning, and testing [54]. VCSELs can also be manufactured in 1-D and 2-D arrays to maximize package density and bandwidth performance. Since the cavity volume of a VCSEL is small, it exhibits low-threshold currents in the sub-mA range at room temperature. The power consumption is therefore low. Long-wavelength VCSELs (1310 nm and 1550 nm) have been demonstrated to transmit directly modulated 10 Gb/s data over distances that are comparable to that of access networks [55].

To further lower the cost of using VCSEL transmitters in ONUs, a WDM-PON scheme that exploits the use of optically injection-locking (OIL) was proposed to potentially achieve uncooled VCSEL operation [56]. Optical injection-locking allows the VCSEL wavelength to be matched to that of the master laser and DWDM grid within a certain detuning range. Together with proper heat sinking, the VCSEL wavelength is locked onto the specific AWG port provided by the CO [57]. Therefore, using injection-locked VCSELs as upstream

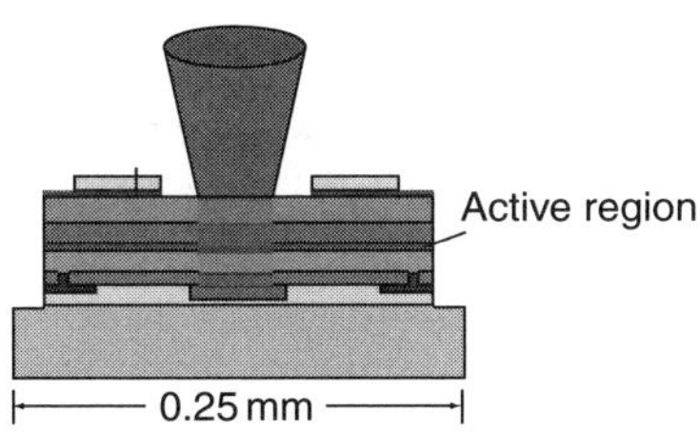

Figure 3.41 Schematic diagram of a vertical-cavity surface-emitting laser (VCSEL) with circular output beam.

transmitters expands the ONU wavelength tolerance and the compatibility with various vendors and systems with slightly different DWDM grid. Optical injection-locking has also been shown to improve laser performance including increasing modulation bandwidth, and reducing optical chirp, laser noise, and nonlinear distortions [58]. Figure 3.42 shows the schematic of a WDM-PON implementing OIL-VCSELs. At the CO, master DFB lasers, either directly or externally modulated with downstream data, are temperature-tuned to emit distinct wavelengths on the DWDM grid. These wavelengths are first multiplexed and then demultiplexed at the remote node for reception at the ONUs. The DFB lasers that carry downstream signals also serve a second function as master lasers to optically injection-lock ONU slave VCSELs onto the WDM grid. At each ONU, an optical splitter divides the optical power of the demultiplexed downstream signal to feed a downstream receiver and to injection-lock a slave VCSEL. The OIL-VCSEL is then directly modulated with upstream data and transmits the data back to the CO. This scheme thereby removes the need for external optical injection-locking sources, external modulators, and potentially wavelength stabilization circuitry. Preliminary experimental results from a unidirectional WDM-PON (refer to Fig. 3.42) show that the OIL-VCSEL favors low-injection powers (<-20 dBm) and responds strongly only to the carrier wavelength, and not the modulated data of the downstream signal [56].

In practice, *bidirectional* optical networks are much preferred due to reduced management and fiber cost savings. However, with optical injection-locking, the

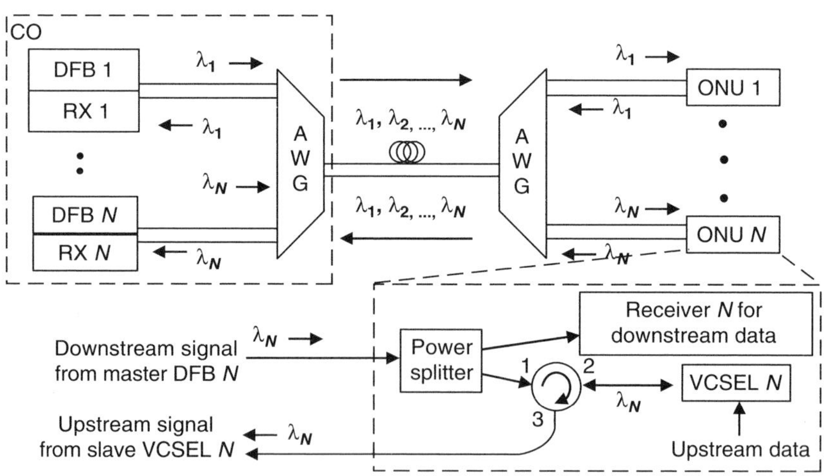

Figure 3.42 WDM-PON using directly modulated OIL-VCSELs as ONU transmitters. Each slave VCSEL, located at the ONU is injection-locked by modulated downstream signal transmitted from a master DFB laser located at the CO. (From Ref [56])

master (downstream) and slave (upstream) wavelengths are identical. The upstream performance at the CO receiver can be potentially degraded by Rayleigh backscatter-induced intensity noise of the master DFB lasers [59]. Compounding the issue is that the master light is not CW but modulated with downstream data. A study was thereby performed on a 25-km bidirectional WDM-PON that utilizes OIL-VCSELs as ONU transmitters. The Rayleigh backscattering effects were analyzed by varying the parameter upstream signal to Rayleigh backscattering ratio (SRR) whilst keeping the extinction ratio (ER) of the downstream and upstream signals constant. Figure 3.43 (a) shows the BER measurements of

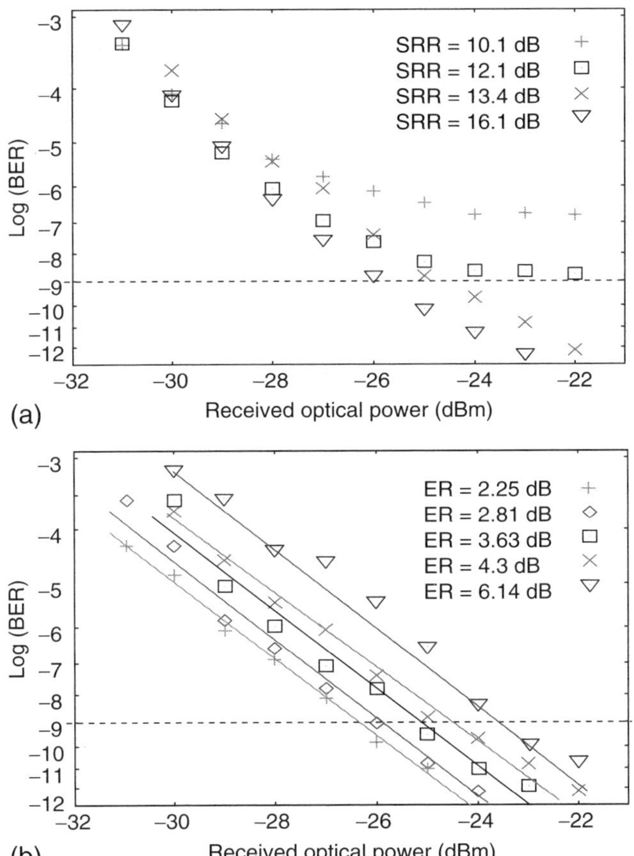

Figure 3.43 Upstream BER curves for different values of (a) upstream signals to Rayleigh backscattering ratio (SRR) at a constant downstream extinction ratio (ER) = 4.5 dB; (b) downstream ER at a constant SRR = 13.4 dB.

2.5 Gb/s upstream signals from the VCSEL optically injection-locked by 2.5 Gb/s downstream signals for different values of SRR. The upstream BER degrades with decreasing SRR, inducing power penalties and error floors. Results also indicate that the influence of Rayleigh backscattering can be drastically reduced to achieve error-free transmission (at BER $<10^{-9}$) if the SRR is higher than 13.4 dB, an easily achievable value in a practical network. Maintaining the SRR and upstream ER at 13.4 dB and 4.5 dB respectively, the downstream ER was then varied with the corresponding upstream BER curves shown in Fig. 3.43 (b). While error-free transmission is achieved in all cases, increase in downstream ER degrades the upstream performance. In the worst case, an increase in downstream extinction ratio from 2.25 dB to 6.14 dB incurs ~3 dB degradation on the upstream BER.

Figure 3.44 plots the minimum power required to achieve BER $<10^{-9}$ as a function of downstream ER for both upstream and downstream signals. The solid and dash lines represent the relative optical ER dependencies of both upstream and downstream signals in a bidirectional network and unidirectional back-to-back setup (i.e. no Rayleigh backscattering or transmission penalty), respectively. Results show a power penalty of <2 dB from 25 km of transmission and Rayleigh backscattering, but more importantly, a linear and small dependence of the upstream signal on the downstream ER as compared to high exponential dependence of the downstream signal on the downstream ER [60].

For colorless operation, identical tunable VCSELs may be placed at each ONU. Tunable VCSELs with continuous tuning in the 1530–1620-nm wavelength

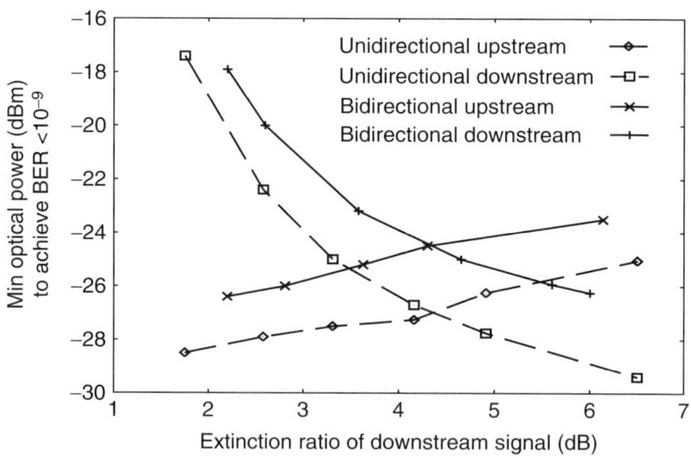

Figure 3.44 Received optical power at BER = 10^{-9} as a function of downstream extinction ratio. Solid line: bidirectional 25-km transmission with Rayleigh backscattering effects. Dash line: unidirectional back-to-back transmission.

regime have been demonstrated [61]. Since the slave VCSEL emission wavelength is typically dependent on its bias current or heat sink temperature, a "training" session may be required to find the lockable wavelength range. This may be done with the assistance of a look-up table, or by forming a feedback loop with measurements of the slave VCSEL reflected power through Port 3 of the optical circulator in Fig. 3.42 or its junction voltage while sweeping the slave VCSEL wavelength. The training session needs to be done only once when the ONU is started up or infrequently as situation becomes necessary, similar to the rebooting of a personal computer.

3.4.3 Colorless ONUs

The emission wavelength of this category of ONUs is nonspecific though it can be selectively determined by external factors such as the filtering properties of the AWG in the remote node or the wavelength of an injection/seeding light into the ONU. This flexibility thereby enables the exact same colorless ONU configuration to be deployed across the network, facilitating mass production. In literature, there exists many schemes that facilitate the colorless feature in ONUs, and these schemes can also be categorized into three main groups described below.

3.4.3.1 Colorless ONUs Based on Spectral-Slicing Techniques

The colorless ONUs in this category benefits from the spectrum-slicing technique to generate unique upstream wavelength channels for transmission. In this technique, the combination of a broadband source and optical filters are exploited to spectrally slice the spectrum of a broadband optical source into narrowband and unique channels for upstream transmission. The earliest proposals were based on implementing a broadband optical source at each ONU with spectral-slicing occurring at the AWG in the remote node [62–64]. Due to spectral slicing, only a narrowband of sliced-spectrum is received at the CO from each ONU, even though upstream data is modulated on the entire broadband spectrum. Figure 3.45 shows the schematic diagram of a WDM-PON with an LED implemented as the broadband optical source in each ONU [62]. Other proposals for broadband optical sources at the ONU include the superluminescent LED (SLED) [63, 64], and erbium doped fiber amplifier (EDFA) as amplified spontaneous emission (ASE) light sources [65], and multimode FP-LDs [66].

In some other proposals, a single centralized broadband source is implemented in the CO instead of distributing broadband optical sources across the

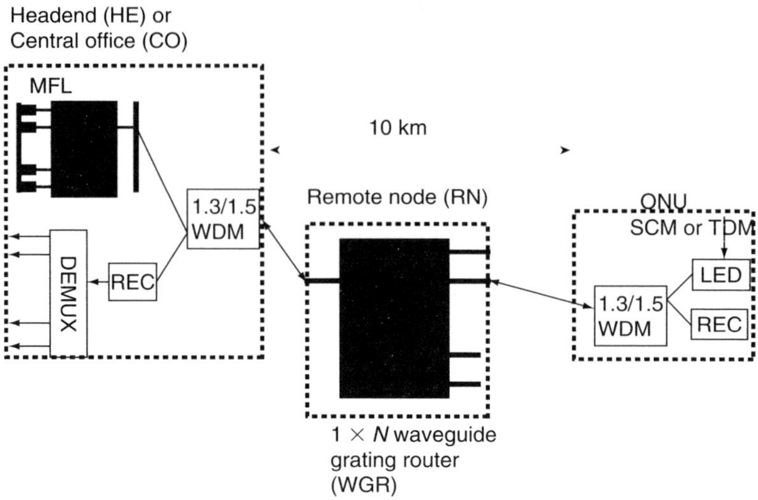

Figure 3.45 Schematic diagram of WDM-PON with broadband optical source (LED) in ONU. (From Ref [62])

network, thus lowering component cost at each ONU [67]. By the same token, the spectrum of the centralized broadband optical source is sliced into multiple narrowband and unique channels at the remote node, with each one distributed toward a separate ONU. At the ONU, the spectrally sliced narrowband wavelength channel is externally modulated with the upstream data to be sent back to the CO. With all the spectral-slicing techniques mentioned above, only a narrow part of the broadband source is effectively utilized for data transport. As a result, the transmission bit-rate and distances are limited by the high spectral-slicing loss, which is further compounded by the low launch powers of LEDs and SLEDs. One can increase the passband of the AWG to reduce spectral slicing losses, but will in turn encounter dispersion issues and will limit the number of users supported in the network. A recent proposal shown in Fig. 3.46 aims at overcoming the power issue by using a centralized supercontinuum broadband light source to achieve 10 Gb/s bidirectional transmission over 40-km distance [67]. The centralized broadband supercontinuum broadband light source consists of an actively mode-locked laser, an EDFA, a polarization controller, a nonlinear photonic crystal fiber, and a C-band band-pass filter to limit the spectral width of the supercontinuum to the C-band for upstream carrier use. Consequently, a C-band supercontinuum is distributed downstream and spectrally sliced at the remote node into multiple upstream carriers which are then modulated with upstream data by the ONUs.

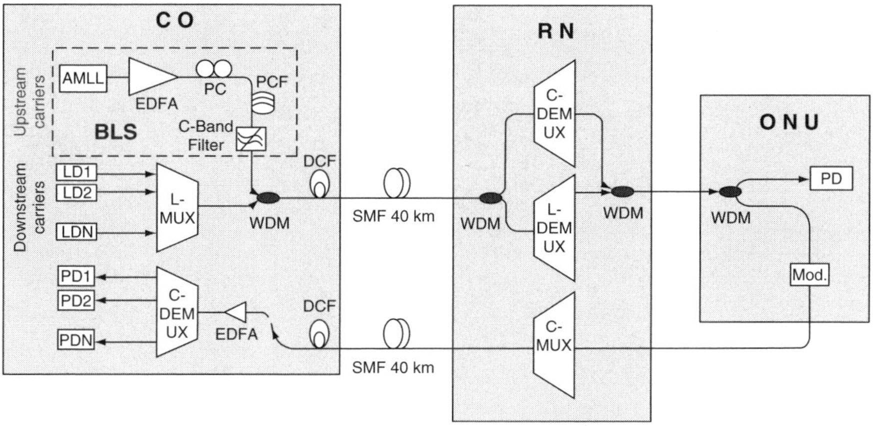

Figure 3.46 Experimental setup of WDM-PON with centralized supercontinuum broadband light source. (From Ref [67])

3.4.3.2 Optically Injection-Locked and Wavelength-Seeded Colorless ONUs

3.4.3.2.1 External Injection-Locking and Wavelength-Seeding

The modulation speeds in colorless ONU configurations based on broadband/multimode sources are limited due to several inherent noise sources such as mode partition noise, intensity noise, and spontaneous–spontaneous beat noise. These effects on the optical signal increase as a function of closer and narrower-bandwidth channels [67, 68]. Alternatively, using an external optical light injected into a multimode laser such an FP-LD can result in the excitation of only one stable mode for high-speed modulation, with all other modes suppressed [69]. Optical injection of multimode lasers (slave lasers) using external optical sources (master lasers) can lock the wavelength of the main mode to that of the master laser, facilitating colorless and uncooled operation without temperature stabilization.

In a WDM-PON, optical injection-locking of FP-LDs has been proposed to function as stable, colorless, and uncooled ONU transmitters [70–73]. Most of the proposed schemes exploit the use of a centralized broadband optical source whose broadband output spectrum is spectrally sliced at the remote node to furnish multiple master injection lights for multiple slave FP-LDs. Aside from facilitating colorless transmitters in ONUs, injection-locked FP-LDs have also been proposed as colorless transmitters in the CO. An example is shown in Fig. 3.47 with CO and ONU bidirectional (BiDi) transceivers that are based on injection-locked FP-LDs and p-i-n photodiodes [72]. The requirements of a

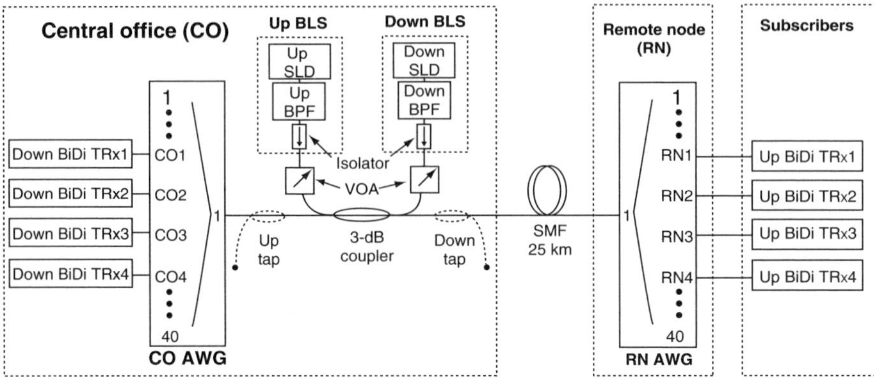

Figure 3.47 Schematic diagram of colorless bidirectional (BiDi) transceivers based on FP-LDs injection-locked by the spectrally sliced light from a broadband super-luminesent LED (SLED). (From Ref [72])

photodiode in a PON are similar to those in any optical network: high sensitivity, low noise, fast response, high reliability, and also cost-effectiveness [74]. In addition, its integration in a burst mode receiver circuit, as discussed in Chap. 4, is important to ensure that these receivers are capable of handling packets of a wide dynamic range of optical power (hence transmission distance), especially in a power-splitting time-shared PON. For cost purposes, p-i-n photodiodes are much preferred especially for the implementation at ONUs. However, avalanche photodiodes (APDs) are gaining wide acceptance in high-speed PONs. A slight drawback to the usage of the APD is the need for a high voltage source to provide a reverse bias voltage of 20–40 V. Nonetheless, the APD has an internal current gain through the impact ionization process, which makes it more tolerable against thermal noise, and is therefore, highly suited for use in an unamplified and thermal noise-limited environment such as PONs.

Referring back to Fig. 3.47, all FP-LDs either located at the CO or ONUs are injection-locked by spectral slices originating from the broadband optical sources (BLSs) located at the CO. Both BLSs for downstream and upstream injection consist of an SLED and an optical band-pass filter to limit the broadband spectrum to the desired wavelength band. An isolator is used to prevent feedback into the SLED. Two distinct bands, the C and E + bands, are allocated for upstream and downstream transmissions, respectively. These particular bands are far enough to allow the usage of WDM filters with relaxed specifications at the CO and ONUs, but also close enough for accurate wavelength filtering using the (FSR) of athermal AWGs. The broadband light from both BLSs are launched into the feeder fiber via a 3-dB coupler. The broadband light to injection lock downstream BiDi transceivers is spectrally sliced by the AWG at the CO, whereas that for the upstream BiDi transceivers is spectrally sliced by the AWG at the RN. As a result of external optical injection, the wavelength of each BiDi transceiver is determined by the wavelength of the incoming spectrally sliced injection light.

Figure 3.48 (a) and (b) show the spectra of the downstream and upstream BLS and the resulting spectrally sliced injection lights measured at the input and output ports of the RN and CO AWGs, respectively. The measurements were performed without connecting any BiDi transceivers to the network. Each Gaussian AWG has 40 passbands (1535–1566 nm) in the C-band, and likewise, 40 passbands (1431–1460 nm) near the E + band. The measured insertion loss of

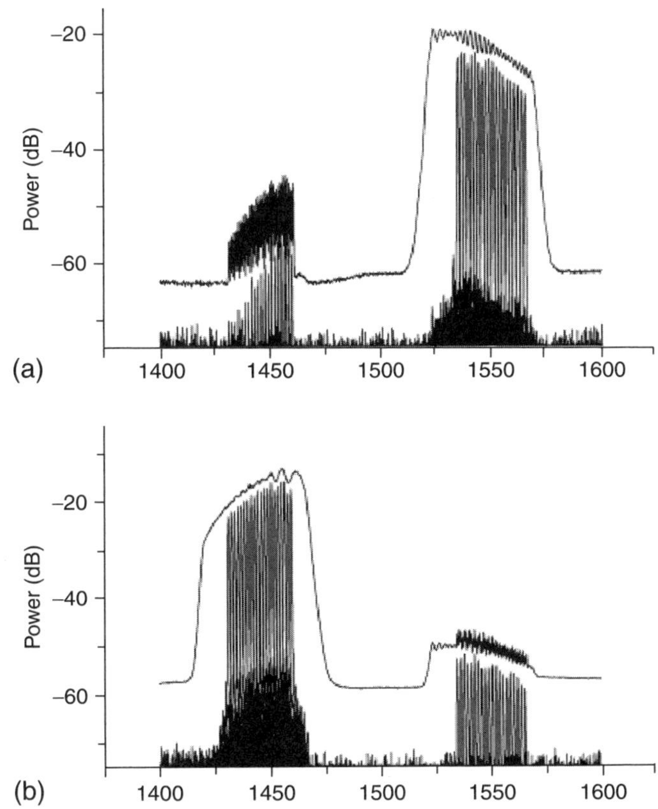

Figure 3.48 Optical spectra of total injection light and spectrum-sliced injection lights measured at the input and output ports of the (a) remote node AWG and (b) central office AWG. (From Ref [72])

both AWGs is 3 dB. Between those two AWGs, the passband misalignment is only <0.02 nm and <0.01 nm in the C and E + bands, respectively. The slight difference between the two bands is due to the intrinsic characteristics of the AWGs. Nonetheless, the penalty arising from this slight misalignment was experimentally proven to be negligible [72]. WDM-PON field trials using such spectral-sliced injection-locked FP-LDs as ONU transmitters have been carried out by Korea Telecom [75], with several broadband equipment providers such as Fujitsu Network Communications [76] and Corecass [77] following suit but with the reflective semiconductor optical amplifier (RSOA) being the colorless ONU transmitter of choice. In both types of colorless ONU transmitters, the efficiency of optical injection and wavelength-seeding can be controlled with careful selection of the laser bias current, modulation index, and optical injection power.

By the same token, injecting an external seeding light into an RSOA enables the generation of a frequency-stable, low-intensity, low-phase noise yet high-power optical source. The emission wavelength of a seeded RSOA follows that of the seeding light [78–81]. An example of deploying wavelength-seeded RSOAs as colorless transmitters in a WDM-PON is shown in Fig. 3.49. At the CO, non-polarized broadband light from an SLED is directed into 10 km of feeder fiber via a 3-dB coupler. The downstream wavelengths and the broadband light (hence the upstream wavelengths), reside on two distinct wavebands, L and C, respectively. The broadband light is spectrum sliced by the AWG at the remote node and is directed into the ONU after traversing through 10 km of distribution fiber. At the ONU, a coarse WDM filter is used to separate the downstream wavelength (in the L band) and spectrally sliced light (in the C-band), with the

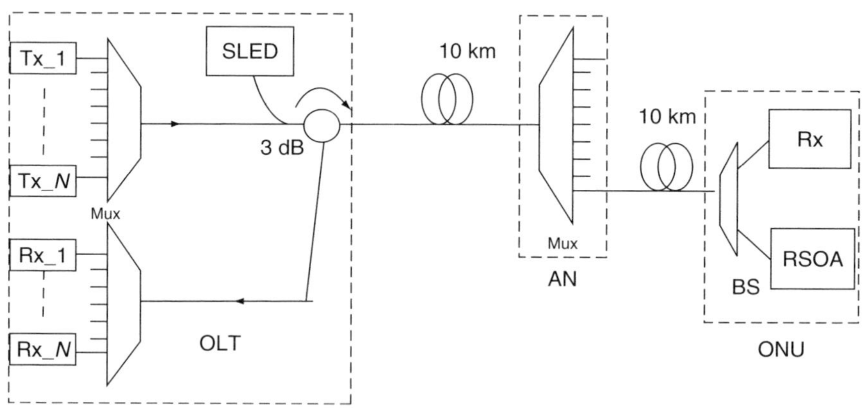

Figure 3.49 Schematic diagram of colorless reflective semiconductor optical amplifier (RSOA) wavelength seeded by spectrally sliced light from a broadband SLED. (From Ref [81])

latter directed into an RSOA. The wavelength-seeded RSOA can then be directly modulated with upstream data that is then sent back upstream to the CO.

Compared to the SOA which has a traveling waveguide structure with two low-reflection facets, a reflective SOA has one front facet that is coated with a very low-reflectivity antireflection coating for light to enter and exit, and a rear facet that is coated with high reflectivity coating to reflect the input light. Figure 3.50 shows an example of a monolithic indium phosphide-based RSOA from CIPhotonics with a waveguide structure that is curved to prevent reflections from the front facet whilst facilitating fairly high reflections from the rear facet [82]. Advances are being made to the structures of RSOAs to increase modulation speeds by reducing junction and parasitics capacitances with multiquantum well [83] and quantum dot technologies [76]. Wavelength-seeded RSOA has the potential to reach high transmission speeds over longer distances than the injection-locked FP-LD due to its high output power and immunity to mode partition noise [83]. More importantly, when operating under the gain saturation condition, RSOAs have an inherent amplitude-squeezing property that suppresses the excess intensity noise of the seeding light (i.e. spectrally sliced ASE light) [83].

3.4.3.2.2 Self-Injection-Locked FP-LDs and Self-seeding Reflective SOAs

Recently, schemes based on self-injection-locking FP-LDs and self-seeding RSOAs were proposed in [84, 85] to remove the requirement for centralized

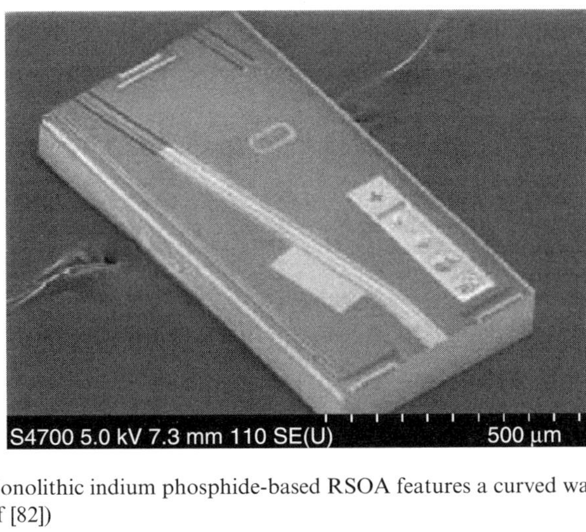

Figure 3.50 Monolithic indium phosphide-based RSOA features a curved waveguide architecture. (From Ref [82])

broadband sources at the CO. In [84], the upstream signal from each FP-LD is reflected back at the remote node to injection lock itself. As shown in Fig. 3.51, the feedback is provided by a fiber Bragg grating (FBG) with a Bragg wavelength centered at the passband of the waveguide grating router port. The temperature coefficient of the FBG is identical to that of the silica-based waveguide grating router, and hence the Bragg wavelength will always track that of the waveguide grating router port. The self-injection-locked FP-LD is directly modulated with a dc-balanced code mark inversion interface to reduce data patterning effects. Further, an optical circulated delay comprising a 1:99 coupler is implemented in front of the FP-LD to remove the coherency between the reflected light and the self-injection-locked upstream signal. Identical FP-LDs can be used at all ONUs since the transmitting wavelength is determined by the FBG and waveguide grating router's spectral characteristics. Figure 3.52 shows the optical

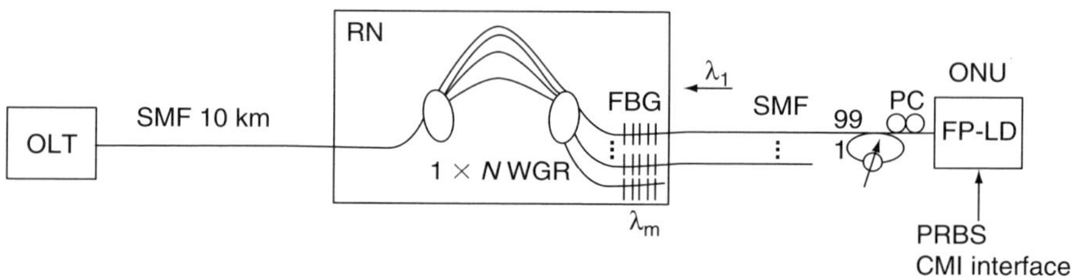

Figure 3.51 WDM-PON with self-injection-locked Fabry–Perot laser diodes (FP-LDs). (From Ref [81])

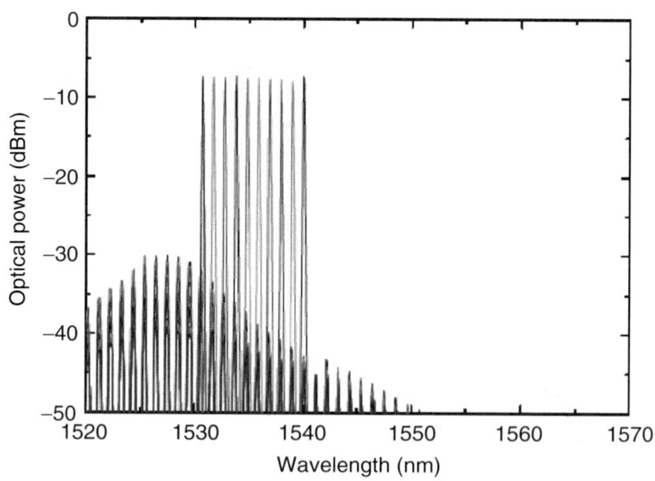

Figure 3.52 Optical spectra of self-injection-locked FP-LD. (From Ref [81])

spectra of an FP-LD that was injection-locked by its own reflected light at different wavelengths (using a tunable FBG), demonstrating colorless operation.

In the self-seeding RSOA scheme proposed in [85], the RSOA at each ONU emits an ASE spectrum in the upstream direction which is then spectrally sliced by the AWG in the remote node. A portion of this light is reflected back into the RSOA via a reflective path comprising a coupler, an optical circulator, and a band-pass filter (BPF), as shown in the schematic diagram of the proposed WDM-PON in Fig. 3.53. The BPF ensures that only one wavelength peak is reflected back to seed the RSOA in the event that the free spectral range of the AWG is smaller than the optical bandwidth of the ASE spectrum. The BPF is also used to reflect the spectrally sliced light of all ONUs, removing the direct correspondence between the AWG and *N* static wavelength filters. An important point to note is that stringent temperature tracking between the AWG and BPF is not required as the AWG's FSR and the passband of the BPF will always coincide to yield one seeding wavelength per output port. Once self-seeding has been established, the RSOA can be directly modulated with the upstream data to be transmitted to the OLT. Identical RSOAs can be used at all ONUs as the seeding and hence transmitting wavelengths are solely determined by the AWG's spectral characteristics. Colorless behavior is demonstrated in Fig. 3.54 which shows the overlapped optical spectra of self-seeded upstream channels. Each channel of the overlapped spectra was measured using the same RSOA connected to a different AWG input port in the upstream direction.

3.4.4 Source-Free ONUs Based on Wavelength Reuse Schemes

In this category of ONUs, no optical sources are actually implemented at the ONU. Instead, the optical carriers from the CO for downstream transmission are

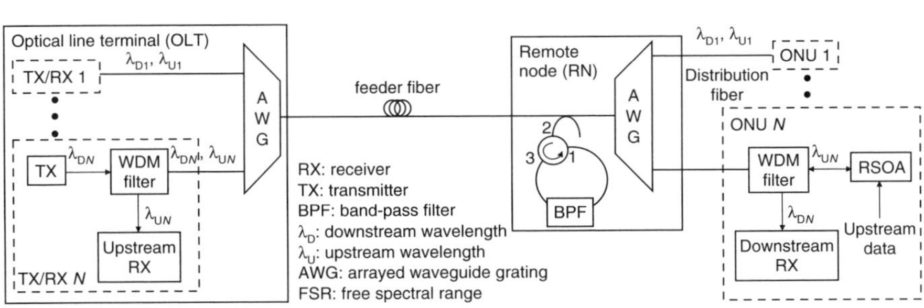

Figure 3.53 Schematic diagram of WDM-PON with self-seeding reflective SOAs. (From Ref [85])

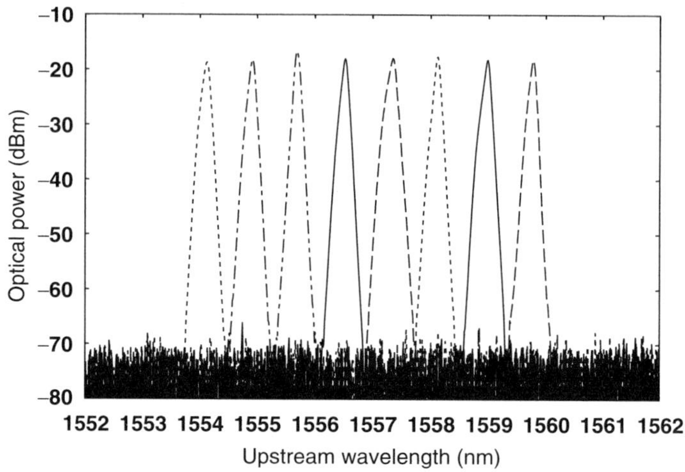

Figure 3.54 Overlapped optical spectra of 1.25 Gbps self-seeded upstream channels. (From Ref [85])

reused for upstream transmission. Such a wavelength reuse scheme to facilitate source-free ONUs was first proposed and experimentally demonstrated in 1994 in the RITE-net architecture (refer to Fig. 3.55) where a tunable laser at the CO steps through a range of wavelengths in time with each wavelength corresponding to a different ONU [86]. Half the time within each packet on a specific wavelength is used to carry downstream data from the CO, whereas the remaining half is CW. At each ONU, an optical power splitter divides the corresponding power of the downstream wavelength into two paths, one for downstream data detection and the other to be remodulated by an external modulator with upstream data, which is subsequently sent back toward the CO. In [86], both downstream and upstream data are subcarrier multiplexed onto the carrier

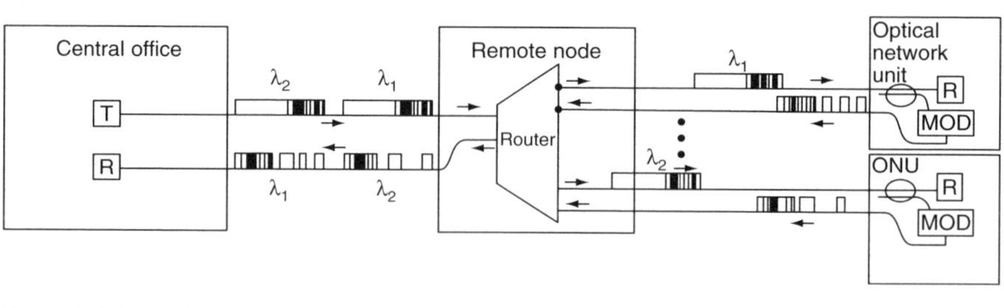

Figure 3.55 RITE-net architecture. The CW portion of each downstream packet is remodulated with upstream data at the optical network unit. (From Ref [86])

wavelength to enable efficient sharing of one CO receiver without packet contention in the time domain.

In wavelength reuse schemes proposed in [87, 88], the CO has N lasers that continuously transmit downstream data on N wavelengths to N ONUs. The downstream signals are encoded in formats that provide constant optical intensity such as optical frequency shift keying (FSK) format [87] and inverse return to zero (I-RZ) [88]. These formats allow remodulation of the downstream carriers with upstream data in amplitude shift keying (ASK) format (i.e. intensity modulation) using an external intensity modulator at the ONU, thereby simplifying the ONU configuration. The constant optical intensity on the downstream carrier minimizes the residual effects of the downstream data on the upstream intensity-modulated data. The scheme proposed in [89] exploits the benefits of a downstream carrier with constant optical intensity to optically injection-lock colorless FP-LDs that can be directly modulated with upstream data. The downstream format used was differential phase shift keying (DPSK). This scheme thereby removes the need for broadband sources at the CO as opposed to the colorless FP-LD scheme described in the preceding section [72].

By comparison, the proposals in [90, 91] choose to implement downstream data in ASK format at a sufficiently low optical ER. Referring to Fig. 3.56, at each ONU, instead of an intensity modulator, a reflective semiconductor optical amplifier that is driven into saturation is used to directly remodulate

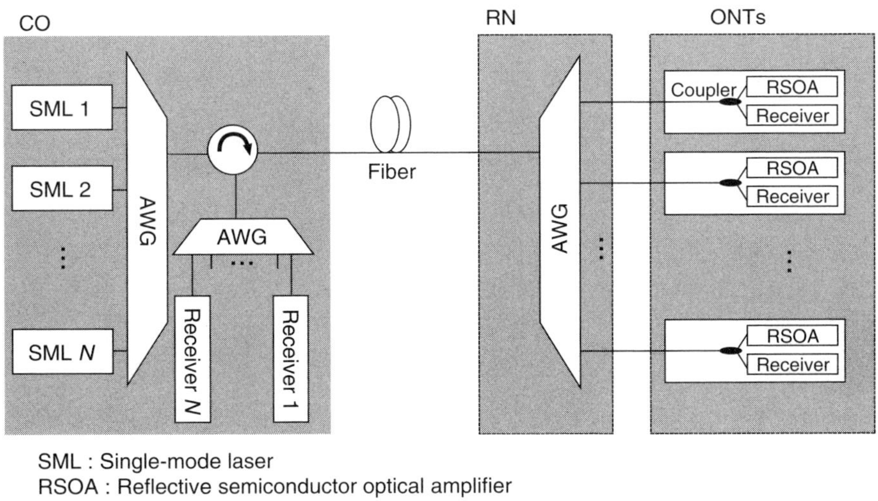

Figure 3.56 WDM-PON using reflective semiconductor optical amplifiers to amplify and remodulate downstream wavelengths carrying downstream signals (ASK format) for upstream transmission. (From Ref [90])

the downstream wavelength with upstream data and at a high optical ER. Compared to schemes using external intensity modulators, the RSOA also provides amplification to compensate for the round-trip loss incurred by the downstream carrier, thereby potentially increasing the distance and coverage of the network. In addition, when driven into saturation, the amplitude squeezing effect of the RSOA enables the modulation of the downstream signal to be suppressed [79]. The magnitude of suppression is strongly dependent on the ER of the downstream signal and also on the input power at the RSOA [90]. That is, if the downstream signal has high ER and is fed into the RSOA at a high-input power, the residual downstream ER will appear as fluctuations on the "1"-level of the upstream data through the remodulation process, thereby degrading the upstream performance at the CO. In this case, the demultiplexer at the CO may be negatively detuned to reduce the residual downstream ER and improve the upstream performance [92]. Figure 3.57 shows the optical spectrum of the downstream DFB laser source at the CO, that is directly modulated with 1.25 Gb/s NRZ data. The optical spectrum shows adiabatic-chirp dominant characteristics with two peaks associated with "1" and "0" levels (12.1-GHz spacing) for the extinction ratio of 6 dB. The passband characteristic of one of the output ports of a negatively detuned demultiplexer (−18.7 GHz from center frequency) is superimposed on the spectrum. The demultiplexer therefore filters out the spectral component associated with the "1" level, thus reducing

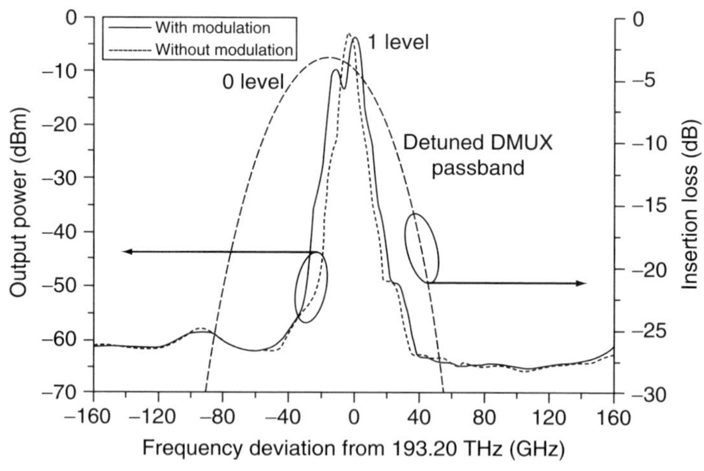

Figure 3.57 Optical spectrum of adiabatic-chirped DFB laser source directly modulated with 1.25 Gbps NRZ signal. Superimposed on the spectrum is the passband characteristic of negatively detuned (−18.7 GHz) demultiplexer. (From Ref [92])

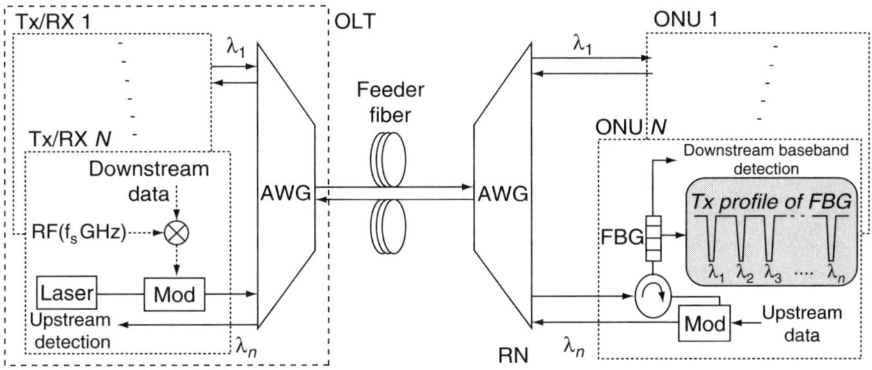

Figure 3.58 Wavelength reuse scheme for WDM-PON using multichannel fiber Bragg grating to filter out downstream carriers for remodulation with upstream data. (From Ref [93])

the spectral power difference between the "0" and "1" levels of the downstream signal, and improving the upstream performance.

Wavelength reuse schemes proposed very recently have revisited the use of subcarriers to distinguish between downstream and upstream signals at the ONU, which was first proposed in 1994. In [93–95], data in ASK format is subcarrier multiplexed onto downstream carriers at the CO. At each ONU, a passive optical notch filter is used to filter out the downstream carrier for base-band remodulation with upstream data. The exact same passive optical notch filter can be used at all ONUs if it is implemented using a multichannel fiber Bragg grating (refer to Fig. 3.58) or a delay interferometer with periodic notch-filtering characteristics. Alternatively, instead of using a filter at each ONU, a delay interferometer may be used at the remote node to simultaneously separate all downstream optical carriers from the RF subcarriers [95]. Compared to [86] which requires the use of additional RF electronics to recover the data on RF subcarriers, using passive optical filtering to separate the RF subcarriers from the base-band carriers enables base-band detection using conventional base-band receivers [93].

3.5 SUMMARY

The access network environment is particularly sensitive to variations in capital and operational costs due to its low cost-sharing characteristic. To date, optical technologies have been developed to reduce the overall capital and operational expenditures for the network, with goals of achieving ease of installation, manageability, upgradeability, customer friendliness, high reliability as well as good performance. Some of the recent and ongoing developments in this area of research have been reviewed in this chapter. Currently, most deployed optical access networks rely on the central element of a passive power splitter at the remote node to broadcast information to all subscribers. PLC technology has significantly improved the performance of optical power splitters, enabling small chip sizes, high reliability, and low cost per port to be achieved. Targeted areas of research are in the manufacturing processes of the fiber array block and PLC chip, primarily to reduce the insertion loss through the reduction of connector losses, and improving the fiber array and splitting ratio uniformity. With the current technology, achievable excess insertion losses of PLC-based splitters are 1–1.5 dB above the ideal theoretical splitting loss with a nonuniformity of within 2 dB over the specified range of operating wavelengths from 1250 nm to 1625 nm.

The ever-increasing demand for bandwidth beyond that provided by the shared capacity of the current power-splitting PON access networks, will necessitate the use of WDM to provide dedicated bandwidths to subscribers. An important optical technology which has a direct cost bearing on the deployment of WDM-PONs is the athermal AWG. As discussed, athermal operation has been achieved using two techniques, namely the incorporation of guiding materials that exhibit negative thermo-optic coefficient with opposite temperature dependence to silica, and the incorporation of a mechanically movable compensation plate in the AWG structure. Athermal AWGs of up to 40 channels at 100/200-GHz channel spacing are now commercially available with good crosstalk performance ($<$30 dB), low polarization-dependent loss ($<$0.4 dB), and low insertion loss ($<$2.0 dB for Gaussian type passband, $<$4.5 dB for semi-flat type passband), and temperature stability over a wide operating range ($<$0.02 nm from 0 °C to 60 °C).

Developments in optical technologies at the installation site of the ONU have also been rapid with the introduction of bendable and angled connectors (for easy and quick connection and disconnection), and customer-tolerant indoor optical fibers. Among these novel optical fibers are the bend-insensitive silica-based single-mode fibers with 15-mm bending radius, the optical curl cord based on hole-assisted fiber technology, and the GI-POF. With the GI-POF, the bending loss has been shown to be easily reduced to 0 dB, though ongoing work is currently focused on reducing its optical attenuation which is still in excess of 25 dB/km at 850 nm and 1300 nm. Developed for indoor fiber installations, these novel fibers enable optical wiring workability with flexible and reliable connections, safety, and cleaner installation with bent sections that are less prone to accidental snagging by customers.

The surge in research of ONU transmitters/configurations to minimize inventory costs has led to the development of colorless ONUs that allow for mass production and deployment across the network. The emission wavelength of each colorless ONU is nonspecific and is determined by the filtering properties of the AWG in the remote node or the wavelength of an injection/seeding light into the ONU. Directly modulated broadband optical sources such as LEDs, which are spectrally sliced at the AWG in the remote node, have been proposed as one form of colorless ONUs. Due to inherently low modulation speeds of these earlier solutions, alternative solutions based on injection-locked FP-LDs and wavelength-seeded RSOAs have been proposed. The emission wavelength of an injection-locked FP-LD or a wavelength-seeded RSOA follows that of the seeding light, which originates from a narrowband slice of light from a broadband optical source located at the CO. More recently, schemes that remove the need for the centralized broadband source have been proposed using self-injection-locking and self-seeding configurations which typically employ an optical feedback element in the remote node. On the other hand, source-free ONU configurations have been proposed to reuse the optical downstream carrier sent from the CO for upstream modulation. To minimize the influence of the downstream modulation on the upstream data, various schemes including the adoption of different modulation formats for downstream and upstream transmissions, as well as the use of subcarriers have been proposed.

It is envisioned that the rapid advancement of optical technologies for passive optical networks will continue to surge ahead to ensure cost-competitiveness with current access technologies. Already, the advancements in the above-mentioned areas have contributed significantly to the success and widespread uptake of power-splitting PONs, especially in Asian countries, with the deployment of the future-proof WDM-PON to follow suit in the near future.

REFERENCES

[1] K. Okamoto, "Planar Lightwave Circuits for FTTH and GMPLS," in Proc. of APOC 2006, paper 6351–28, Kwangju, Korea. Sept. 3–7, 2006.

[2] I.-B. Sohn, M.-S. Lee, J.-Y. Chung, "Fabricatiom of optical splitter and passive alignment technique with a femtosecond laser," *IEEE PTL*, vol.17, no.11, pp2349–2351, Nov., 2005.

[3] E. Edmon, K.G. McCammon, R. Estes, J. Lorentzen, "Chapter 2: Today's broadband fiber access technologies and deployment considerations at SBC," *Broadband optical access and FTTH*, ed. Chin-Lon Lin, Chap. 3, p17, (John Wiley & Sons, 2006).

[4] Alliance Fiber Optic Products, Inc., PLC Splitter data sheet, available from http://www.afop.com/pdf/Couplers,%20WDMs,%20Splitters/PLC%20Splitter%20Premium.pdf

[5] R. Gritters, H. Taguchi, and A. Geraci, "Low Insertion Loss PLC Splitters for Improved PON Performance," White paper from Seikoh Gaiken USA, Fiber optic product news http://www.fpnmag.com/scripts/ShowPR.asp?PUBCODE = 025&ACCT = 0004717&ISSUE = 0312&RELTYPE = PR&PRODCODE = 0000&PRODLETT = A

[6] K.A. McGreer, H. Zu, C. Ho, N. Kheraj, Q. Zhu, M. Stiller, and J. Lam, "Planar lightwave circuits for PON applications" Proc. of OFC, paper NWD4, Mar., 2006.

[7] M.K. Smit, "Progress in AWG design and technology" Proc. of 2005 IEEE IEEE/LEOS Workshop on Fibers and Optical Passive Components, pp26–31, June, 2005.
[8] M.K. Smit, "New focusing and dispersive planr component based on an optical phased array," *Elect. Lett.*, vol.24, no.7, pp385–386, Mar., 1988.
[9] H. Takahashi, S. Suzuki, K. Kato, and I. Nishi,"Arrayed-waveguide grating for wavelength division multi/demultiplexer with nanometer resolution," *Elect. Lett.*, vol.26, no.2, pp87–88, Jan., 1990.
[10] C. Dragone, "An NxN optical multiplexer using a planar arrangement of two star couplers," *IEEE Phot. Technol. Lett.*, vol.3, no.9, pp812–815, Sept., 1991.
[11] Y. Barbarin, X.J.M. Leijtens, E.A.J.M. Bente, C.M. Louzao, J.R. Kooiman, and M.K., Smit, "Extremely small AWG demultiplexer fabricated on InP by using a double-etch Process," *IEEE Phot. Technol. Lett.*, vol.16, no.11, pp2478–2480, Nov., 2004.
[12] A. Kaneko, T. Goh, H. Yamada, T. Tanaka, and L. Ogawa, "Design and applications of silica-based planar lightwave circuits," *IEEE Journal of Selected Topics in Quantum Electronics*, vol.5, no.5, pp1227–1236, Sept./Oct. 1999.
[13] Y. Inoue, A. Himeno, K. Moriwaki, and M. Kawachi, "Silica-based arrayed-waveguide grating circuit as optical splitter/router," *Elect. Lett.*, vol.31, no.9, pp726–727, April 27, 1995.
[14] Y.P. Li, L.G. Cohen, C.H. Henry, E.J. Laskowski, and M.A. Cappuzzo, "Demonstration and application of a monolithic Two-PONs-In-One device," Optical Communication, 1996. ECOC '96. 22nd European Conference on, vol.2, pp123–126, Sept. 15–19, 1996.
[15] Y. Inoue, A. Kaneko, F. Hanaw, H. Takahashi, K. Hattori, and S. Sumida, "Athermal silica-based arrayed-waveguide grating (AWG) multiplexer," Proc. of 11th International Conference on Integrated Optics and Optical Fibre Communications, and 23rd European Conference on Optical Communications, vol.5, pp33–36, Sept. 1997.
[16] Y. Inoue, "Athermal silica-based arrayed-waveguide grating (AWG) multiplexer," *Elect. Lett.*, vol.33, no.23, pp1945–1947, Nov., 1997.
[17] A. Kaneko, S. Kamei, Y. Inoue, H. Takahashi, and A. Sugita, "Athermal silica-based arrayed-waveguide grating (AWG) multi/demultiplexers with new low loss groove design," *Elect. Lett.*, vol.36, no.4, pp318–319, Feb. 2000.
[18] Y. Inoue, Y. Ohmori, M. Kawachi, S. Ando, T., Sawada., and H. Takahashi, "Polarization mode converter with polyimide half waveplate in silica-based PLCs," *IEEE Phot. Technol. Lett.*, vol.6, no.5, pp626–628, May, 1994.
[19] K. Maru, Y. Abe, M. Ito, H. Ishikawa, S. Himi, H. Uetsuka, and T. Mizumoto, "2.5%-Δ silica-based athermal arrayed waveguide grating employing spot-size converters based on segmented core," *IEEE Phot. Technol. Lett.*, vol.17, no.11, pp2325–2327, Nov., 2005.
[20] K. Maru, K. Matsui, H. Ishikawa, Y. Abe, S. Kashimura, S. Himi, and H. Uetsuka, "Super-high-Δ athermal arrayed waveguide grating with resin-filled trenches in slab region," *Elect. Lett.*, vol.40, no.6, pp374–375, Mar. 2004.
[21] S. Kamei, K. Iemura, A. Kaneko, Y. Inoue, T. Shibata, H. Takahashi, and A. Sugita, "1.5%-Δ athermal arrayed-waveguide grating multi/demultiplexer with very low loss groove design, vol.17, no.3, pp588–590, Mar. 2005.
[22] D. Kim, Y. Han, J. Shin, S. Park, Y. Park, H. Sung, S. Lee, and Y. Lee, "Suppression of temperature and polarization dependence by polymer overcladding in silica-based AWG multiplexer," Optical Fiber Communications Conference, 2003. OFC 2003 vol.1, pp61–62, Mar. 23–28, 2003.
[23] N. Keil, H.H. Yao, C. Zawadzki, J. Bauer, M. Bauer, C. Dreyer, and J. Schneider, "Athermal all-polymer arrayed-waveguide grating multiplexer," *Elect. Lett.*, vol.37, 9, pp579–580, Apr. 26, 2001.

[24] J. Hasegawa, and K. Nara, "Low loss (/spl sim/1.4 dB) 200 GHz-16ch athermal AWG compact module for metro/access network," Optical Fiber Communication Conference, 2005. *Technical Digest*. OFC/NFOEC vol.2, p3, Mar. 6–11, 2005.
[25] L. Leick, M. Boulanger, J.G. Nielsen, H. Imam, and J. Ingenhoff, "Athermal AWGs for colorless WDM-PON with −40 °C to +70 °C and underwater operation," Proc. of OFC, Postdeadline paper PDP31, Mar., 2006.
[26] N. Ooba, Y. Hibino, Y. Inoue, and A. Sugita, "Athermal silica-based arrayed-waveguide grating multiplexer using bimetal plate temperature compensator," *Elect. Lett*., vol.36, no.21, pp1800–1801, Oct. 2000.
[27] NEL, NTT Electronics Corporation website: http://www.nel-world.com/products/photonics/ather_awg.html
[28] Furukawa Electric Co., Ltd. website: http://www.fitel.com/news/060301.htm
[29] Hitachi Cable, Ltd. http://www.hitachi-cable.co.jp/en/product/news/20020320.html
[30] NeoPhotonics Corporation website: http://www.neophotonics.com/products/nei.php?proid = 42
[31] Gemfire Corporation website: http://www.gemfirecorp.com/Resources/ATHERMAL%2003–2005_output.pdf
[32] H. Hayashida, M. Yasunaga, T. Ema, K. Nakazawa, Y. Hoshino, H. Tanaka, and Y. Oda, "Reducing costs for first one mile FTTH lines," Proc. of OFC, paper WP2, Mar. 2005.
[33] H. Shinohara, "Broadband expansion in Japan," Proc. of OFC, Plenery Talk, Mar. 2005.
[34] "Connectors in FTTH Networks," White paper, available from www.adc.com/Library/Literature/103178AE.pdf
[35] S.C. Zimmel, "PON deployment of angle polished connectors," White paper, available from http://www.adc.com/Library/Literature/1326454.pdf
[36] StockerYale, Bend Insensitive Fiber BIF1550-L2, BIF-RC-1550-L2, Data Sheet available from http://www.stockeryale.com/o/fiber/products/bif-1550-l2_bif-rc-1550-l2.pdf
[37] H. Sakabe, H. Ishikawa, Y. Tanji, M. Terasawa, M. Itou, and T. Ueda, "Performance of Bend-insensitive single mode fiber and its application to optical access networks," *SEI Technical Review*, no.59, pp32–36, Jan. 2005.
[38] K. Nakajima, K. Hogari, J. Zhou, K. Tajima, and I. Sankawa, "Hole-assited fiber design for small bending and splice losses," *IEEE Phot. Technol. Lett*, vol.15, no.12, pp1737–1739, Dec. 2003.
[39] K. Hiramatsu, S. Tomita, T. Kurashima, and E. Araki, "Flexible optical fiber curl cord," *NTT Technical review*, vol.2, no.5, pp57–61, May, 2005.
[40] A. Fujioka, K. Osono, Y. Endo, and H. Akahoshi, "Key Materials for Electronic Information and Communication Equipment," *Hitachi Review*, vol.55, no.1, pp21–25, 2006, available from http://www.hitachi.com/ICSFiles/afieldfile/2006/04/25/r2006_01_101.pdf
[41] Y. Koike, E. Nihei, and T. Ishigure, "High bandwidth (2 GHz km) large core (0.5–1.0 mm) GI polymer optical fiber," Lasers and Electro-Optics Society Annual Meeting, 1993. LEOS '93 Conference Proceedings. *IEEE*, pp349–350, Nov. 15–18, 1993.
[42] Y. Koike, and E. Nihei, "Polymer optical fibers," Optical Fiber Communications, 1996. OFC '96, pp59–60, Feb. 25–Mar. 1, 1996.
[43] Y. Koike, T. Ishigure, and E. Nihei, "High-bandwidth graded-index polymer optical fiber," *J. Lightwave Technol*., vol.13., pp1475–1489, July, 1995.
[44] T. Ishigure, A. Horibe, E. Nihei, and Y. Koike, "High-bandwidth, high-numerical aperture graded-index polymer optical fiber," *J. Lightwave Technol*., vol.13, no.8, pp1686–1691, Aug., 1995.
[45] T. Ishigure, M. Satoh, O. Takanashi, E. Nihei, T. Nyu, S. Yamazaki, and Y. Koike, "Formation of the refractive index profile in the graded index polymer optical fiber for gigabit data transmission," *J. Lightwave Technol*., vol.15., pp2095–2100, Nov. 1997.

[46] Y. Watanabe, Y. Takano, T. Tsukamoto, Y. Matsuyama, Y. Hioki, and T. Hisano, "The new optical wiring system for LAN in premises and residential area using perfluorinated GI-POF," Optical Fiber Communication Conference and Exhibit, 2002. OFC 2002 pp629–630, Mar. 17–22, 2002.
[47] K. Makino, T. Ishigure, and Y. Koike, "Waveguide parameter design of graded index POFs for bending loss reduction," *J. Lightwave Technol.*, vol.24, no.5, pp2106–2114, May 2005.
[48] W.R. White, L.L., Jr. Blyler, R. Ratnagiri, and M. Park, "Manufacture of perfluorinated plastic optical fibers," Optical Fiber Communication Conference, 2004. OFC 2004, paper THI1.
[49] G.P. Agrawal, Semiconductor Lasers and Amplifiers from Lightwave Technology: Components and Devices, Chap. 5, p179 (Wiley-Interscience, 2004).
[50] U. Troppenz, J. Kreissl, W. Rehbein, C. Bornholdt, T. Gaertner, M. Radziunas, A. Glitzky, U. Bandelow, and M. Wolfrum, "40 Gb/s directly modulated InGaAsP Passive Feedback DFB Laser," Proc. of ECOC, Postdeadline Paper Th4.5.5, 2006.
[51] Furukawa Electric Co., Ltd., Full C-band thermally tunable DFB laser module with integrated wavelength monitor, available from http://www.furukawa.co.jp/fitel/eng/active/pdf/Cooled/OD-YAH46001C.pdf., Dec. 2004.
[52] C. Chang-Hasnain, "Progress and prospects of long-wavelength VCSELSs," *IEEE Opt. Commun. Mag.*, vol.41, ppS30–S32, 2003.
[53] VERTILAS GmbH, "Introduction to VCSELs," available from http://www.vertilas.com/technologie_einfuhrung.php
[54] VERTILAS GmbH, "VCSELs for Data/Telecommunication Applications," available from http://www.vertilas.com/produkte_daten_.php
[55] W. Hofmann, G. Gohm, M. Ortsiefer, E. Wong, and M.C. Amann, "Uncooled High-Speed 1.55 um VCSELs for CWDM Access Networks," IEE European Conference on Optical Communications (ECOC), postdeadline paper Th4.5.4, Cannes, France, Sept. 2006.
[56] E. Wong, X. Zhao, C.J. Chang-Hasnain, W. Hofmann, and M.C. Amann, "Optically Injection-Locked 1.55 um VCSEL as upstream transmitters in WDM-PONs," *IEEE Phot. Technol. Lett.*, vol.18 (22), pp2377–2373, Nov. 2006.
[57] L. Chrostowski, C. Chang, and C.J. Chang-Hasnain, "Uncooled injection-locked 1.55 μm tunable VCSELs as DWDM transmitter," in Proc. Optical Fiber Communication Conference, vol.2, pp753–754, Mar. 2003.
[58] L. Chrostowski, X. Zhao, and C.J. Chang-Hasnain, "Microwave performance of optically injection-locked VCSELs," *IEEE Trans. Microwave. Theory Tech.*, Special Issue on Microwave Photonics, vol.54, pp788–795, Feb. 2006.
[59] W.S. Jang, H.C. Kwon, and S.K. Han, "Suppression of Rayleigh backscattering in a bidirectional WDM optical link using clipped direct modulation," *IEE Proc. Optoelectron.*, 151 (4), pp219–222, 2004.
[60] E. Wong, X. Zhao, C.J. Chang-Hasnain, W. Hofmann, and M.C. Amann, "Extinction ratio dependence of 1.55 um VCSELs injection-locked by downstream signals in a WDM-PON," Proc. in ECOC, paper We3.P.161, Cannes, France, Sept. 2006.
[61] W. Yuen, G.S. Li, R.F. Nabiev, M. Jansen, D. Davis, and C.J. Chang-Hasnain, "Electrically-pumped directly-modulated tunable VCSEL for metro DWDM applications," *GaAS IC Symposium Tech. Dig.*, pp51–52, Oct. 2001.
[62] M. Zirngibl, C.H. Joyner, L.W. Stulz, C. Dragone, H.M. Presby, and I.P. Kaminow, "LARnet, a local access router network," *IEEE Phot. Technol. Lett.*, vol.7, no.2, pp215–217, Feb. 1995.
[63] K.-Y. Liou, U. Koren, E.C. Burrows, J.L. Zyskind, K. Dreyer, "A WDM access system architecture based on spectral slicing of an amplified LED and delay-line multiplexing and

encoding of eight wavelength channels for 64 subscribers," *IEEE Phot. Technol. Lett.*, vol.9, no.4, pp517–519, April 1997.
[64] S.S. Wagner, and T.E. Chapuran, "Broadband high-density WDM transmission using superluminescent diodes," *Elect. Lett.*, vol.26, no.11, pp696–697, May 24, 1990.
[65] W.T. Holloway, A.J. Keating, and D.D. Sampson, "Multiwavelength source for spectrum-sliced WDM access networks and LAN's," *IEEE Phot. Technol. Lett.*, vol.9, no.7, pp1014–1016, July 1997.
[66] S.L. Woodward, P.P. Iannone, K.C. Reichmann, and N.J. Frigo, "A spectrally sliced PON employing Fabry–Perot lasers," *IEEE Phot. Technol. Lett.*, vol.10, no.9, pp1337–1339, Sept. 1998.
[67] Bo Zhang, Chinlon Lin, Li Huo, Zhaoxin Wang, and Chun-Kit Chan, "A simple high-speed WDM PON utilizing a centralized supercontinuum broadband light source for colorless ONUs," Optical Fiber Communication Conference, 2006 and the 2006 National Fiber Optic Engineers Conference, p3, Mar. 5–10, 2006.
[68] C.F. Lam, M.D. Feuer, and N.J. Frigo, "Performance of pin and APD receivers in high-speed WDM data transmission systems employing spectrally sliced spontaneous emission sources," *Elect. Lett.*, vol.36, no.18, pp1572–1574, Aug. 31, 2000.
[69] H. Sanjoh, H. Yasaka, Y. Sakai, K. Sato, H. Ishii, and Y. Yoshikuni, "Multiwavelength light source with precise frequency spacing using a mode-locked semiconductor laser and an arrayed waveguide grating filter," *IEEE Phot. Technol. Lett.*, vol.9, no.6, pp818–820, June 1997.
[70] H.D. Kim, S.G. Kang, and C.H. Lee, "A low-cost WDM source with an ASE injected Fabry–Perot semiconductor lasers," *IEEE Phot. Technol. Lett.*, vol.12, pp1067–1069, Aug. 2000.
[71] D.J. Shin, D.K. Jung, J.K. Lee, J.H. Lee, Y.H. Choi, Y.C. Bang, H.S. Shin, J. Lee, S.T. Hwang, and Y.J. Oh, "155 Mbit/s transmission using ASE-injection Fabry-Perot laser diode in WDM-PON over 70°C temperature range," *IEE Elect. Lett.*, vol.39, pp1331–1332, 2003.
[72] D.J. Shin, Y.C. Keh, J.W. Kwon, E.H. Lee, J.K. Lee, M.K. Park, J.W. Park, K.Y.Oh, S.W. Kim, I.K. Yun, H.C. Shin, D. Heo, J.S. Lee, H.S. Shin, H.S. Kim, S.B. Park, D.K. Jung, S. Hwang, Y.J. Oh, D.H. Jang, and C.S. Shim, "Low-cost WDM-PON with colorless bidirectional transceivers," *J. Lightwave Technol.*, vol.24, pp158–165, Jan. 2006.
[73] Soo-Jin Park, Chang-Hee Lee, Ki-Tae Jeong, Hyung-Jin Park, Jeong-Gyun Ahn, and Kil-Ho Song, "Fiber-to-the-home services based on wavelength-division-multiplexing passive optical network," *J. Lightwave Technol.*, vol.22, no.11, pp2582–2591, Nov. 2004.
[74] G.P. Agrawal, *Chapter 7 Photodetectors* from *Lightwave Technology: Components and Devices*, Chap. 7, p253 (Wiley-Interscience, 2004).
[75] G.Y. Kim, J. Seo, M.J. Kang, S.B. Koh, H.K. Seo, S.T. Lee, S.G. Jong, and Jeong, Ki-Tae, "FTTH field trial of injection locking based WDM-PON system," Proc. of SPIE, vol. 6354, 2006.
[76] Meghan Fuller, *Lightwave* Oct., 2006. "SOAs still awaiting big chance," available from http://lw.pennnet.com/display_article/275513/13/ARTCL/none/none/SOAs_sti
[77] Corecess, Inc., "Corecess Intros DWDM-PON", Sept. 13, 2006, available from http://www.lightreading.com/document.asp?doc_id=103644, http://www.corecess.com/eng/main.asp
[78] S.-B. Park, D.K. Jung, D.J. Shin, H.S. Shin, I.K. Yun, J.S. Lee, Y.K. Oh, and Y.J. Oh, "Demonstration of WDM-PON with 50 GHz channel spacing employing spectrum-sliced reflective semiconductor optical amplifiers," *Elect. Lett.*, vol.42, no.20, pp1172–1173, Sept. 28, 2006.

[79] P. Healey, P. Townsend, C. Ford, L. Johnston, P. Townley, I. Lealman, L. Rivers, S. Perrin, and R. Moore, "Spectral slicing WDM-PON using wavelength-seeded reflective SOAs," *IEE Elect. Lett.*, vol.37, pp1181–1182, 2001.
[80] S.H. Shin, O.K. Jung, D.J. Shin, S.B. Park, J.S. Lee, L.K. Yun, S.W. Kim, Y.H. Oh, and C.S. Shim, "16 /spl times/ 1.25 Gbit/s WDM-PON based on ASE-injected R-SOAs in 60/ spl deg/C temperature range," Optical Fiber Communication Conference, 2006 and the 2006 National Fiber Optic Engineers Conference, p3, Mar. 2006.
[81] F. Payoux, P. Chanclou, M. Moignard, and R. Brenot, "Gigabit optical access using WDM PON based on spectrum slicing and reflective SOA," in Proc. European Conference of Communications (ECOC), vol.3, pp455–456, Sept. 2005.
[82] CIPhotonics, Inc., Data sheet: "SOA-RL-OEC-1550 – 1.55 um reflective semiconductor optical amplifier (SOA)," available from http://www.ciphotonics.com/PDFs_29Aug06/SOA_RL_OEC_1550_G.pdf
[83] H.C. Shin, J.S. Lee, I.K. Yun, S.W. Kim, H.I. Kim, H.S. Shin, S.T. Hwang, Y.J. Oh, K. Oh, and C.S. Shim, "Reflective SOAs optimized for 1.25 Gbit/s WDM-PONs," Lasers and Electro-Optics Society, 2004. LEOS 2004. The 17th Annual Meeting of the IEEE, vol.1, Nov. 7–11, pp308–309, 2004.
[84] S. Hann, Tae-Young Kim, and Chang-Soo Park, "Direct-modulated upstream signal transmission using a self-injection locked F-P LD for WDM-PON," Optical Communication, 2005. ECOC 2005. 31st European Conference on, vol.3, pp451–452, Sept. 25–29, 2005.
[85] E. Wong, K.-L. Lee, and T. Anderson, "Low-Cost WDM Passive Optical Network With Directly-Modulated Self-Seeding Reflective SOA," *Elect. Lett.*, vol.42(5), pp299–300, Mar. 2006.
[86] N.J. Frigo, P.P. Iannone, P.D. Magill, T.E. Darcie, M.M. Downs, B.N. Desai, U. Koren, T.L. Koch, C. Dragone, H.M. Presby, and G.E. Bodeep, "A wavelength-division multiplexed passive optical network with cost-shared components," *IEEE Phot. Technol. Lett.*, vol.6, no.11, pp1365–1367, Nov. 1994.
[87] Ning Deng, Chun-Kit Chan, Lian-Kuan Chen, and F. Tong, "Data remodulation on downstream OFSK signal for upstream transmission in WDM passive optical network," *Elect. Lett.*, vol.39, no.24, pp1741–1743, Nov. 27, 2003.
[88] Guo-Wei Lu, Ning Deng, Chun-Kit Chan, and Lian-Kuan Chen, "Use of downstream inverse-RZ signal for upstream data re-modulation in a WDM passive optical network," Optical Fiber Communication Conference, 2005. Technical Digest. *OFC/NFOEC*, vol.5, p3, Mar. 6–11, 2005.
[89] Wai Hung, Chun-Kit Chan, Lian-Kuan Chen, and F. Tong, "An optical network unit for WDM access networks with downstream DPSK and upstream remodulated OOK data using injection-locked FP laser," *IEEE Phot. Technol. Lett.*, vol.15, no.10, pp1476–1478, Oct. 2003.
[90] W.R. Lee, M.Y. Park, S.H. Cho, J. Lee, C. Kim, G. Jeong, and B.W. Kim Hung, "Bidirectional WDM PON based on gain saturated reflective SOAs," *IEEE Phot. Technol. Lett.*, vol.17, pp2460–2462, 2005.
[91] J.J. Koponen and M.J. Soderlund, "A duplex WDM passive optical network with 1:16 power split using reflective SOA remoduator at ONU," Proc. in OFC, MF99, OFC, 2003.
[92] W. Lee, S.H. Cho, M.Y. Park, J.H. Lee, C. Kim, G. Jeong, and B.W. Kim, "Wavelength filter detuning for improved carrier reuse in loop-back WDM-PON," *Elect. Lett.*, vol.42, no.10, pp596–597, May 11, 2006.

[93] M. Attygalle, N. Nadarajah, and A. Nirmalathas, "A novel technique for wavelength reuse in WDM-PON," Lasers and Electro-Optics Society, 2005. LEOS 2005. The 18th Annual Meeting of the IEEE, pp224–225, Oct. 2005.

[94] M. Attygalle, N. Nadarajah, A. Nirmalathas, "Wavelength Reuse Scheme for Source Free Optical Network Units in WDM Passive Optical Networks," Proc. of 31st European Conference on Optical Communication, paper TU3.5.3, 2006.

[95] Z. Xu, Y.J. Wen, M. Attygalle, X. Cheng, W.-D. Zhong, Y. Wang, and C. Lu, "Multiple channel carrier-reused WDM Passive Optical Networks, Proc. of 31st European Conference on Optical Communication, paper, Th4.3.2, 2006.

Chapter 4

Transceivers for Passive Optical Networks

Yongmao Frank Chang and Badri Gomatam

Vitesse Semiconductor, Camarillo, California

4.1 INTRODUCTION

Optical access networks have received considerable attention for their potential to alleviate bandwidth constraints in the last mile [1–5]. It has already been recognized as an important technology for high-bandwidth deployments. Recently, true fiber-to-the-home (FTTH) installments in various forms of passive optical networks (PONs) have finally been realized in many Asian countries. It was reported that there were over 100,000 new FTTH subscribers every month in Japan alone. This technology is also quickly spurring new markets in North America and Europe, Middle East, and Africa (EMEA) regions [6].

Optical transceivers are key to optical communications since they perform the fundamental task of optical-to-electrical signal conversion. Optical transceivers rely on the use of semiconductor lasers and detectors, in conjunction with integrated circuits (ICs), and novel packaging technologies to provide cost-effective means for transmitting and receiving analog and digital signals over fiber-optic cables.

Transceivers (transmitters and receiver) are perceived as significant components of the cost in FTTH networks. For these price-sensitive systems, there is tremendous pressure to develop low-cost, high-performance, and compact transceivers that operate under real-world system constraints. This chapter describes the building blocks, their associated technologies, system evaluations, and related issues of such transmitters and receivers, or simply transceivers.

4.1.1 Historical Background

The recent tremendous growth in IP traffic has accentuated the aggravating lag of access network capacity, rendering the "last mile" the bottleneck between local area and backbone networks. It is currently believed that copper-based access networks (cable modem and many kinds of digital subscriber lines or DSLs) do not provide either the minimum bandwidth or the required transmission distance for delivering future services of voice, data, and video programs such as IPTV and HDTV. To resolve this problem, PONs are identified as a promising, cost-effective, and future-safe access solution [1]. Full-service access PON networks can offer efficient gigabit transport with guaranteed Quality of Service (QoS) to customers [7], through different fiber-to-the-premises/curb/node/business/home/users (FTTx) scenarios.

Nowadays PON systems are prevalent in Asia (Japan and Korea). The total number of FTTH users in Japan alone has exceeded 7 million with accelerated growth to reach over a 100,000 users per month. PON transceivers have entered into the stage of low-cost and high-volume mass production. Evolutions of new optical technologies, and the ability to integrate more functionality into a single IC, both for digital and analog needs, are two directions to further drive down the cost.

Figure 4.1 illustrates a typical PON system, consisting of an optical line terminal (OLT), multiple optical network units (ONUs) and a passive optical splitter connecting the OLT to the ONUs. In the downstream direction, the OLT broadcasts continuous downstream packets to each of the ONUs in the 1480- to

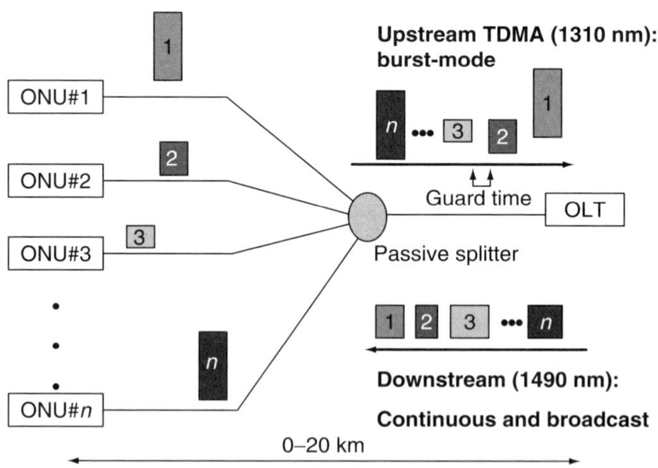

Figure 4.1 The general PON network architecture for the FTTx scenario.

1500-nm wavelength band. Because each node receives the entire original signal, each downstream packet contains a header identifying the ONU that should receive the packet. The downstream traffic may use encryption to restrict each ONU to decode only the part it is supposed to receive.

Each ONU gets synchronized with the OLT by extracting the clock from downstream traffic. Some PONs can be configured to change bandwidth allocations to individual ONUs on the fly, depending on the users' needs. This enables carriers to better utilize their systems and also provide premium users with more bandwidth. As explained in Chap. 2, some systems use the 1550-nm-wavelength window in downstream for CATV video broadcast-service overlay.

The more challenging aspect is the upstream traffic, where one OLT collects all upstream traffic from multiple ONUs in the 1310-nm window in a "multipoint-to-point" fashion. In order to avoid data collisions between ONUs, upstream traffic is managed by a time-division-multiple-access (TDMA) scheme. The OLT allocates transmission time slots, i.e. Gates, to each ONU. When an ONU receives the Gate frame, the ONU will transmit its MAC frames with the designated speed during the allocated time slots. The whole system is synchronized so that the data from different ONUs will not interfere with each other. As a result, the OLT always receives a bursty stream of packets separated by some guard time during which no signal is transmitted by any ONU.

4.1.2 PON Transceiver Evolution

Fiber entered access networks around the 1990s [1]. Initially, the bit rate was lower than 100 Mbps. This was later upgraded to 150 Mbps and 600 Mbps along with the hierarchy of SDH. The original multiplexing scheme was based on asynchronous transfer mode (ATM) cells, which was later replaced by Ethernet packets. The first standard specifications of PON were established as the G.983 series recommendations by ITU-T in 1998 [8]. Almost simultaneously, independent PON systems at 100 Mbps were developed and put into commercial use, albeit in limited regions, as cost-effective systems.

In the last 3 or 4 years, the PON data rate has exceeded 1 Gbps and the IEEE 802.3 committee established 802.3ah (EPON Standard) in June 2004 [9]. Several operators in Japan (in particular NTT) introduced systems complying with the standard as gigabit Ethernet-PON (GE-PON) for real field applications. About the same time, ITU-T established the G.984 series G-PON standards with up to 1.244-Gbps upstream speed and up to 2.488-Gbps downstream speed [10]. Optical transceivers complying with these standards have been released by chip and module vendors and further developments of optical transmission devices in accordance with standardized specifications have greatly boosted FTTH applications. Detailed descriptions of various PON architectures can be found in Chap. 2.

4.2 PON SYSTEM REQUIREMENTS

The performance characteristics of a conventional TDM-PON system is defined by the feeder bandwidth it provides, its bandwidth efficiency, splitter ratio, and the differential reach it supports. This section discusses the physical layer specifications and requirements that PON transceivers need to comply with.

4.2.1 PON System Power Budgets and Specifications

Table 4.1 lists the split ratios and power budgets for three PON types. BPON supports 16 or 32 ONUs and dynamic upstream bandwidth allocation [8]. Both BPON and G-PON adopted the three classes (A, B, C) of ITU-T G.982-specified power budgets, i.e. 20, 25, and 30 dB between the OLT and an ONU. G-PON upgrades BPON with higher data rates shared by 16–64 or even 128 nodes. Its higher speed enable carriers to split the bandwidth to serve a greater number of ONUs than BPON, but also puts a higher strain on the power budgets.

The IEEE 802.3ah standard specifies 10-km and 20-km transmission distances [9, 11–13]. These physical-layer specifications are referred to as

Table 4.1

PON power budgets[a]

Type & protocal	Split ratio	Power budget
BPON	32 max	Channel loss (only[b])
		Class A optics: 20 dB
		Class B optics: 25 dB
		Class C optics: 30 dB
GPON	64 max	Channel loss (only[b])
		Class A optics: 20 dB
		Class B optics: 25 dB
		Class C optics: 30 dB
EPON	16 nominal	PX10 US: 23 dB
	32 permitted	PX10 DS: 21 dB
		PX20 US: 26 dB
		PX20 DS: 26 dB

[a] There also exists a class B+ budget in field deployments. It is midway between class B and C with an overall loss between a minimum of 10 dB and maximum of 28 dB.

[b] The dispersion and noise related optical path penalty is specified to be less than 1 dB for both BPON and G-PON.

1000BASE-PX10 and 1000BASE-PX20. EPON supports at least 16 fiber splits, a performance exceeding this minimum boundary is acceptable and the maximum PON split-ratio was not specified because it depends on actual fiber-link losses and physical-layer capabilities. In reality, most EPONs are deployed with a 1:32 split ratio, with some trials done for 1:64 or even 1:128 split ratios depending on the available link budgets and fiber-plant configurations [14].

Optical link budgets are determined by individual vendors' active components —lasers, receivers, and PON chips. Traditional BPON equipment has typically used class B optics, but it was determined that some of the PON networks of 20-km transmission lengths were actually stretching the budget to the limits, forcing active equipment manufacturers to increase link budgets to 26.5 dB. These increased budgets, coupled with a possible requirement to increase the split ratios of G-PON, resulted in an increase in the class B budget to allow for a 28-dB loss budget—thus, establishing the class B+ optics category [15, 16].

With class A optics typically associated with fiber-to-the-curb (FTTC) applications, class B and B+ optics provide today's FTTP PONs with the best reach and split ratios. Nevertheless, despite the increase in capability, the class B+ optics have not escalated to the price level of higher class C while supporting better PON loss budgets, thanks to low-noise trans-impedance ICs, which results in the recent improvements in the receiver sensitivities. In fact, EPON deployments have also taken B+ equivalent optics into account. In the future, however, the need to transport greater distances (30 km or 40 km) and even higher split ratios (1:128) could eventually force equipment manufacturers to use class-C equivalent optics.

4.2.2 Physical-Layer Specifications

Physical-layer specifications are mainly related to the physical medium-dependent (PMD) layer, which includes optical/electrical converters and clock and data recovery (CDR). The physical layer passes data traffic from physical medium to layer 2 and vice versa. The PMD layer specifies optical transmitters and receivers with transmission power and receiver sensitivity values for each power budget and transmission speed. To keep optical transceiver costs low, forward error correction (FEC) is allowed as an option, albeit at a slight loss of data throughput (Chap. 2).

Table 4.2 summarizes the ITU-T G.983.2 BPON standards [8]. Since G-PON is expected to replace BPON with high rates, its requirements are set to be somewhat consistent with those of BPON [17–19].

While G-PON aims at accommodating full services, EPON primarily aims at transmitting Ethernet frames. ITU-T tried to have a common physical-layer specification with EPON, but the commonality of the specification is imperfect

Table 4.2
ITU-T G.983.3 BPON standards

Items		Downstream	Upstream
Technology		ATM with WDM	ATM with WDM
Bit rate (Mbps)		155/622	155
Wavelength (nm)	Basic band	1,480–1,500	1,260–1,360
	Enhance band (op1)	1,539–1,565	1,260–1,360
	Enhance band (op2)	1,550–1,560	1,260–1,360
Output power (dBm)	Class A	−3 to −7.5	−7.5 to 0
	Class B	−2.5 to +2	−5.5 to +2
	Class C	−0.5 to +4	−3.5 to +4
Extinction ratio (dB)		>10	>10
MLM Laser max RMS width (nm) (for basic band)		<1.8	<5.8
Receiver sensitivity (dBm)	Class A	−28.5	−28.5
	Class B	−28.5	−31.5
	Class C	−31.5	−34.5
CID immunity		>72 digits	>72 digits

because IEEE was considering using only the existing Ethernet transceivers. Table 4.3 compares the physical layer requirements of ITU-T G-PON with IEEE EPON Standards [9, 10, 19].

4.2.3 PON Burst-Mode Timing Requirements

Burst-mode transmission in the PON system upstream direction imposes stringent timing requirements on both ONU and OLT. Figure 4.2 provides an outline of the timing definitions for the physical layer burst-mode upstream transmission from an ONU transmitter to the OLT receiver [19].

The multiple upstream access and burst-mode behavior require that ONU transmitters must generate time-gated bursty signals within the allocated time slots assigned by the MAC layer. This means laser power levels and extinction ratio must be stabilized within a short period of time, and then remain unchanged before the time slot completes. During the time slots allocated to other ONUs, an ONU transmitter should remain shut off and must not send upstream light or even low-level noise. Otherwise this would create cross talks and disturb the operational upstream services. This requires fast ONU transmitter switching, which should be normally within subnanosecond rise and fall time, after a power up or power down or a first connection to the network.

Table 4.3

Physical-layer requirements of ITU-T G-PON and IEEE EPON standards

Item		FSAN / ITU-T G-PON	IEEE EPON
MAC Layer	Service	Full services (Ethernet, TDM, POTS)	Ethernet data
	Frame	GEM frame	Ethernet frame
PHY Layer	Distance	10/20 km (Logical: 60 km)	10/(20) km
	Branches	64 (Logical: 128)	16 or over
	Bit rate	Up: 156 M, 622 M, 1.25 Gbit/s Down: 1.25 G, 2.5 Gbit/s	1.25 Gbit/s (Up and Down)
	Bandwidth	Scrambled NRZ	1 Gbit/s (8B10B coding)
	Opt. loss	15/20/25 dB	15/20 dB
	Wavelength	Down: 1480–1500 nm Up: 1260–1360 nm (Available to video signals overlay)	Down: 1480–1500 nm Up: 1260–1360 nm (Available to video signals overlay)
	Upstream burst timing	Guard: 25.6 ns Preamble: 35.2 ns (typical) Delimiter: 16.0 ns (typical)	Laser turn on/off: 512 ns (max) AGC setting and CDR lock: 400 ns (max)

To summarize, there are two basic requirements for an ONU transmitter. Firstly, fast switching and initialization is very important in high split ratios, particularly for quick recovery of the traffic after a network failure which simultaneously affects a number of ONUs. Secondly, fast tracking of slow power drift is also needed, which is important for long packet transmission.

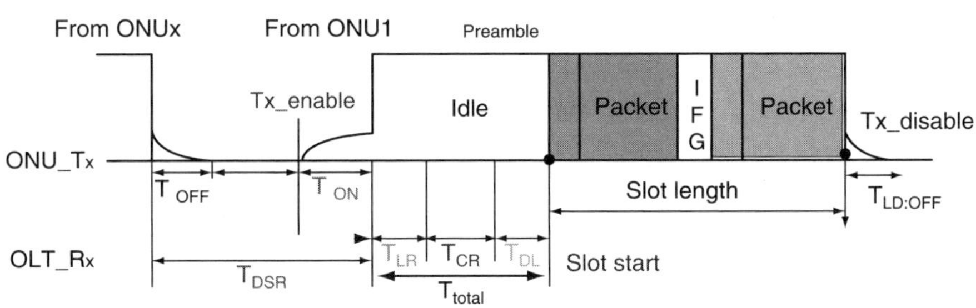

Figure 4.2 Physical layer burst-mode timing definition. T_{ON} – Tx turn-on time; T_{OFF} – Tx turn-off time; T_{DSR} – Rx dynamic sensitivity recovery time; T_{LR} – Rx level recovery time; T_{CR} – Rx clock recovery time; T_{DL} – Rx delimiter time.

The OLT receiver is required to achieve a high sensitivity (i.e. low received power for error-free operation), a wide dynamic range and fast response for the ease and flexibility in network deployment. As shown in Fig. 4.1, every ONU could have a very different transmission loss to the OLT, depending on how far it is separated from the OLT. So the OLT could receive bursty signals in fast succession with large signal-level variations and phase differences from burst to burst. The worst-case performance scenario occurs when a weak burst follows a strong one or vice versa [20]. The OLT receiver must have a fast response to amplitude variation and a short settling time, i.e. a burst-mode operation to quickly recover the bursty signal.

Table 4.4 summarizes the burst-mode timing specifications for BPON, G-PON, and EPON. As shown, both BPON and G-PON rely on very strict synchronization requirements. In BPON, an upstream frame consists of 53 time slots, each consisting of a 3-byte physical layer preamble (PLP) and 1-byte cyclic redundancy check (CRC) in the overhead [21]. When two consecutive time slots are given to different ONUs, these 4 bytes, or approximately 205.8 ns, of the overhead should be sufficient to disable the laser in the first ONU, turn on the laser in the second ONU, and perform receiver gain and clock resynchronization at the OLT.

G-PON specifies similar tight timing parameters [19]. For example, in G-PON with 1.244-Gbps rate, 32-bit guard times (25.6 ns) are allotted for laser on and laser off, 44-bit preamble times (35.4 ns) are allocated in the overhead for the gain control and clock recovery. Table 4.5 lists key specifications of class B

Table 4.4

G-PON and EPON burst-mode timing comparison

Upstream Data rate (Mbps)	Tx Enable (bits)	Tx Disable (bits)	Total (g+p+d) (bits)	Guard time (Tg) (bits)	Preamble time(Tp) (bits)	Delimiter time(Td) (bits)
155.52	2	2	32	6	10(64 ns)	16
622.08	8	8	64	16	28(44.8 ns)	20
1,244.16	16(12.9 ns)	16(12.9 ns)	96(77.2 ns)	32	44(35.2 ns)	20
(2,488.32)	32	32	192	64	108(43.2 ns)	20

Standard EPON	Tx Enable (ns)	Tx Disable (ns)	Tagc (ns) Receiver AGC setting	Tcdr (ns) CDR locking	Tcga (bits) Code group aligning
PX10	512	512	400		32
PX20	512	512	400		32

Table 4.5

Key PMD parameters of G-PON class B 1.244 Gbps upstream [10, 20]

Items	Unit	Single fiber
		ONU TX (optical interface O_{ru})[a]
Bit rate	Mbit/s	1,244.16
Wavelength	nm	1,260–1,360
ODN[b]	dB	Class B: attenuation range 10–25
Mean launched power	dBm	min −2, max +3
Maximum Tx enable	bit	16
Maximum Tx disable	bit	16
Extinction ratio	dB	>10
		OLT RX (optical interface O_{lu})[a]
Minimum sensitivity	dBm	−28 @ BER = 10^{-10}
Minimum overload	dBm	−7
Minimum overload	dBm	−13[c]
RX dynamic range	dB	21[d]
RX dynamic range with PLM	dB	15
Consecutive identical digit immunity	bit	More than 72
Overhead length	bytes	12
Guard time	bytes	4

[a] Optical interface O_{ru} and O_{lu} are specified in ITU-T G.983.1.
[b] Optical distribution network (ODN), class A: 5–20 dB, ODN class C: 15–30 dB.
[c] Using power-leveling mechanism (PLM) at ONU transmitter.
[d] Including 1 dB of optical path penalty over the ODN for worst case.

G-PON PMD layer in the upstream direction as defined in ITU-T Recommendation G984.2 [10].

In many cases, the dynamic range of the signal arriving from different ONUs demands a longer receiver settling time than the allotted guard time. To reduce the range of necessary gain adjustments, BPON and G-PON perform a power-leveling mechanism (PLM) in which the OLT instructs individual ONUs to adjust their transmission powers so that the signal levels received by the OLT from different ONUs are close enough.

For the sake of completeness, Table 4.6 lists the key optical PMD interface parameters as well as timing and signal issues in the 20-km EPON physical-layer upstream specifications.

1. Fabry–Perot (F-P) laser is assumed. The allowed maximum RMS spectral line widths are listed in 802.3ah Table 60.4 [9].
2. Optical distribution network (ODN) PX10: 0.5–10 km; 5–20 dB. PX20: 0.5–20 km; 10–24 dB.
3. Including the chromatic dispersion penalty which is expected to be below 1.5 dB when all link parameters are simultaneously at worst-case values. The

Table 4.6

Key PMD parameters of IEEE 802.3ah EPON 20-km 1.25 Gbps upstream [9, 22]

Description	Unit	Single fiber
		ONU Tx (optical interface TP2)
Bit rates	Gbd	1.25
Center wavelength range (a)	nm	1,260–1,360
Average lauching power	dBm	min −1; max +4
Extinction ratio	dB	>6
Launching OMA	dBm	>−0.22
Max Tx enable	ns	512
Max Tx disable	ns	512
Max TDP	dB	1.8
		ODN uplink: 0.5 m to 20 km (b)
Attenuation range	dB	min 10; max 24
		OLT Rx (optical interface TP3)
BER		1E−12
Average Rx power	dBm	max −6
Max Rx sensitivity (c)	dBm	<−27
Max sensitivity in OMA	dBm	−26.2
Signal detect threshold	dBm	−45
Stress Rx sensitivity	dBm	<−24.4
$T_{receiver\_settling}$	ns	<400
Rx dynamic range (d)	dB	>21

chromatic dispersion penalty is a component of transmit and dispersion penalty (TDP).

4. Rx dynamic range is also known as loud/soft ratio.

It should be mentioned that EPON took a different approach to select physical-layer timing parameters. Initially in the IEEE 802.3ah task force, several alternatives for burst-mode timing specification have been considered, including the proposal to use very short laser-on, laser-off, AGC, and CDR resettling times, similar to the G-PON specifications. After extensive analysis and debates, the IEEE 802.3ah standard relaxed timing parameters, and specified the following parameters: laser-on time of 512 ns, laser-off time of 512 ns, and receiver settling time of less than 400 ns (negotiable). The reasoning was that the ONUs, being mass-deployed devices, must be as simple and inexpensive as possible. Therefore, PMD components should have high yield (which would lower the cost), and should not mandate implementation of digital interfaces, which otherwise would be mandatory if ONUs were required to negotiate laser on/off times. The OLT device can be more expensive as only a single device is used per EPON network. Therefore, the OLT is allowed to negotiate and adjust its receiver parameters such as the receiver settling time.

Time has shown that the relaxed physical specification of EPON facilitates its widespread deployments. There are many suppliers for EPON optics. The performance and yield of EPON transceivers are increasing while the cost is decreasing. At the same time, suppliers of G-PON transceivers are struggling with the more demanding optical requirements on the ITU-T specification [14, 23, 24]. The much tighter timing parameters and a wider dynamic range make the G-PON burst-mode electronics technically more challenging than their BPON and EPON counterparts. So far only a few G-PON 1.244-Gbps upstream OLT chipsets are commercially available. Nevertheless, PON is a robust technology option for meeting customers' demands for high bandwidth. While G-PON systems were available as early as 2006, the FSAN operators consensus on key G-PON requirements in 2006 facilitated G-PON equipment being available from multiple vendors and ready for deployment in North America starting in early 2007.

4.3 TRANSCEIVER TECHNOLOGIES

Optical transceivers are classified by wavelengths, data rates, reaches, packaging types, electrical and optical interfaces, temperature ranges, etc. The manufacturing follows a process from discrete optics to optical subassembly and integrated chips to module assembly and test. Figure 4.3 shows various optical transceivers currently used with optical access systems. Conventional transceiver technologies based on discrete optoelectronic chips, components, and coaxial packaging [19] are still the key enabler for the industry. The high cost of transceivers is one of the major barriers to mass deployment.

PON transceivers are bidirectional devices that use different wavelengths to transmit and receive signals between an OLT at a central office (CO) and the ONTs at end users' premises. There are currently two standard types of transceivers: the diplexer and the triplexer transceivers, respectively. For the diplexer

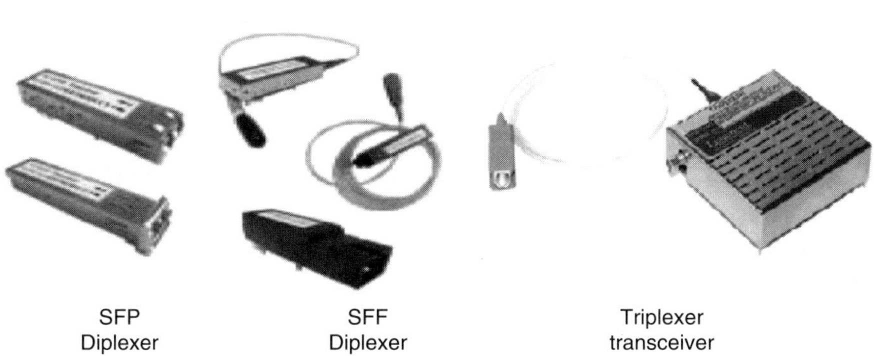

Figure 4.3 Various PON diplexer and triplexer transceivers for optical access systems, courtesy of LuminentOIC Inc.

transceivers, the wavelengths are designated in accordance with industry standards, namely, 1310 nm for the upstream and 1490 nm for the downstream wavelengths. For triplexer transceivers, the 1550 nm wavelength is allocated for analog broadcast video overlay in the downstream direction (Chap. 2). It is also possible that digital video signals are carried on the downstream 1490-nm wavelength using video over IP technologies.

For rapid adoption of FTTH, it is important to reduce the cost of optical transceivers. In particular, the overall PON system cost is weighted more toward the ONU, as the OLT cost is shared among the number of FTTH users. So the ONU transceiver should be the main target for cost reduction whenever possible. In summary, the technological challenges in the optical transceivers for PON systems exist in the following areas:

- High-output-optical-power and high-sensitivity OLT at the CO to compensate for the losses introduced by the optical splitter and the transmission fibers connecting subscribers' premises.
- Burst-mode optical transmission technologies for the upstream link.
- Cost-effective packaging of optical devices.
- Integration of more digital and analog functions into a single IC.

4.3.1 Transceiver Building Blocks

Figure 4.4 illustrates the functional building blocks of PON transceivers in terms of the physical-layer chipsets for both continuous-mode and burst-mode operations [20, 22]. This is also the basis of various PON transmitters, receivers,

Figure 4.4 Feature function blocks of optical transceivers showing both downstream and upstream directions. In the upstream direction, an ONU transmitter consists of a burst-mode laser driver (BM-LDD), and an F-P laser in the form of transmit optical subassembly (TOSA); the OLT receiver consists of a PIN or avalanche photodiode (PIN or APD), BM-trans-impedance (BM-TIA) ROSA (receive optical subassembly), a limiting post-amplifier (post-amp) and a burst-mode clock/data recovery (BM-CDR) unit.

and subsystems. The ONU side consists of a downstream receiver (Rx) and an upstream transmitter (Tx), while the OLT side consists of a downstream Tx and an upstream receiver Rx. The upstream Tx contains a BM laser driver (BM-LDD) and an F-P laser in TOSA transmit optical subassembly (TOSA) form. The upstream Rx contains a PIN or APD/TIA in receive optical subassembly (ROSA) format, a limiting amplifier (post-amp) and a burst-mode clock/data recovery (BM-CDR) unit. For both the ONU and the OLT, the transmitter and receiver sections are combined onto a single optical fiber though a wavelength-division-multiplexing (WDM) coupler.

Three burst-mode chips are identified as key enablers for PON upstream burst-mode transmission: a BM-LDD with fast and accurate power-level control, a high sensitivity and wide dynamic range PIN or APD/TIA with receiver front-end post-amp, and a BM-CDR for fast recovery of received data signal. These PON chipset are critical and technically challenging parts to achieve the optimum system performance.

Considering Fig. 4.4 in another way, the transceiver blocks can be grouped into two main blocks, one is the optical block and the other is the electrical block [25]. The optical block consists of two TO-CAN-type LD and PD (Fig. 4.7), a WDM filter and a metal housing, which is also called the bidirectional optical subassembly (BOSA), as it is popular that these devices are assembled and tested altogether. The WDM filter is 45 degrees tilted to the incident light beam so that it can separate (or combine) the upstream (1310 nm) and downstream (1490 nm or 1550 nm) signals (Fig. 4.9). The electronic block consists of analog frond-end ICs. There are trends to integrate continuous limiting post-amplifier with the burst-mode laser driver on a single IC. Higher-level integration will combine the CDR and SerDes (serializer/deserializer) functions with the PON MAC processor.

4.3.2 Optical Transmit and Receive Devices

Optical transmitter devices typically consist of distributed-feedback (DFB) or F-P laser diodes [26–28]. DFB lasers are capable of single-mode lasing with the help of a distributed grating on top of the active regions. DFB lasers exhibit high efficiency over a wide temperature range, this device structure is particularly useful for PON systems particularly in the OLT side. DFB lasers (typically uncooled) feature a low threshold current, a wide operating temperature range, a high side-mode suppression ratio (40 dB typical), and high-speed response (0.12-ns rise and fall time, 20–80%).

F-P lasers are used to reduce the cost of optical modules particularly on the ONU side. For PON systems, since the ONU transmitter also needs high-optical output power (so that a cheaper and less sensitive PIN receiver can be used at the

OLT), the output power of an F-P laser can be higher compared with DFB. To increase the optical output power launched from the front facet of the laser, the F-P laser design can be optimized without employing an isolator and an efficiency of 0.45 W/A is achievable with an optical output power of 20 mW at 25 °C. The capacitance of the device can be optimized to permit high-speed modulation at 1.25 Gbps. Figure 4.5 shows the L-I curve of typical DFB and F-P lasers, respectively.

High-sensitivity avalanche photodiodes (APDs) are used in optical receivers at OLTs because ONUs are usually equipped with lower-power optical sources in order to control the cost. When an APD is biased near its breakdown voltage, photocurrent amplification takes place, making it possible to obtain better optical sensitivity. Typically, InGaAs detector is adopted because of sufficient responsivity in the wavelength range from 1.0 μm to 1.6 μm for the absorption region and an InGaAs-InP planar structure is employed which applies InP for the avalanche region. A typical APD for gigabit PON applications has an active diameter around 35 μm, responsivity of 0.9A/W at the 1310-nm wavelength and a frequency bandwidth of 2.5 GHz. Figure 4.6 shows the V-I curve of such an APD. The breakdown voltage is about 60 V, and a multiplication factor of 10 or greater at an incident optical power of 0.3 μW is obtainable.

Currently, optical transmit and receive devices still rely on stand-alone TO-CANs or based on coaxial packaging [27]. TO packaging techniques have made significant progresses recently. These packaging techniques have drawn more and more attention for transmitters and receivers due to their attractive

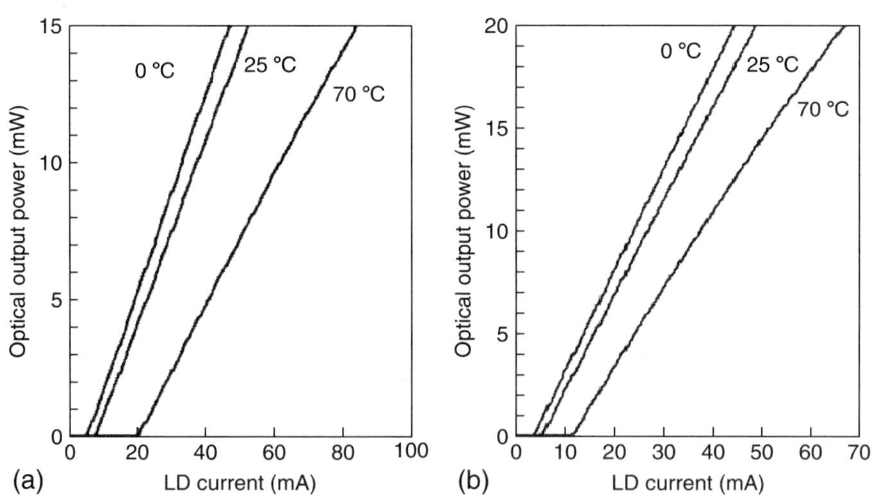

Figure 4.5 Typical L-I curves of (a) DFB-LD and (b) F-P LD.

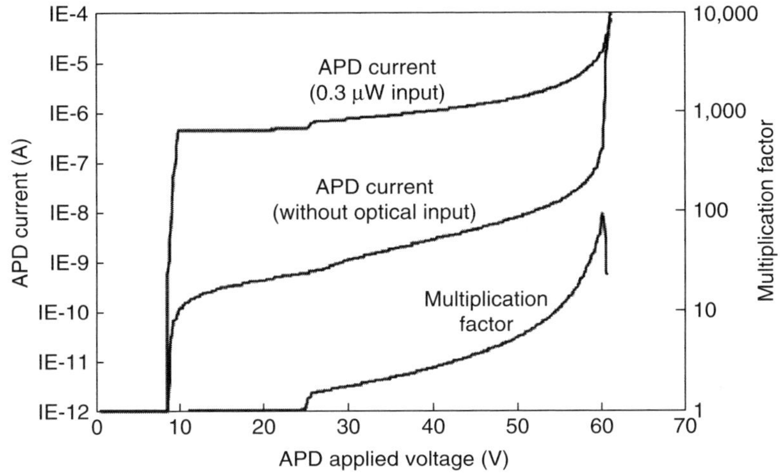

Figure 4.6 APD V-I curve and multiplication factor.

advantages of low cost, small size, and ease of usage. TO laser modules are widely used instead of butterfly laser modules in high-speed transmitters, transceivers, transponders, and small-form-factor (SFF) modules.

Figure 4.7 (a) shows the TO-CAN package of a DFB laser. The aspherical lens helps to achieve an excellent optical coupling efficiency of about 65%. A monitor photo diode (PD) is mounted at the back of the DFB laser in order to detect optical output power from the DFB laser and to control the DFB laser so as to keep the optical output power constant in case of laser temperature change. An optical isolator is employed to suppress the noise generated due to optical feedback. The size and cost of the optical isolator have been reduced by mounting it on the optical input end of the optical fiber where the diameter of the beam launched from the DFB laser becomes minimal.

Figure 4.7 (b) shows the TO-CAN structure of an APD with the preamplifier IC embedded. Because the APD has a large active diameter, an optical coupling efficiency of almost 100% is achievable and optical alignment is simpler compared with a laser diode (LD).

Another important optical device is the WDM coupler shown in Fig. 4.4, which is a bulk optical component based on multiplayer interference coating (thin-film filter) technology. Its function is to separate the upstream and downstream signals in a PON system. Conventionally, the WDM filter is deposited onto a substrate such as a slab of glass, cut into square chips with size of several millimeters, and mounted on an optical subassembly (OSA) (Fig. 4.8). A new technology was recently reported [29] whereby a WDM filter is deposited on the facet of a polished fiber at 30 degrees angle directly in order to minimize the filter

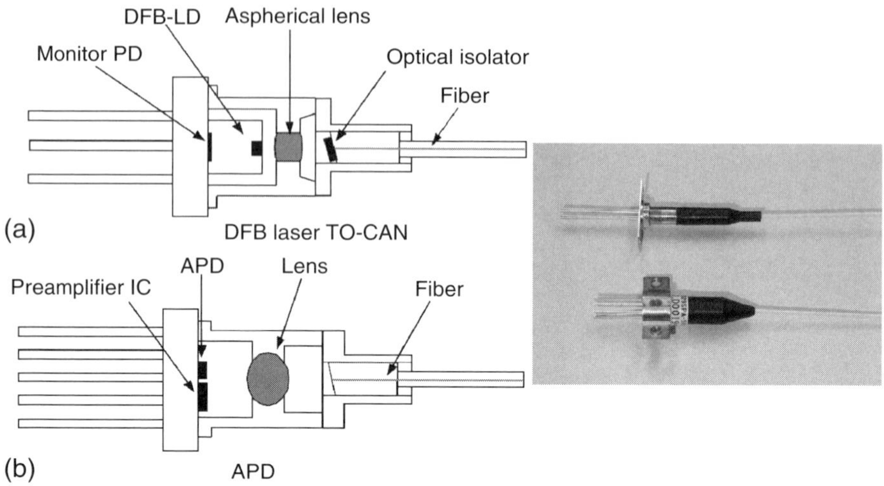

Figure 4.7 Schematics structures of TO-CAN packaging for DFB (a) and APD (b). Their photographs are shown on the right.

size. In this way, the WDM filter, which accounts for the major portion of the OSA cost, was realized simply by the fiber assembly process within a package. Fiber coupling with the LD has been done within the omega groove in the plastic molding by butt-coupling a fiber polished at 30 degrees. Figure 4.8 shows the transmission wavelength spectrum characteristic and a photograph of the WDM filter formed of the fiber facet. It provides 20-dB isolation between the maximum wavelength of 1360 nm in the upstream band and the minimum wavelength of 1480 nm in the downstream band. Furthermore, in order to achieve both the low cross talk from the 1310-nm transmitter wavelength, and high isolation between

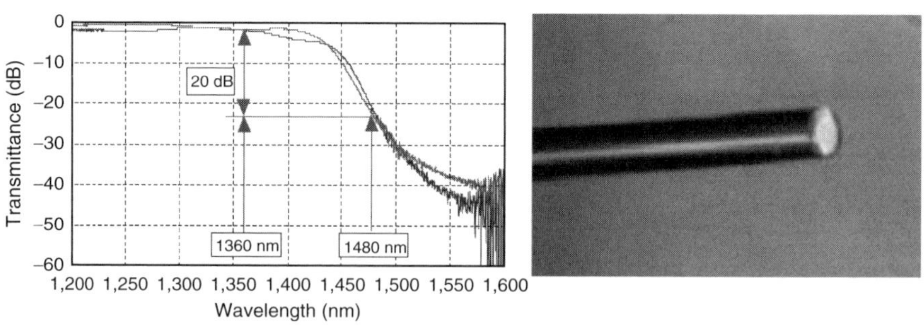

Figure 4.8 Transmission spectrum characteristic (left) and photograph (right) of the WDM filter formed on a fiber facet [29]. The two curves represent two different light polarizations.

the downstream digital light and the downstream video signal of 1550 nm to the required 40 dB or more, a band-pass filter with a pass band from 1480 nm to 1500 nm can be inserted in front of the PD in addition to the fiber end filter.

4.3.3 Bidirectional Optical Subassembly (BOSA)

The bidirectional optical subassembly (BOSA) takes up a major portion of the total cost of PON transceivers itself. The incumbent technology used for today's BOSA development is based on bulk-optic assembly technology using thin-film filters (TFFs) [30–33], which relies on discrete TO-CAN packaged components.

Now this technology is facing serious competition from new advanced technologies based on highly integrated planar lightwave circuits (PLCs) and automated passive assembly [34–39]. One of the challenges faced by bulk-optic TFF diplexer/triplexer technology is that the alignment of lasers, detectors, and filters often requires an active process, in which all components must be powered up during assembly. This adds complexity and cost. However, PLCs, with its promise for passive assembly, better yield and cost, continue to challenge this conventional technology.

4.3.3.1 Bulk-Optic Assembly Technology

Conventional two-TO-CAN-type BOSA consists of an LD CAN, a PD CAN, and a WDM filter. In this two-TO-CAN-type BOSA configuration, LD and PD are actively aligned with an SMF independently. This two-step alignment process is time-consuming and makes the BOSA expensive.

Therefore one cost-reduction approach is to simplify the assembly process by realizing the BOSA function into a single TO-CAN package [30–33]. Both the bare LD and PD dies are integrated into the BOSA chip using a surface-mount silicon microlens on a silicon substrate. This single TO-CAN-type BOSA reduces the total part counts and the corresponding part cost compared with the traditional two-TO-CAN type. One-time-only active alignment between the BOSA chip and an SMF is required. This simplified alignment process is another cost advantage.

Figure 4.9 shows the optical system schematics of a BOSA package. The precise alignment of the Si microlens on the substrate is realized by physically contacting the microlens sidewall on the Si V-groove (not shown in the figure). The emitted light (1310 nm) from the LD chip is collimated by a nearby Si microlens. The collimated beam passes through the WDM filter and is focused onto the SMF by the external ball lens on the sealing cap. The downstream

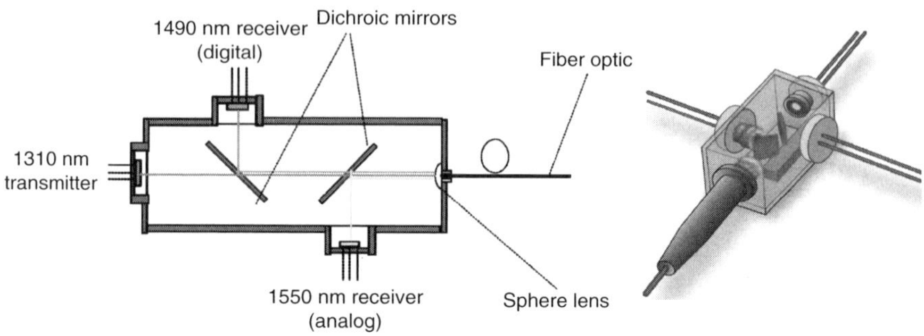

Figure 4.9 BOSA schematics for three wavelengths based on thin-film technology.

optical beam of 1490 nm from the SMF is collimated by the ball lens, and reflected by the WDM filter. The PD chip is placed on the silicon substrate facing downwards. The collimated beam is focused by another off-axis Si microlens and redirected to the upper PD chip by the mirror at the end facet of the V-groove. The BOSA chip is expanded to a triplexer by adding a second WDM filter, a third Si microlens, and a second PD for the video signal.

The BOSA TO-CAN can be fabricated sufficiently small to fit into a conventional SFF transceiver package. In developing such a compact BOSA, two things have to be taken into account. Fabrication and placement misalignment of the optical elements may cause overall coupling efficiency variation. This is compensated by the single active alignment process. The ball lens on the sealing cap serves as an imaging lens between LD and SMF, and SMF and PD. This shared imaging lens configuration and the precise parts alignment based on surface mount technology (SMT) realized one-time LD and PD optical paths alignment by actively monitoring the LD output power into SMF without monitoring the PD current.

Another consideration is to take great care of optical and electrical cross talks between the transmitter and receiver. To suppress optical cross talk, a TO-CAN cap with ball lens is fabricated whose inside surface is finished with low-reflection coating. LD emission from rear facet would be absorbed by surface mount monitor PD on a Si substrate. When the monitor PD is not used by LD driving circuit, putting black resin at the back-facet efficiently absorbs the scattered light. For electrical cross talk, pin assignment of such a BOSA may not be perfect, because the pins for transmitter and receiver are aligned in parallel. But, those pins can be separated by ground layer in circuit board so that electrical cross talk can be suppressed efficiently. To suppress electrical cross talk inside the BOSA, wire-bonding route of LD to pin, preamplifier to pin, and PD to preamplifier are designed to make a right angle to each other with as far a separation as possible. Figure 4.10 shows a BOSA in TO-CAN with driving

Figure 4.10 The BOSA in TO-CAN package with driving circuit [32].

circuit [32]. The minimum receiver sensitivity is found to be −28.5 dBm at a bit-error rate of 10^{-12}, which is very good. The cost of this BOSA design is reduced by reducing the part count used and by simplifying the assembly process.

4.3.3.2 Planar Lightwave Circuits (PLCs)

PLCs have been the subject of active R&D for many years [40]. PLCs are fabricated with the same processing technology used for making electronic ICs. Unlike traditional bulk-optic assemblies, where light is guided through a series of lenses and filters in free space, in the PLC approach, optical signals pass through waveguides on a chip, much in the same way electrical signals are routed through an electronic IC. With this promise, PLCs are now emerging to compete with bulk-optics in BOSA diplexers and triplxers for FTTH [34].

Traditional PLC components have been widely used to fabricate arrayed waveguide grating (AWG) and optical splitter devices [40]. Although AWGs have proved themselves in DWDM applications that require a large number of tightly spaced wavelengths, conventional AWGs have shown themselves quite unsuitable for FTTH applications. The large chip size of an AWG makes it prohibitively expensive for FTTH applications, and the free-spectral range of an AWG is typically much too small to cover the full PON wavelength range (1260–1565 nm). These shortcomings have required the development of new PLC filter technologies, such as dispersion bridge gratings [35].

In general, there are two competing types of PLC approaches for the FTTH market: the external-filter PLCs and the embedded-filter PLCs such as those that feature dispersion bridge gratings [35].

In the external-filter PLC, the chip contains waveguides only for routing light to different parts of the chip, and has no embedded wavelength-filtering capabilities (see Fig. 4.11). Instead, deep pits are etched into the chip, into which

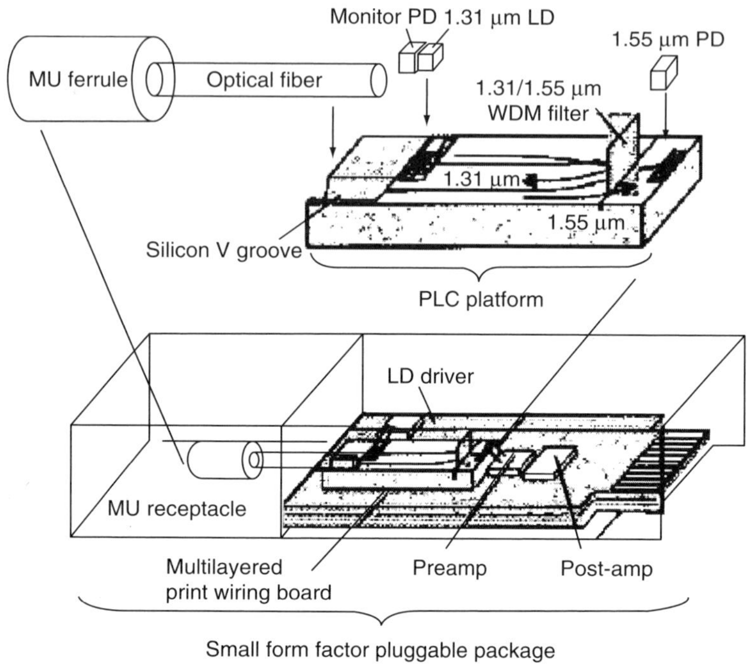

Figure 4.11 Optical circuit configurations with external filters.

TFFs are accurately dropped, aligned, and bonded in place. These TFFs perform all of the WDM functions of splitting/combining wavelengths. In essence, the PLC platform acts as a new packaging technology for simplifying the alignment and assembly of TFFs. This approach, coupled with an efficient means for mounting lasers and detectors onto the same chip, provides a high-volume approach to manufacturing FTTH transceiver chips.

Figure 4.11 shows one optical circuit configuration with external filters [42]. It includes a WDM circuit on a PLC substrate with a dielectric thin film filter installed in a cross-branch optical waveguide circuit. In addition bare chips of a 1310-nm transmitter LD and a 1550-nm receiver PD are mounted on opposite sides of the substrate since the uncoupled light from the LD can be prevented from entering the PD by positioning a WDM filter between them. The MU (a small form factor connector) ferrule is connected to the PLC with a short fiber fixed in a V-groove fabricated on the substrate. The LD, PD, and MU ferrule are mounted on the PLC substrate simply from above and this is suited to mass production. Bare dies or the ICs and the PLC substrate are mounted on a printed circuit board (PCB) designed to isolate the LD and PD electrically and thus achieve compact integration. This technology was applied to a Gbps single-fiber SFP module.

This external-filter PLC technology has matured in recent years, and products based on the approach are now generally available. However, the main challenge in this architecture remains yield, and therefore cost. Although passive alignment in this approach simplifies the assembly process, the LD and PD chips used tend to be expensive because custom structures are required for efficient optical coupling with the waveguide. Thus this waveguide-based approach is still questionable for its cost structure. The embedded-filter PLC approach takes integration to the next level by embedding the wavelength-filtering technology [35, 43] directly into the optical chip. Advanced WDM filtering technology, such as Bragg gratings (as shown in Fig. 4.12), can be fabricated on the chip itself by incorporating into the regular processing steps in wafer manufacturing. This eliminates the need for any external TFFs, greatly simplifying the subsequent assembly and packaging steps. The result is a highly integrated PLC design that requires no external lenses or filters of any kind. The low cost and high efficiency of this approach are greatly compounded by the fact that all chips are made in a wafer form, where a single 6-inch silicon wafer can contain over 500 triplexers based on Dispersion Bridge Grating [34].

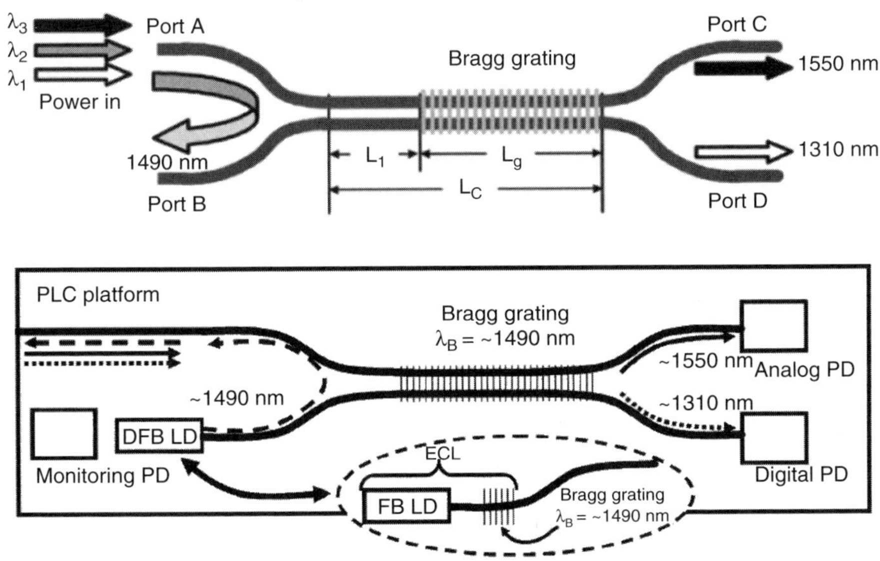

Figure 4.12 The feasibility of grating assisted WDM filter on a PLC platform (top) and conceptual schematic (bottom) of a triplexer using this structure [43].

4.3.4 PON Transceiver Modules

PON transceivers still represent a very active area in industry research and development due to the huge market opportunity. Because of the PON point-to-multipoint nature, PON burst-mode transceiver must face several unique characteristics to meet the requirements. Despite widespread interest, there is still no related PON transceiver multisource agreement (MSA) defined to support this application. Optical module vendors normally collaborate with system vendors on a case-by-case basis in specifications for diverse systems. Because of the EPON influence, SFF and SFP MSAs [44] gained some popularity, while other form factors like GBIC [45] and proprietary triplexer designs still exist for this application [46].

From the EPON/G-PON system compatibility and reusability viewpoint, transceiver modules can be split into blocks of BOSA, electrical subassembly (ESA), housing, and thermal management. The BOSA module must satisfy transmitter optical power, and receiver sensitivity, etc. Considerations on the ESA side include burst-mode physical-media-dependant (PMD) driving performance, system control signal acceptable, high receiver sensitivity, and receiver power dynamic range. The cross-talk effects between transmitter and receiver circuits also need to be examined. One most important aspect is the specific testing items such as transmitter ON/OFF time and receiver settling time. It is of key importance for system designers to maintain system performance and utilize system data processing ability. On the mechanical side, many critical issues need to be considered, such as thermal, EMI, temperature, humidity effect, etc.

Figure 4.13 is an exemplary block diagram and photograph of a diplexer transceiver module developed for ONU usage. Since the ONU transceiver module was to be incorporated in the ONT apparatus installed at a subscriber's premise, it was designed in conformity with the cheaper SFF standard. The LC or SC type pigtails were normally adopted as the optical connector termination. And the module plugs into a 2×5 or 2×7 pins connector for electrical connection on a circuit board operating from a power supply of 3.3 V $\pm$ 5%.

The ONU transceiver module shown in Fig. 4.13 consists of a case, an LC or SC connector receptacle and a printed circuit board, an OSA and a driver circuit. The OSA is micro-optics-based, and contains an optical band-pass filter, a photo detector (PD), an electric trans-impedance amplifier (TIA) IC on the receiver side, and an F-P laser diode and monitor PD on the transmission side. A WDM multiplexes the transmitter and receiver on a single fiber. Using a custom-designed IC, which integrates a trans-impedance preamplifier and a limiting amplifier, this device realizes the 2R function by the OSA alone, and achieves reduced component count and lower power consumption of about 0.8 W.

Figure 4.14 shows an exemplary block diagram and photograph of a triplexer transceiver module for the ONU. Based on the need for installing high-density

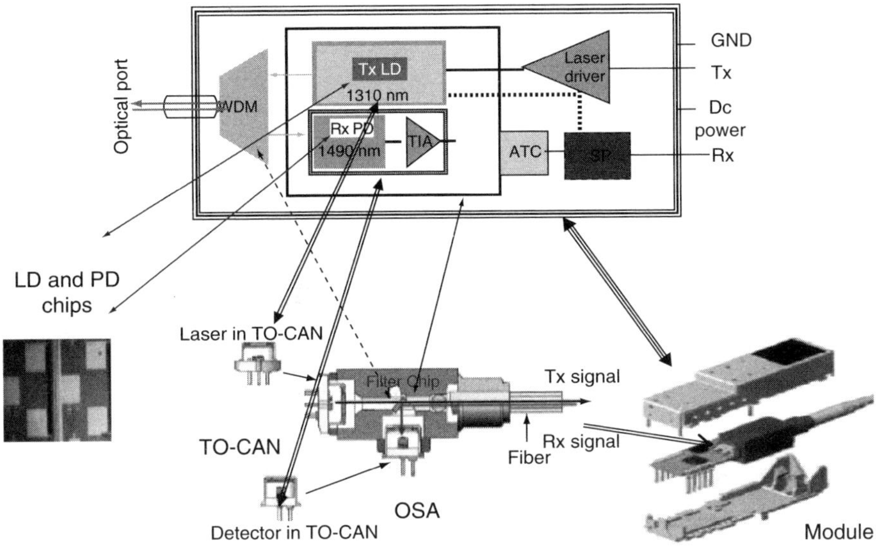

Figure 4.13 Bidirectional ONU diplexer transceiver module.

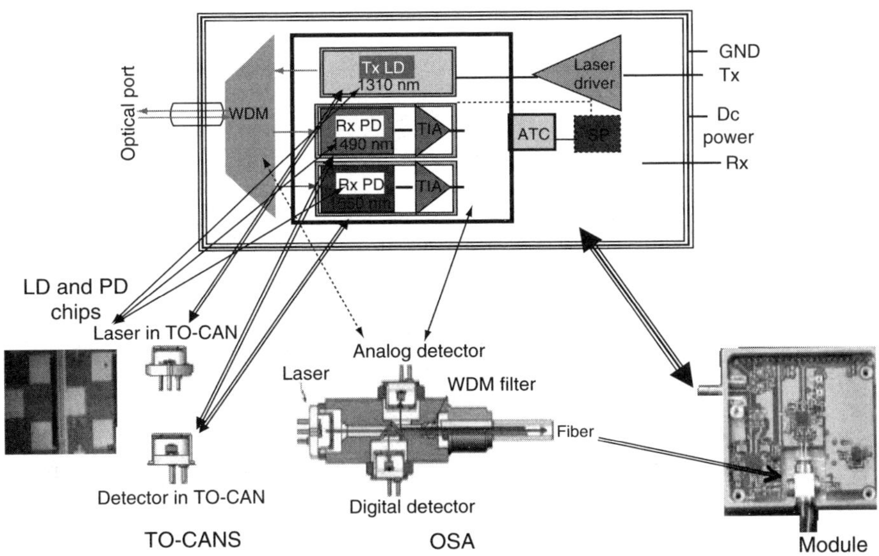

Figure 4.14 Bidirectional ONU triplexer transceiver module.

boards inside the transmission equipment rack the height of the transceiver module for the OLT had to be 8.5 mm or less, conforming to the transceiver size for the standard full-service access network ATM-PON. The LC- or SC-type pigtail was adopted as the optical connector termination.

In GE-PON, the OLT needs a DFB laser having a 1490-nm single longitudinal mode wavelength with high output power. For this reason, in the OSA a TO-CAN package was adopted as the transmitting laser and a high coupling efficiency design was adopted using an aspheric lens. The receiver side had the same structure as the OSA for ONU, having a fiber with WDM filter, receiving PD, and receiving amplifier with a passive alignment assembly in a plastic package. The receiving amplifier for the OLT consists of a custom-designed IC implementing 2R functional burst receiver with a trans-impedance preamplifier and limiting amplifier.

4.4 BURST-MODE ELECTRONICS

PON systems require special burst-mode transmission and reception [47] as one of the key technologies for the upstream direction (from the subscriber to the central office). This is necessary because multiple subscribers share the same optical fiber using TDMA methodology. The key components for such a PON system are the burst-mode transmitter inside the ONU at the subscriber end and the burst-mode receiver inside the OLT at CO. Burst-mode transmitters and receivers are normally realized with ICs by taking advantage of low-cost and low-power CMOS process.

4.4.1 Conventional vs Burst-Mode Data

Figure 4.15 shows three kinds of data signal formats in digital communications. Figure 4.15 (a) is the well-known continuous mode data. A binary sequence is continuously sent with an approximately balanced ratio of 1s to 0s, and the interval between any two logic symbols is strictly limited. Examples are the 8B10B and 64B66B line codes commonly used in point-to-point data link applications such as gigabit and 10-gigabit Ethernet systems. Figure 4.15 (b) is known as the burst-mode data, where the ratio of 1s to 0s and the intervals between the logic transitions are not constrained. The sequences have the same amplitude for the same logic symbols. In Fig. 4.15 (c), the signal amplitude may vary from burst packet to burst packet, and a guard time is usually used between different bursts. The latter two kinds of patterns usually appear in burst-mode multiple-access networks such as PON. The task for a burst-mode receiver is to recover pattern (b) or (c) correctly and instantaneously.

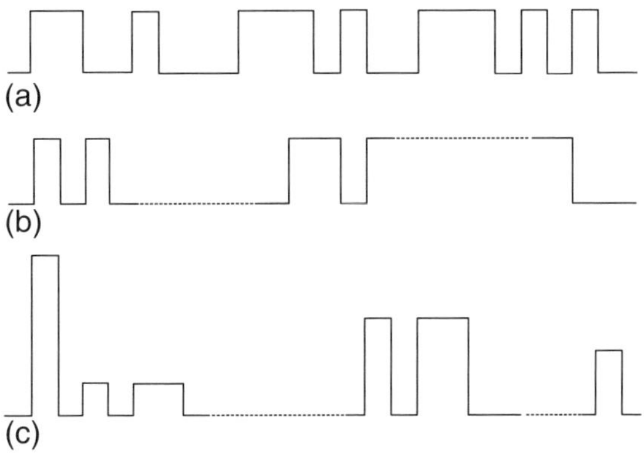

Figure 4.15 Data formats in digital communication: (a) continuous-mode data, (b) burst-mode data, (c) burst packet data [47].

Conventional transmitters and receivers are only suitable for continuous mode data transmission as AC coupling is normally used. AC-coupled circuitry can provide high receiver sensitivity more easily. However, because of the charging and discharging time of the capacitors associated with the AC-coupled signal path, the rate of average amplitude change in the received data is conventionally limited to timescales on the order of millisecond (ms) to microsecond (μs), and may not be allowed to vary rapidly with time.

The operation of a burst-mode receiver is very different from that of a conventional receiver. The main difference is that the burst-mode receiver is typically DC-coupled and the threshold setting of the receiver circuitry must adapt to the amplitude of the received signal in a very short time. Secondly, clock and phase recovery by a burst-mode receiver must be performed very quickly in nanosecond (ns) timescale (typically within a small fraction of a packet transmission time).

4.4.2 Burst-Mode Transmitter (BM Tx)

The BM-Tx at an ONU contains two critical blocks [20, 48–53]: the laser diode residing in a TOSA or BOSA package and the laser diode driver IC (LDD) which provides the required DC bias and modulated currents to the laser diode. The driver IC has to be low in power consumption since it must run from a backup battery during power outages, low cost in order to be competitive with regard to the copper-based infrastructure, and stable over a wide temperature range because it could be located outdoors.

The burst nature of upstream transmission raises special requirements for the laser driver circuitry. In a TDMA scheme, the BM-LDD must generate time-gated bias currents and driver currents between at least 1 mA and 160 mA within sub-ns rise and fall time [48]. Time-gated bias current provided by BM-LDD prevents the laser diode from emitting significant residual power during its idle time. Otherwise, if some residual power was emitted from all inactive ONUs, it would build up at the OLT receiver input and result in undesirable shrinking of Rx dynamic range. The laser power extinction during the idle state must be at least −25 dB to −30 dB below the power during the data burst. For fast switching consideration, however, the laser current cannot be completely turned off during the idle state due to turn-on delay effects and duty-cycle distortion. The prevailing solution is to hold the laser biasing very close to its threshold current. In this way high extinction is obtained and turn-on delay can be avoided at the same time [54]. Higher power biasing is undesirable because it reduces extinction ratio, while lower may push laser into spontaneous emission regime (the so-called subthreshold biasing), which leads to distortion in optical modulation.

BM-LDD is required to control laser diode to stably maintain average optical power and extinction ratio over a wide temperature range (e.g. −40 to 80 °C). Figure 4.16 shows the temperature characteristics of a typical F-P laser in terms of L-I curve (i.e. optical intensity vs laser current). As shown in Fig. 4.16, the optical output power depends on the drive current, threshold, and slope efficiency of the laser, which is a strong function of laser structure and operating temperature. To compensate laser power and extinction ratio variation due to temperature, a fast automatic power control (APC) circuitry is required. The APC counts on the reference feedback from a monitor photodiode (PD) in the LD package (usually located on the back facet of the laser). The APC algorithm

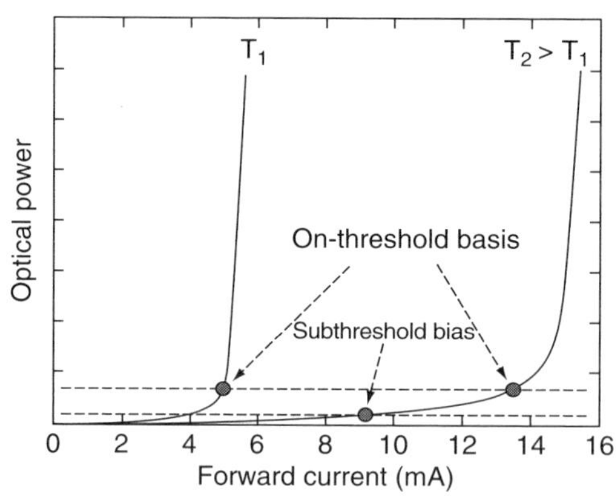

Figure 4.16 Temperature characteristics of a typical F-P laser as L-I curves.

performs two kinds of functions to actively stabilize both the high and low optical levels (the so-called dual-mode laser power control). In the laser idle state (no data burst being transmitted), the peak detector circuit monitors the idle power; and a feedback loop is established to hold the idle power with laser bias close to its threshold. In the laser active state during the transmission of data bursts, the peak detector circuit switches to monitor average output optical power of the burst via the back-facet monitoring PD; and the feedback loop is established to hold the laser bias and modulation current to the desirable reference level, thus a reliable and constant optical power is transmitted.

Figure 4.17 illustrates the block diagram of a typical burst-mode laser driver IC [49]. The input signals can be applied in low-voltage CMOS (LVCMOS) or low-voltage positive-emitter-coupled logic (PECL) format. In this design, the data signal is predistorted in the turn-on delay compensation (TODC) block before driving the laser diode with the laser driver stage (LDS) so that it compensates for the turn-on delay of the laser diode, which can then be operated without the bias current. The signal from the monitor photodiode is fed back into the peak comparator (PC), which compares the photodiode peak current with a reference value. The digital section (DIG) controls the on-state current of the LDS and also provides the end-of-life (EOL) alarm. The interface between

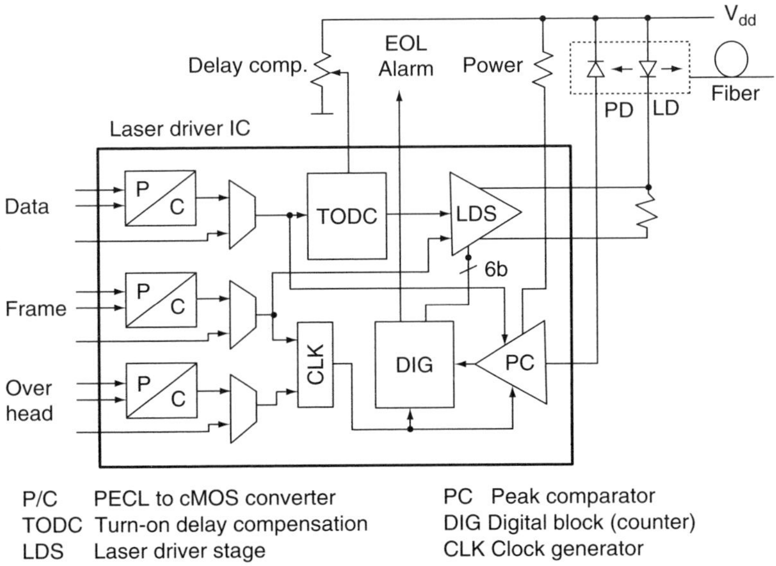

Figure 4.17 Block diagram of a typical burst-mode laser driver IC [49].

the BM-LDD and the laser diode is DC-coupled to avoid the slow response introduced by AC-coupling.

Figure 4.18 compares the LDS circuits for continuous and burst-mode operations. Continuous mode LDS circuits as shown in Fig. 4.18 (a) are designed to maintain a constant bias current, i.e. constant average optical output power. And there is no need to be able to quickly change the bias. That is why most continuous driver cannot handle burst data. However, LDS has circuits built to modulate the laser diode at the data rate, which is very fast and typically at Gbps rate. Similar circuit can be applied to modulate the bias current at the same speed. As shown in Fig. 4.18, laser driver circuits can be designed to have short T_{ON} and T_{OFF} performance in the range of several ns. From this point of view, burst-mode driver does not have much complexity premium over continuous mode driver.

Figure 4.19 shows two standard implementations of the APC function in a burst-mode laser driver. APC circuit based on feedback from the monitor photodiode is necessary. In the first approach (top) the wideband I/V converter and peak detector typically consume a significant amount of power. Furthermore, the analog peak detector has a limited hold time which may lead to incorrect power levels at the beginning of the burst [49].

Figure 4.19 (bottom) shows an alternative APC circuit with less power consumption and also eliminates the problem regarding the limited hold time. This APC circuit operates on a burst-by-burst rather than a bit-by-bit basis and therefore does not require fast and power-hungry circuits. During the first data

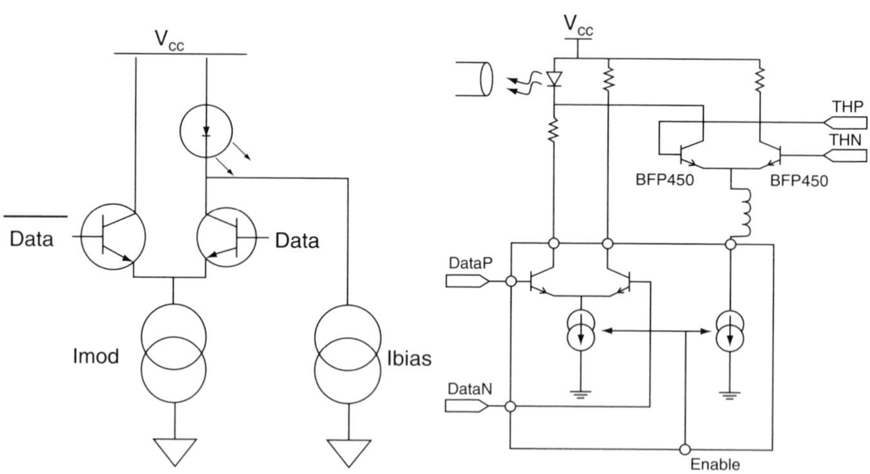

Figure 4.18 Comparison of (a) continuous and (b) burst-mode laser driver stage.

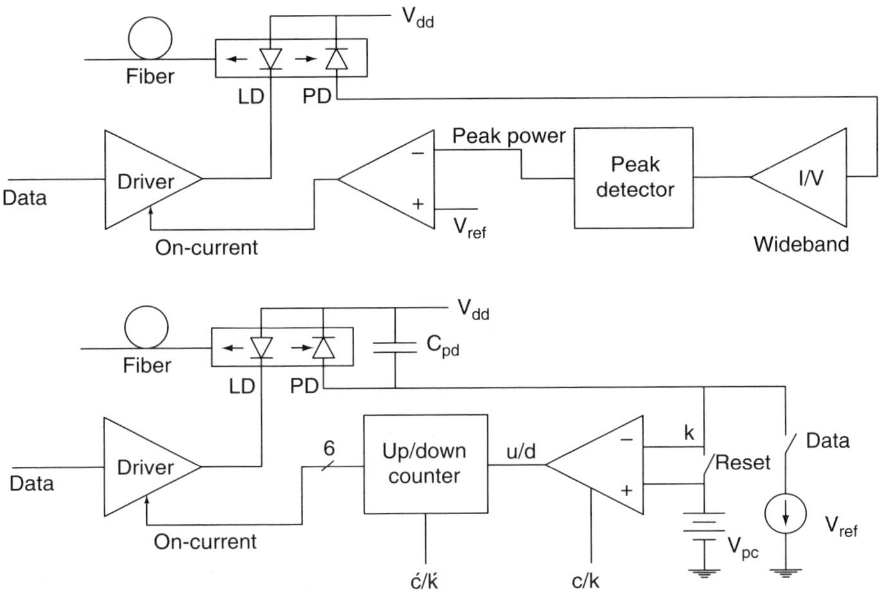

Figure 4.19 Block diagrams of two typical APC circuits.

burst, the photodiode capacitance, Cpd**,** is precharged to a known voltage. Then, during the data burst, this capacitance is charged up by the current from the photodiode and simultaneously discharged by current pulses generated by the data and the reference current source. At the end of the burst, the voltage on the capacitance is compared to the precharge voltage source, with a clocked comparator. Depending on the result, a counter controlling the laser output power is stepped up or down. Because the selected power level is stored by a digital up/down counter, its hold time can be infinite.

Compared to the conventional bit-by-bit power control method, the burst-by-burst method adapts more slowly. But even this "slow" adaptation time is only a few milliseconds, which is very fast compared to the timescale of temperature changes and the lifetime of the laser. After system power-up, the first few packets are transmitted at a power level that is too low because the counter has not yet adjusted itself to the correct value. But this happens only the first time the ONU is powered up and can be taken care of at installation time by transmitting a short series of dummy bursts (during power outages the ONU is powered by a backup battery and the counter value remains valid).

The optical and timing performance of a realized BM-Tx for EPON [22] is shown in Fig. 4.20, where a filtered eye diagram, bursty packet pattern, laser

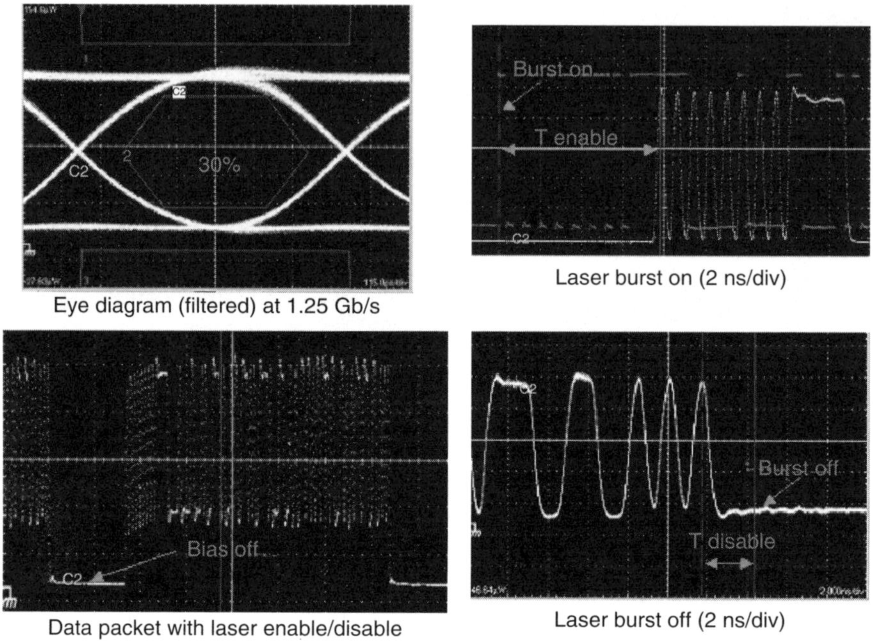

Figure 4.20 Optical and timing performance of an EPON BM-Tx: (top left) Optical eye diagram using 4th-order Thompson filter at 1.25 Gbps; numbers inside box are GbE eye-mask margin. (Bottom left) Optical bursty signal with Laser bias on/off. The measured laser burst-on and burst-off times are shown on the right side [22].

burst-on and burst-off patterns are shown. The eye diagram is measured with PRBS $2^7 - 1$ data with over 30% eye margin. The measured laser idle power is measured to be <-47 dBm. Laser burst-off time is less than 2 ns, while burst-on time is estimated to be less than 8 ns (note: the time delay due to fiber pigtail was not calibrated).

Figure 4.21 shows exemplary measurements of eye diagrams over a wide temperature range (−40 to 80 °C). Because of the APC circuits, a stable transmitted optical power with at least 25% eye margin and above 10-dB extinction ratio was maintained. Note the slightly double tracing for eye diagram at −40 °C is due to the slight impedance mismatch between BM-LD and F-P laser.

Because of its fast response time in the ns range, today's off-the-shelf burst-mode laser driver ICs typically can be built for multi-rates from 155 Mbps to 1.25 Gbps as universal part to cover the requirement of various PON flavors such as BPON, EPON, and G-PON. The laser driver can independently control the bias and modulation currents via the APC loop that maintains the target

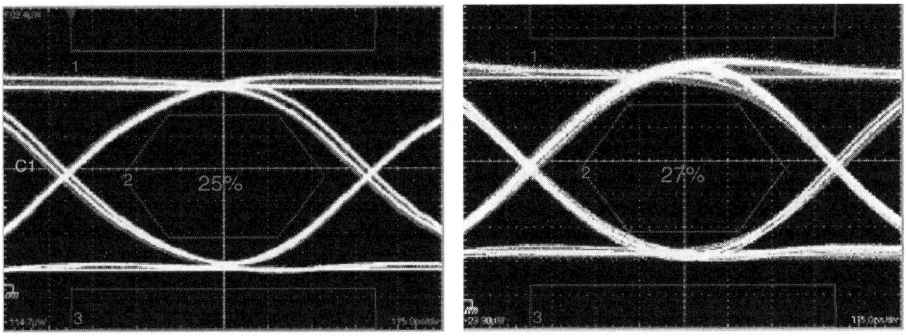

Figure 4.21 Optical eye diagrams at −40 °C and 80 °C case temperature with optical power above 0 dBm and ER above 10 dB. The numbers inside box are GbE eye-mask margin.

extinction ratio by compensating for laser aging, temperature and voltage changes. Laser driver ICs are intended to drive DFB and F-P lasers, and capable of independently driving 100 mA of modulation current and 80 mA of bias current. To reduce the ONU module cost, continuous limiting amplifiers for the downstream are normally integrated with the burst-mode laser driver to form a single chip.

4.4.3 Burst-Mode Receiver (BM Rx)

The challenge of BM receiver is to quickly restore the logic levels to that of the individual signal bursts. The general principle is to use DC-coupled methodology so that when each packet burst arrives, the circuit first measures its power levels; then adjusts the threshold accordingly. Burst-mode receivers can be divided into two categories [47] according to its structure: (1) feedforward configuration [55], and (2) feedback configuration [56] as shown in Fig. 4.22.

In the feed-forward implementation, a conventional DC-coupled preamplifier can be used. The received signal is first amplified by this preamplifier and then split into two branches. The first branch of the output from the preamplifier is DC-coupled to a differential amplifier. The second branch is feed-forwarded into a peak detection circuitry to extract the amplitude information of received packets. From the output of the peak detector, the proper threshold level can be set adaptively in front of the differential amplifier. At the output of the differential amplifier, amplitude-recovered data packets are ready for further processing.

In the feedback configuration, signal amplitude recovery is done in the preamplifier stage. A differential input/output trans-impedance amplifier with

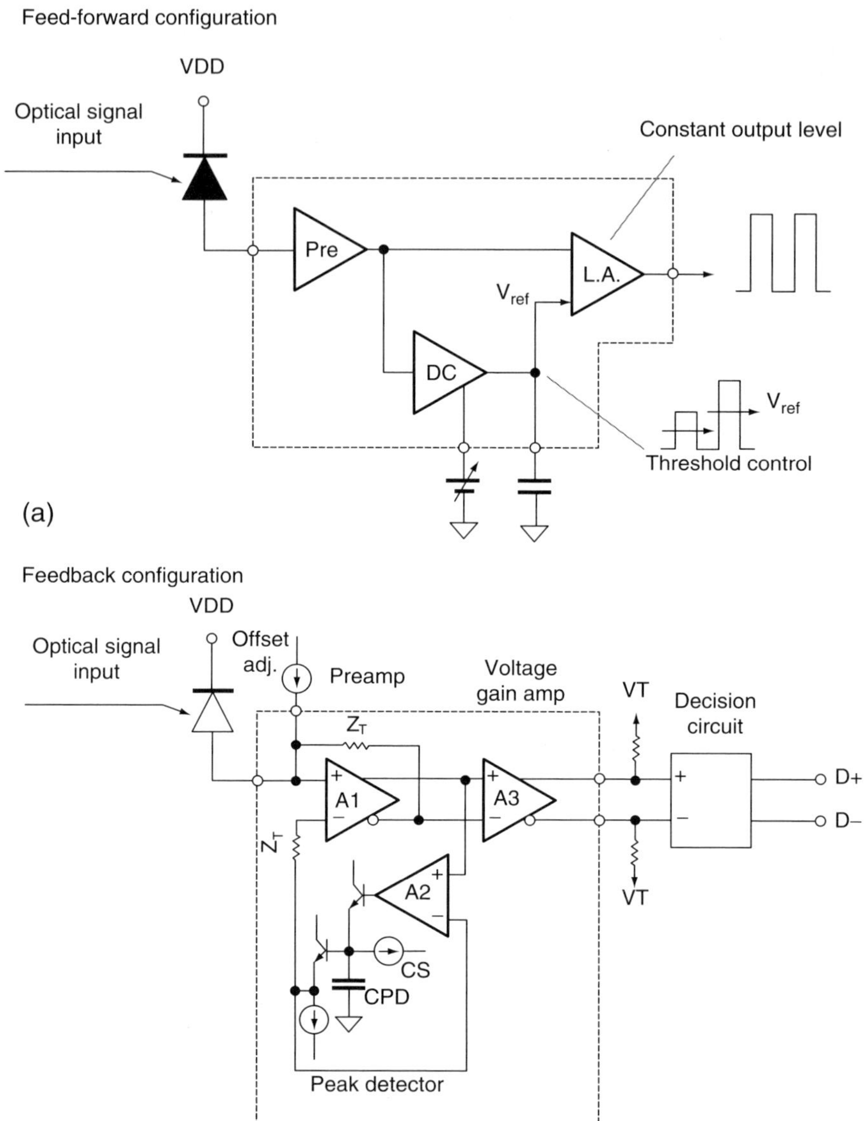

Figure 4.22 Feed-forward and feedback implementations of optical burst-mode receivers.

a peak detection circuit is used to form a feedback loop. The peak detector circuit determines the instantaneous detection threshold for the incoming signal. The output of the preamplifier is DC-coupled to a differential post-amplifier for further amplification. As far as the hardware implementation is concerned, the operation for feedback configuration is more stable than feed-forward type, since a feedback loop enables the receiver to work more reliably, but a differential input/output preamplifier is needed. In the feed-forward configuration, a conventional DC-coupled preamplifier can be used in the receiver; however, the circuitry needs to be carefully designed to prevent oscillation in the receiver.

Burst-mode receivers have also been classified in literatures according to the way the threshold is set. In the first approach, the BM-receiver threshold is adaptively determined according to the input signal [47]. Thus it is also called automatic threshold control (ATC) method. In the second approach, the receiver threshold is determined completely from the preamble field through an automatic gain control (AGC) technique, and held constant in the payload field.

Figure 4.23 shows the block diagram [57] of a burst-mode amplifier IC, which includes a limiter amplifier, an output buffer, and an ATC circuit. It can operate on a +3.3 V single-voltage supply. The ATC circuit consists of a peak detector, a DC-feedback circuit, a 1/2 circuit, and a reset circuit. As shown in the ATC circuit response in Fig. 4.23, the peak detector senses a "1" level of the input signal, while the DC-feedback circuit holds a "0" level. The 1/2 circuit generates a threshold level at the center of these two levels. To clear the peak detector output, the reset circuit quickly discharges a peak-hold capacitor with an external reset signal. It is obvious that the ATC circuit relies upon a high-precision peak detector to achieve both high sensitivity and fast response.

Burst-mode receivers are required to operate over wide dynamic ranges, have fast response from the first bit of a packet, and accept signals of ultra-low

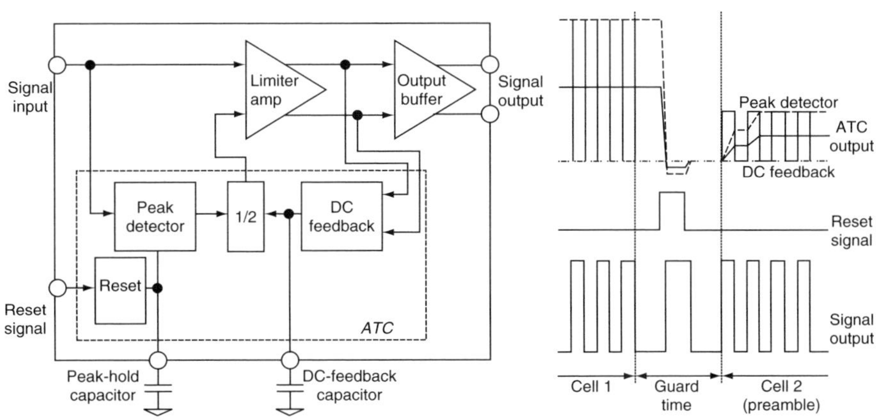

Figure 4.23 Block diagram of a burst-mode preamplifier IC employing ATC (left) and the response of ATC circuit (right) [56].

extinction ratios [58–65]. To achieve this operation, the preamplifier must realize burst-by-burst gain control, that is, high gain for small signal, and low gain for large signal. At the same time, it is required that a burst-mode preamplifier realizes acceptability of low extinction ratio signals, and high sensitivity.

If a large signal with a low extinction ratio is input to a conventional AGC preamplifier, the output waveform will have a large bias, as shown in Fig. 4.24 (a). The amplitude of the output signal is also squeezed, leading to difficulties in discriminating between "0" and "1" properly. To solve the problem of degrading the extinction ratio in output signal, a cell-AGC approach [60] is proposed that controls trans-impedance gain cell by cell according to the amplitude of the input signal. (Here a cell has the same meaning as a packet burst.) Figure 4.24 (b) shows the response of a cell-AGC for burst signals with low extinction ratios. Line G1 in the graph represents the trans-impedance gain for a large-signal cell; line G2 for a small-signal cell, which has higher gain than that for a large-signal cell. By maintaining the gain during the same cell, the "0" level is not as large as that in a conventional preamplifier. This improves the circuit's capability to properly discriminate between "0" and "1." Therefore, this method enables receivers to recover burst signals even with low extinction ratios.

Figure 4.25 (a) shows the cell-AGC preamplifier configuration. This circuit consists of a bottom-level detector (BLD), a gain-control circuit (GCC), a reset circuit, and a FET (Field Effect Transistor) connected in parallel to a feedback resistor. Figure 4.25 (b) shows the response of each block for burst-signal inputs. The BLD quickly detects the bottom level of Amp3, and a hold circuit in the

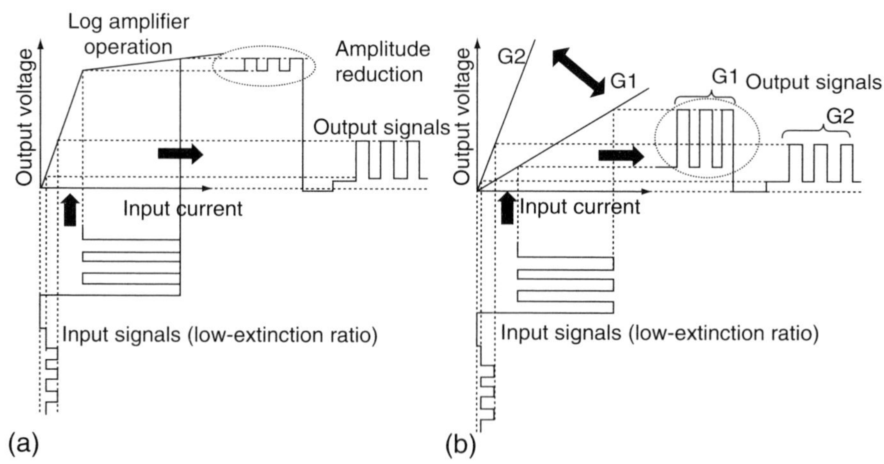

Figure 4.24 The comparison of (a) a conventional AGC and (b) a cell-AGC for burst-mode inputs with low extinction ratio [60].

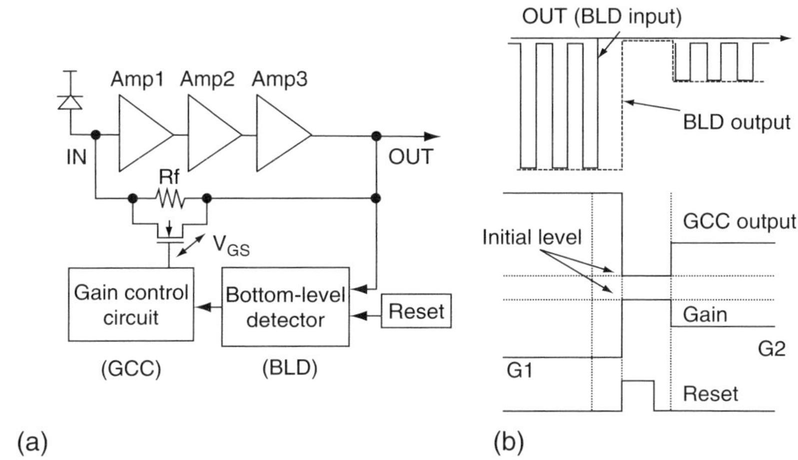

Figure 4.25 Configuration and operation principles of cell-AGC based preamplifier IC.

BLD keeps this level. Depending on this level, the GCC generates a constant voltage during operation in the same cell, and determines the base voltage of the FET, which is connected in parallel with the feedback resistor to reduce the total feedback resistance. As the IC input current increases (due to higher input optical power), the base voltage of the FET also increases, resulting in lower trans-impedance. So, the trans-impedance can be changed by the input current. At the change of cells, the reset signal is launched into the BLD, the output of the GCC and trans-impedance gain return to the initial level. As a result, a high dynamic range is achieved with low extinction ratio signals.

It should be mentioned that burst-mode receivers described above are all based on DC-coupled implementations. This approach has very low burst-mode penalty. Such devices are very suitable for BPON and G-PON which specify very tight timing parameters. For example, in G-PON with 1.244-Gbps data rate, 32-bit guard times (25.6 ns) are allotted for laser on and laser off, 44-bit preamble times (35.4 ns) are allocated in the overhead for the gain control and clock recovery. As is well known, EPON has specified a relaxed receiver settling time of 400 ns, allowing AC-coupling [25, 66] to be implemented in practical applications with short enough time constants. When an AC-coupled circuit with a small time constant is applied to the receiver, fast response can be obtained. Figure 4.26 depicts the block diagram of AC-coupled versus DC-coupled burst-mode receivers. In Fig. 4.26, the AC-coupling circuit consists of a single interstage capacitor, an output impedance-matched preamplifier, and an input impedance-matched limiting amplifier.

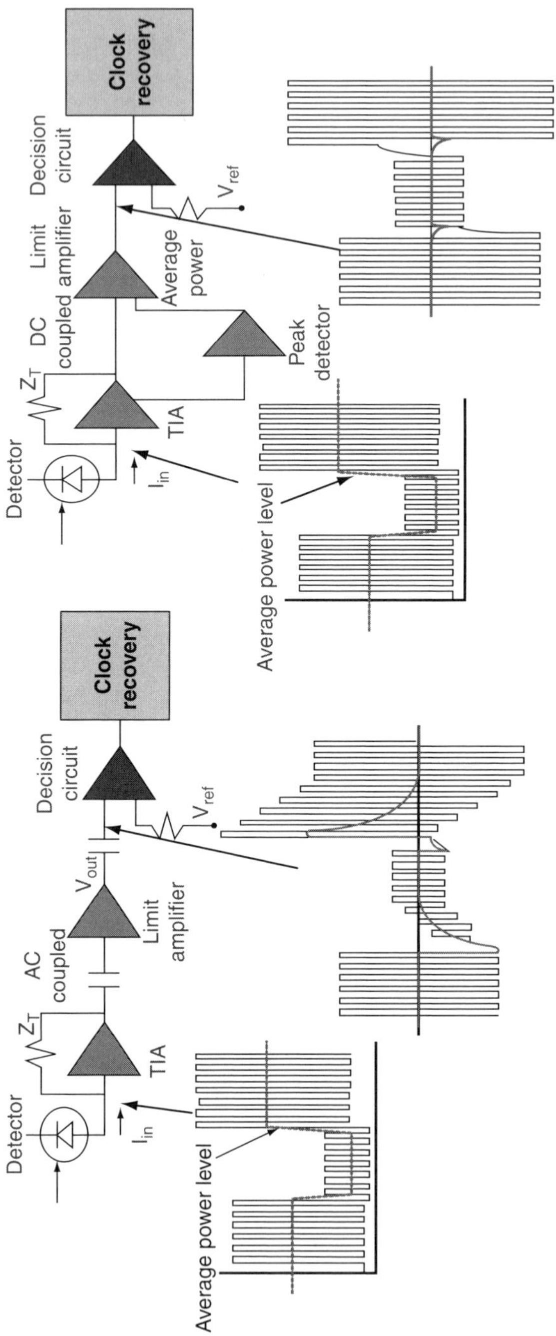

Figure 4.26 Comparison of AC-coupled (left) versus DC-coupled (right) burst-mode receivers.

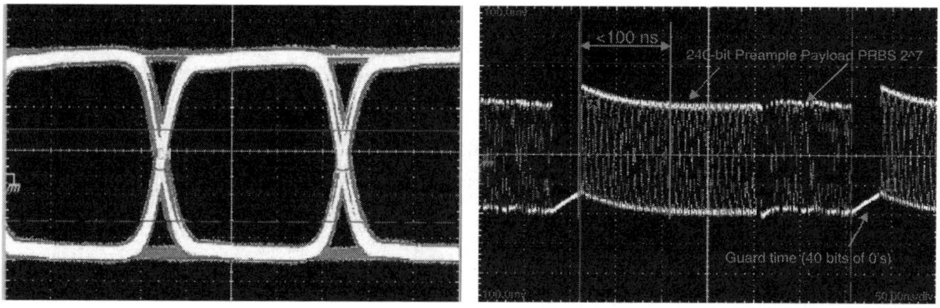

Figure 4.27 An example of 1.25 Gbps BM APD/TIA output eye diagram (left) and settling time measurements (right) [22].

Figure 4.27 shows an example of eye diagrams and settling time for a realized 1.25-Gbps BM APD-TIA. This BM APD/TIA inside a ROSA package has a high sensitivity of −36.0 dBm, with overload of over −1.5 dBm for measured BER at 10^{-12}. Its settling time is measured to be less than 100 ns, well below the required of 400 ns.

4.5 TRANSCEIVER DIGITAL-DIAGNOSTIC MONITORING

Recently, there have been renewed interests in the industry in performance monitoring of optical-access networks. The aim of this optical supervision request is to provide a cost-effective tool toward the lowest downtime of the network, by enabling operators to have the best diagnostics and fewer truck rolls.

Most of the designs of optoelectronic transceivers follow the requirements in SFP [44] and GBIC [45] module multisource agreements (MSAs) (although there is still no PON transceiver MSA at the time of writing), which recently added enhanced digital-diagnostic-monitoring interface specifications as specified in the document SFF-8472, *Digital Diagnostic Monitoring Interface for Optical Transceivers*, revision 9.5 [67]. This interface allows real-time access to device-operating parameters and conditions, and it includes a sophisticated system of alarm and warning flags, which alerts end users when particular operating parameters are outside of factory-set normal ranges. Such digital-diagnostic-monitoring function [68–70] provides a user with critical information concerning the status of the transmitted and received signals, allowing better fault isolation and error detection.

Transceiver digital diagnostics typically monitors module temperature, receiver power, transmitter bias current, and transmitter power. Beyond these module parameters, there is also a need to monitor fiber link failures, degradations, and bad connections without disturbing the many ongoing services. Fiber fault can be detected by using optical time domain reflectometry (OTDR) [71–75]. Recently nonintrusive embedded OTDR approaches have been proposed [76, 77] to remotely monitor the health of access networks, which will be described in this section.

4.5.1 Module Parameter Monitoring

Existing SFP and GBIC MSA specifications define a 256-byte memory map in EEPROM, which is accessible over a 2-wire serial interface at the 8-bit address 1010000X (A0h). The new digital-diagnostic-monitoring interface makes use of the 8-bit address 1010001X (A2h), so the originally defined serial ID memory map remains unchanged. The interface is identical to, and is thus fully backward compatible with both the GBIC Specification and the SFP MSA. Operating and diagnostics information is monitored and reported by a digital diagnostics transceiver controller (DDTC), which is accessed via a 2-wire serial bus. The details of the complete interface and its physical characteristics can be referred to the MSA specs [44, 45].

To make the digital-diagnostic-monitoring information more meaningful and to ensure consistency from vendor to vendor, the MSAs define the minimum accuracy requirements of monitored parameters. System designers must ensure that they meet these requirements over the defined operating power supply and temperature ranges, while minimizing the cost impact.

Usually, the output of the physical value of each parameter is an analog voltage or current from the trans-impedance amplifier, the laser driver, or the post-amplifier. Engineers use analog to digital converters (ADCs) to digitize those physical values. With the digitized values, a microcontroller can then either process data as part of a control loop, trigger an alarm, or just record the data into a register. The ability to provide accurate data to the microcontroller depends on the accuracy of measured analog data, the noise source, along with the ADC accuracy.

Because PON transceivers have limited board space and are price-sensitive, there are always trade-offs in choosing a high-resolution and high-precision ADC and a stable voltage reference. These constraints force designers to determine the minimum requirements for the ADC and the voltage reference, given the known incoming-signal characteristics, to meet the accuracy requirement.

4.5.1.1 Temperature Monitoring

Existing SFP and GBIC MSAs require accurate monitoring of the absolute temperature in an environment with no active temperature-control scheme, with wide dynamic range, and with better than 3 °C overall variation. Whatever sensors to choose, the accuracy, stability, and repeatability of the sensing result over the temperature and power-supply variations are the most important factors. If the monitoring-voltage versus temperature relationship is nonlinear, the accuracy levels refer to the accuracy under the least accurate measurement corner.

As an analysis example [70], assume a linear relationship between the absolute temperature and the sensing voltage of the ADC, with an accuracy of Δ_{TEMP}. The incoming signal voltage of 500 mV to 1 V represents the temperature of 0 °C and 100 °C, respectively. One can assume that $\Delta_{TEMP} = \pm 1\%$ (relative to the incoming voltage), and $\Delta_{VREF} = \pm 1\%$. An ADC with 10-bit resolution improves the temperature monitoring accuracy by more than 1 °C over an 8-bit ADC (Fig. 4.28). Such a change is significant enough to make a difference for an accuracy requirement of 3 °C in the SFP MSA document. The difference between the 10- and the 12-bit ADC, on the other hand, is within 0.25 °C. The improvement from an oversampling of nine times is 0.5 °C. When the accuracy of the voltage-reference variation is within $\pm 0.25\%$, the accuracy

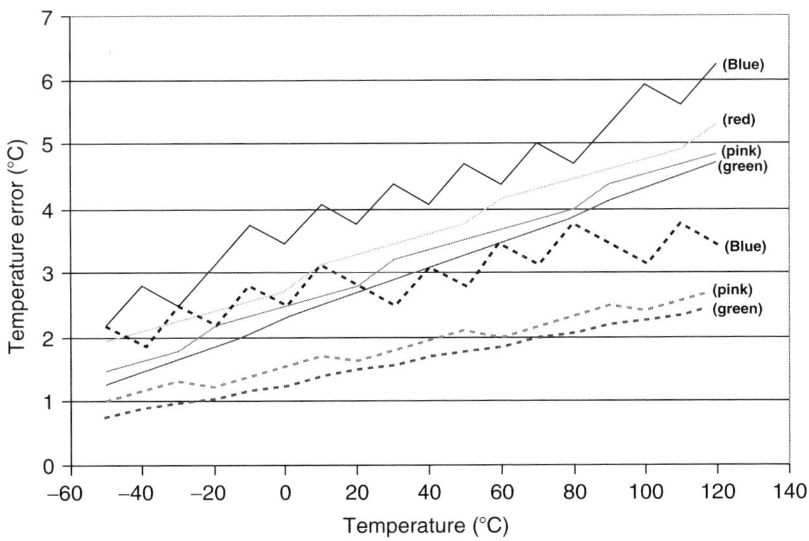

Figure 4.28 Examining temperature error at different ADC resolutions. The *y*-axis shows the absolute value of the maximum error. The overall error for the temperature monitoring is $\pm y$ °C. The blue, pink, and green lines represent the 8-, 10-, and 12-bit ADCs, respectively. The solid lines represent the cases in which the accuracy level of both the temperature sensor and the voltage reference is within $\pm 1\%$, and the dashed lines represent a $\pm 0.5\%$ accuracy level. The red solid line is a 10-bit ADC without oversampling [70].

improvement for a 10-bit ADC is from 0.7 °C at low temperature to 1.65 °C at high temperature. Such improvement is equivalent to changing the resolution of the ADC from 8 to 10 bits. Thus an ADC with 10-bit or higher resolution and a stable voltage reference are equally important factors for accurate temperature monitoring. When the incoming signal's stability is within ±0.5%, one can relax the stability requirement for the voltage reference to ±0.5% as well, and can still achieve a 3 °C accuracy across the entire operating temperature range.

4.5.1.2 Receiver-Power Monitoring

SFP and GBIC MSAs require that the receiver-power-monitoring accuracy be within ±3 dB. There are many factors that impact the accuracy of receiver optical power monitoring. The most important factors are the variation of the photodetector responsivity over temperature, supply voltage, and aging.

Assuming a detector with a responsivity of 0.9 A/W to perform this analysis, analysis [70] shows that the error introduced by an 8-bit ADC can be as high as 3.83 dB. The ADC resolution needs to be 10 bits or higher to achieve the necessary accuracy of ±3 dB. The oversampling by nine times for a 10-bit ADC can improve the accuracy by 1 dB, which is important when the receiver is operating at the sensitivity level. When monitoring the receiver input power, the reference-voltage variation is less important than the variation of the random dark current from the device at the sensitivity level [70].

4.5.1.3 Transmitter Bias Current Monitoring

The current MSA (SFP or GBIC) requirement for transmitter-bias-current monitoring is an accuracy of better than ±10%. Transmitter-bias-current monitoring is usually more accurate than temperature and receiver-power monitoring, because it does not cover as wide a dynamic range. However, other sources of error have to be considered, such as the fluctuation of the bias current itself when the bias-control voltage is static. Offset also exists between the control voltage and the monitoring voltage, because one must infer the actual bias rather than directly measure it to prevent interference with the performance of the semiconductor laser.

If we assume that the default overall variation from the actual current flowing through the laser chip to the monitoring voltage is 3% and the ratio of $V_{BIAS\ MON}$ to I_{BIAS} is 1 V to 100 mA, analysis [70] shows the bias-current monitoring accuracy from an 8-bit ADC is marginal at high currents and becomes dramatically worse at low currents. The accuracy at low bias current is worse because the absolute value of the signal is smaller. Hence, the relative noise contribution increases. Such a condition may be present in uncooled lasers operating at low environmental temperature. In general, the oversampling technique helps as much as 5% under low-current conditions. Nonetheless, the approximate bias current for edge-emitting lasers is less than 10 mA, close to its threshold current, so the accuracy is not a major concern. An improvement on the reference voltage by 5% contributes only 0.5% of the monitoring accuracy.

4.5.1.4 Transmitter Power Monitoring

The current MSA requirement for the transmitter-power monitoring is ±1 dB. Most semiconductor lasers today still rely on the back-facet monitoring. Figure 4.29 shows the typical characteristics between the laser driving current and the monitoring current and optical output power for 1310-nm and 1550-nm laser diodes [78]. It can be seen that the transmitter power is easier to monitor than the three previously mentioned parameters if the only concern is the back-facet-monitoring current.

The transmitter-power-monitoring voltage usually ranges from several hundred millivolts to several volts. If the module is running under APC, the APC loop itself introduces some error. An extra error may arise from the thermal misalignment of the back-facet photodetector, or the responsivity change in the photodetector over temperature and supply voltage. Assuming a default source error of 2%, error-analysis results from transmitter-bias-current monitoring are similar to those for the transmitter-back-facet-power monitoring. Although an 8-bit ADC may seem good enough for transmitter-power monitoring, a 10-bit ADC works better because it allows a 0.5-dB margin for the tracking error [70].

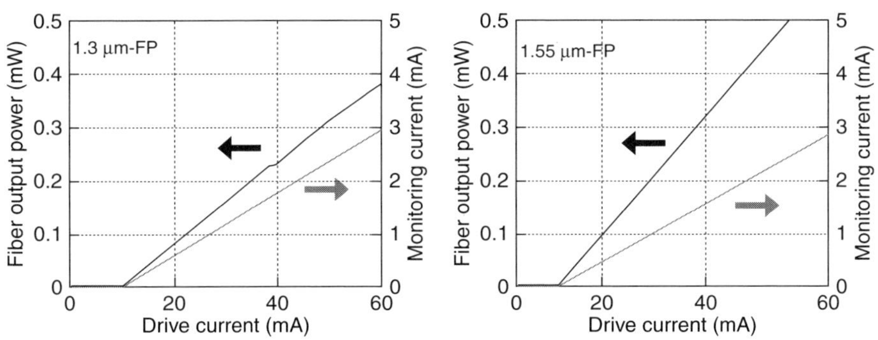

Figure 4.29 A Laser output power and monitoring current versus driving current.

4.5.2 Fiber OTDR Monitoring

Conventionally, fiber monitoring is performed by OTDR measurements at the OLT side in an intrusive way [71–75]. OLT OTDR data, however, suffer from reduced sensitivity due to high splitting losses and from ambiguity due to the superposition of OTDR traces originating from different fiber branches. This subsection introduces the OTDR principles and describes a novel embedded OTDR approach [76, 77, 79], which integrates inexpensive fiber monitoring functionality into an ONU transceiver module.

4.5.2.1 OTDR Operation Principle

As an example, Fig. 4.30 shows an OTDR system in the dotted box in conjunction with a typical WDM-PON architecture [74] that is based on either DFB LDs or wavelength locked F-P LDs. The proposed scheme consists of a broadband light source (BLS), a tunable band-pass filter (BPF), an F-P LD, an optical circulator, an OTDR receiver, and other electrical parts. An F-P LD is

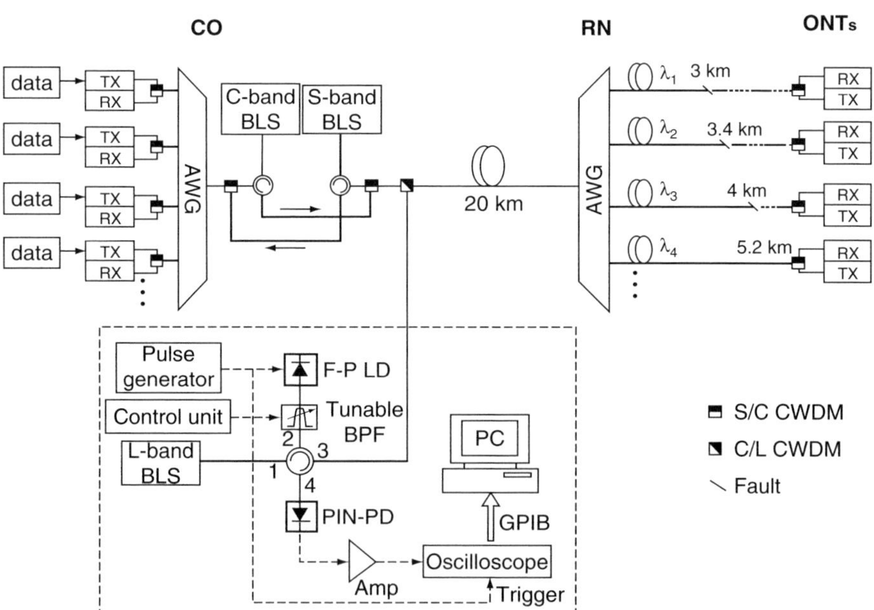

Figure 4.30 A tunable OTDR example for in-service monitoring of the fiber fault in WDM-PON [74].

AR-coated to enhance the ASE injection efficiency. The laser has a cavity length of 600 μm. The AWG has 100 GHz channel spacing with a Gaussian type pass band. The tunable BPF has a 3-dB bandwidth of 0.56 nm. The OTDR signal was coupled into the feeder fiber of the WDM-PON close to OLT site through a C/L WDM filter that passes C and S bands and reflects L band (for the inputs at the common port).

In the experiment, OTDR pulse wavelength was assigned in L-band, while the upstream signal was in C-band and the downstream signal in S-band. However, a single AWG was used for the remote node, since it has periodic transmission characteristics. A different OTDR band from the signal bands enables in-service monitoring with negligible cross talk. The operation principle can be explained as follows. The control unit adjusted the tunable BPF to the target wavelength. The light from the L-band BLS was then spectrum-sliced by the tunable BPF and injected into the F-P LD. We then obtained a quasi-single-mode output from the F-P LD that was directly modulated by a series of electrical pulses. The F-P LD output was filtered by the tunable BPF to suppress residual side modes and coupled into the feeder fiber. The backscattered light from the transmission fiber was detected by the PIN photodetector and processed to find fault positions, as in a conventional OTDR. An oscilloscope and a computer were used as an A/D converter and a signal processor, respectively.

Figure 4.31 shows the fiber fault detection results from OTDR pulses. Clear indications of faults at different locations were observed. The calculated fault positions match the lengths of the drop fibers, which are 3 km, 3.4 km, 4 km, and 5.2 km from the remote node, respectively. The peak at 20 km was resulted from reflection at the AWG.

4.5.2.2 ONU-Embedded OTDR

The large scale of modern PON systems makes cost-efficient fiber monitoring from the ONU side desirable, to reduce the operation and maintenance costs. By integrating OTDR functionality into optical transceiver modules, embedded OTDR becomes an integral part of the network, which can be accessed by the PON management system to monitor and test the quality of the physical layer.

The block diagram of an ONU burst-mode transmitter (BM-TX) with embedded nonintrusive OTDR functionality is shown in Fig. 4.32. It contains the traffic data input (DataIn), a burst transmission enable (BEN) signal, and an OTDR trace output. A 1310 nm laser module with back-facet monitor PD (MPD) generates data bursts and the MPD monitors the emitted optical power. When the BEN signal indicates that no data bursts are transmitted upstream (idle window), the optical front-end is switched from transmit mode into OTDR mode, and the LD (switched from forward-bias to low or zero-bias) and/or the MPD act

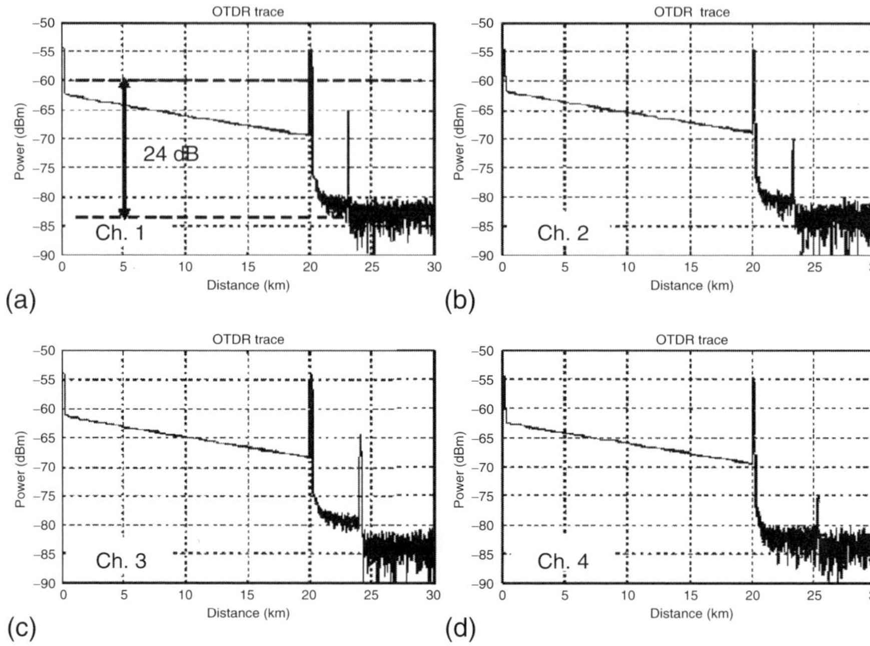

Figure 4.31 Fiber fault detection results for fault locations at (a) 3 km, (b) 3.4 km, (c) 4 km, and (d) 5.2 km from the remote node using the setup from Fig. 4.30.

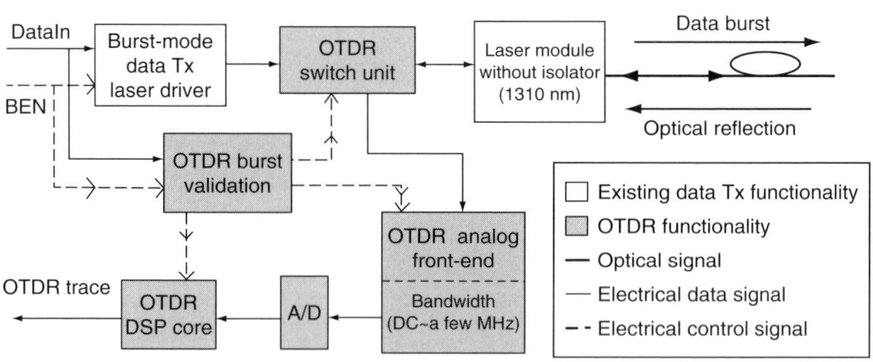

Figure 4.32 Embedded OTDR into ONU burst-mode transmitter. OTDR analog front-end bandwidth is limited to 5 MHz [76].

as an OTDR photodetector. So the backscattering caused by the data burst can be acquired with maximal reuse of components and at minimal cost. The data bursts transmitted by the ONU have a variable length and a limited optical power compliant with G-PON specifications, and cannot be used as an excitation for classic impulse-response OTDR. However, from known parameters, the data burst length (μs scale) and the length of the fiber to be tested (km), the BM-TX can detect whether a data burst is long enough as excitation for step-response OTDR. If this happens, the OTDR measurement only depends on the fiber status and the ONU excitation power, and not on the data-burst length. This negative step response (NSR) approach provides a novel way to reuse traffic data bursts to perform truly nonintrusive fiber monitoring [76].

Figure 4.33 illustrates three nonconventional low-cost implementations to integrate the OTDR functionality inside ONU blocks [77]. The first approach uses a 10/90 optic coupler to separate reflection from transmitted light, but this method requires additional dedicated optics (coupler) and reduces link budget. The second approach uses the laser diode as a photodectector to detect the

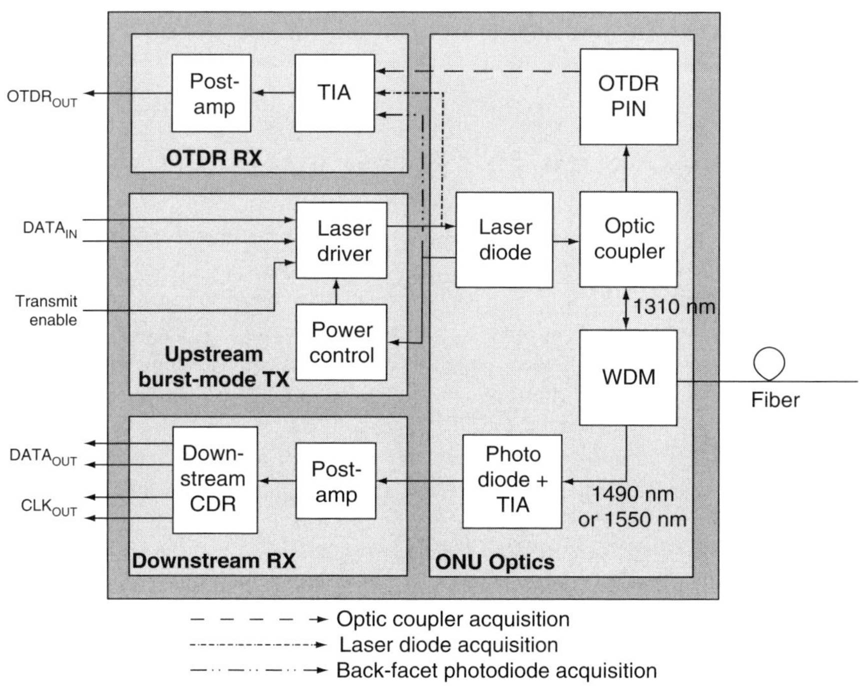

Figure 4.33 ONU block diagram with OTDR functionality integrated [77].

reflection (echo) in case the laser has no isolation. This approach is economical but requires fasting switching of the laser driver and the OTDR Rx. Similarly, the third method measured the reflection using the back-facet PD connected to the laser diode.

It is still questionable how robust this method will be. Simulations and experimental results show that this OTDR method can yield as good precision for locating an abrupt change in attenuation as the classic impulse-response backscattering, without the disadvantages of an intrusive technique. The advantages of this method are obvious:

1. Embedded nonintrusive fiber plant monitoring within the ONU transmitter.
2. No penalty on network performance and no interference with the MAC.
3. Use F-P or VCSEL transmitters without optical isolators. Do not need additional optical components, but need limited amount of high-speed electronics to be integrated in the laser driver chip at marginal cost.
4. No transmission of specific OTDR signals is required, and the required signal processing can be handled during idle time by an off-chip microcontroller.
5. Information on the state of the fiber plant can be obtained continuously, and used to prove fiber plant performance, or to trigger preventive maintenance or repair actions.
6. Fiber plant degradation can be detected long before transmission errors occur or services fail.

4.6 PON TRANSCEIVER SYSTEM EVALUATION

For a PON access system, the transmission and reception of different optical wavelengths bidirectionally over a single fiber complicates optical diagnostics and measurements of PON transceiver modules. As described in Chap. 2, traditional PON architecture adopts a point-to-multipoint configuration. Since many ONU users are connected to one OLT, a shared time-domain multiplexing (TDM) channel is used in the downstream direction. In the upstream, TDMA approach is used to allow each ONU to transmit its traffic during its allocated time slots. This is significantly different from conventional metro or long-haul optical networks, which are typically based on point-to-point wavelength connections and therefore there is a clear distinction between the different network layers. For those networks, the physical-layer performance of conventional links can be tested using standard PRBS patterns regardless of higher layer protocols due to their continuous-mode nature.

For this reason, and other economic considerations, there is still the lack of fully conformed PON test equipment. Access system vendors are forced to develop their own PON-specific diagnostic procedures [20, 22, 62, 80–83].

In reality, PON transceiver system tests have to take into account the bursty nature of the data and the disparity in distances from each ONU to OLT. During the process of OLT receiver sensitivity characterization, the signal amplitude and clock phase can differ from packet to packet. Therefore, successive packets have to be considered with a large difference in optical power and clock phase alignment in order to catch the worst-case scenario. In other words, the OLT receiver is regarded as part of a PON system in conjunction with several ONUs and not only as a stand-alone point-to-point link receiver.

Figure 4.34 shows a typical system setup for PON transceiver module tests, which resembles an actual field implementation of a PON system. It includes two ONUs configured in the worst-case scenario. One is placed near the OLT and the second one far down the link. The OLT receiver has to capture a weak upstream burst that follows a much stronger one. The OLT sensitivity can be evaluated in three scenarios: continuous-mode measurement, single ONU burst-mode measurement, and dual ONU burst-mode measurement. The most important parameters in a burst-mode receiver evaluation are sensitivity, dynamic range, and response time. The system dynamic range is defined as the ratio of the strongest to the weakest optical signals that can be received by the OLT burst-mode receiver with guaranteed performance specification. The burst-mode penalties [20, 22, 46, 66, 84] in terms of the sensitivity difference among these three test scenarios are considered key figures of merit for verifying the system performance of PON transceivers.

4.6.1 G-PON Transceiver System Evaluation

The ITU-T G984.2 [10, 20] recommended allocation of the physical-layer test pattern and overhead at 1.244 Gbps are illustrated in Fig. 4.35. The burst length can be as long as 125 μs, normally taken as PRBS payload. The mandatory total length of overhead at 1244 Mb/s is 96 bits consisting of guard time (mandatory 32 bits), preamble time (44 bits), and delimiter time (20 bits). The length of the guard time is determined by the laser turn-on/turn-off time, time shifts caused by: (1) slight variations in the fiber delay; (2) the fiber propagation delay equalization granularity determined by the ranging process; and (3) the APD and transistor discharge time. The preamble can be split into two parts, amplitude recovery for threshold determination and clock phase alignment for clock phase recovery. As explained earlier, quick extraction of the decision threshold and clock phase from a short preamble at the start of each packet results in a sensitivity penalty, the so-called burst-mode penalty.

The challenge in G-PON burst-mode receiver is to operate over a wide dynamic range while recovering the data in a very short time. In case a weak packet follows a strong packet with the same timing as in Fig. 4.35, bit errors are

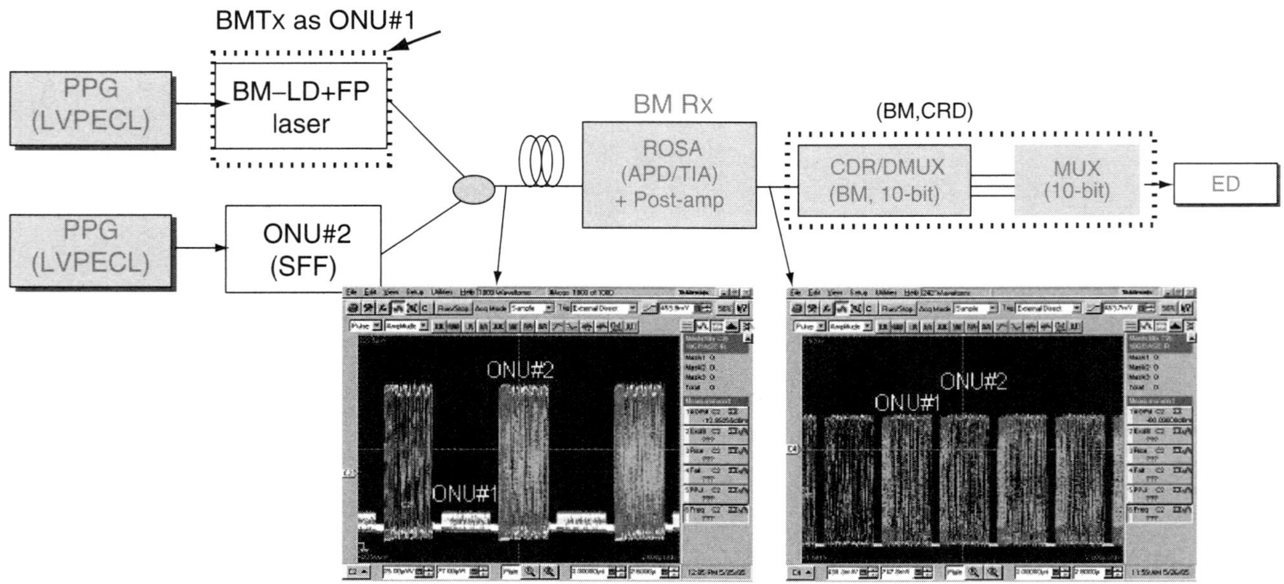

Figure 4.34 Typical burst-mode test setup configuration consisting of two ONUs. A weak packet from BM-Tx as ONU#1 followed by a strong packet from commercial SFF module Tx as ONU#2 emulates the worst-case condition [22].

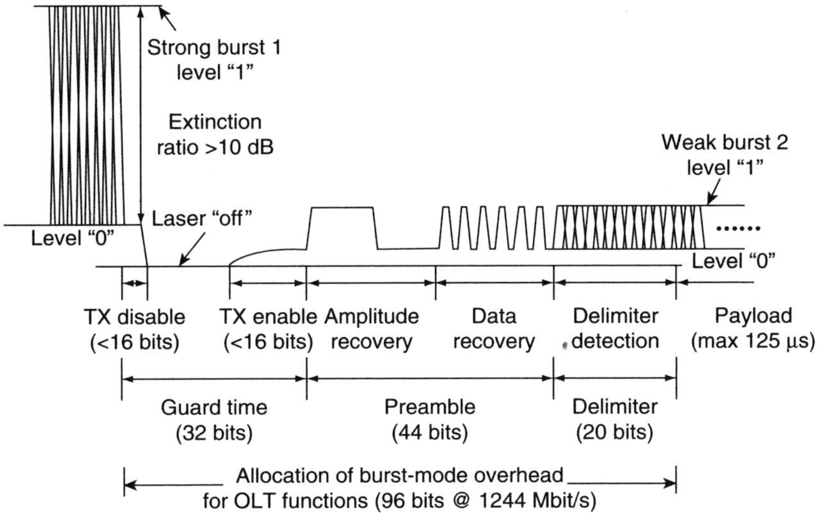

Figure 4.35 A typical G-PON upstream burst-mode test pattern when strong packets from BM-Tx as ONU#1 are followed by the weak packets from ONU#2. The overhead guard time consists of the mandatory 32 bits, preamble 44 bits, and delimiter time 16–20 bits.

counted solely for the payload part of the weak packet as a function of the optical power of the strong packet into the OLT receiver. Figure 4.36 shows the measurements of burst-mode BER and sensitivity penalty performed for a DC-coupled G-PON APD receiver at 1.244 Gbps [20]. It is obvious that a sensitivity penalty is incurred. The major cause of this burst-mode sensitivity penalty is the turn-off tail from the APD and TIA chip conflicting with the strict requirements of short guard time and preamble length. The worst-case measured sensitivity is about −30.5 dBm at BER of 10^{-10} when the optical power of the strong packet reaches −9 dBm, which is close to the overload power of the APD. This represents about 2-dB worst-case burst-mode receiver sensitivity penalty and a dynamic range of about 21.5 dB. The effect of guard time on the burst-mode penalty is also shown in Fig. 4.36 as a function of the optical power of the preceding packet. Sensitivity penalty of ~1.5 dB due to the tail of a strong packet at −12 dBm was observed with a guard time of 32 bits (25.6 ns). An improvement of 1.4 dB in Rx sensitivity is possible when the guard time is relaxed to 64 bits (51.2 ns).

It should be mentioned that G-PON takes advantage of the implementation of PLM (power-leveling mechanism) to relax the 21-dB dynamic range specification of the OLT receiver. The PLM allows ONUs to operate in three discrete output power modes. For example, for a 1.244-Gbps uplink, the following mean launching powers have been defined: (1) Normal mode: minimum/

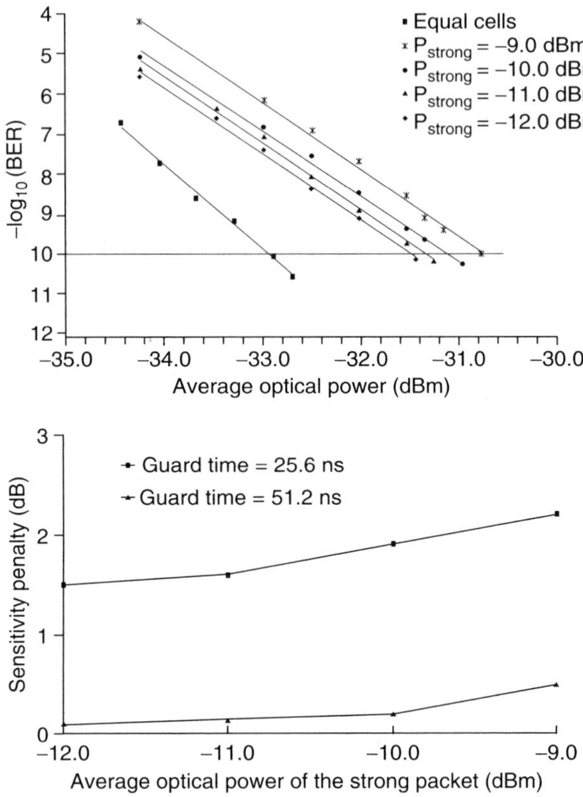

Figure 4.36 Typical measurements of burst-mode BER (top) and sensitivity penalty (bottom) performed for a DC-coupled G-PON OLT Rx at 1.244 Gbps.

maximum = −2/+3 dBm, as stated in Table 4.7; (2) Mode 1 = Normal −3 dB; and (3) Mode 2 = Normal −6 dB. The PLM mode can be set locally in the ONU transceiver via a serial peripheral interface (SPI) according to the control signals coming from the MAC layer. For the detailed PLM procedure, one can refer to G984.2, Appendix II [10]. In summary, the PLM works as follows. The OLT Rx first measures the received average power and compares it with two threshold voltages. It then decides whether the incoming optical signal is too low or too high or within the range. When an ONU receives the message from OLT to switch from one mode to another, it sets its emitted power within the range of the new mode and then resumes sending upstream data.

One advantage of employing PLM is to reduce the dynamic range requirement by 5–6 dB (from 21 dB to 15 dB) for the OLT receiver. Another advantage

Table 4.7

State-of-the-art G-PON transceiver performance parameters for the upstream link at 1.244 Gbps, in comparison with the ITU-T G.984.2 specifications [10]

Parameter	Measured (without PLM)	Spec.	Unit
Bit rate	1.244	1.244	Gbit/s
ODN (Class B)	>15 dB differential range	min 10 max 25	dB
Mean launched power (without PLM)	1 dB tolerance over −40 °C to 80 °C	min −2 max +3	dBm
Extinction ratio	>10	10	dB
Consecutive identical digit immunity	72	72	bit
Receiver sensitivity	−31.6 (w/o WDM)	−28 (with WDM)	dBm
Receiver overload	−4.7 (w/o WDM)	−7	dBm
Dynamic range	26.9	21	dB
Overhead	12	12	byte
Guard time	4	4	byte

is that it increases the laser lifetime and reduces the power consumption of ONUs when working in Mode 1 and/or Mode 2. It also reduces possibly strong optical reflections due to nearby ONUs. Figure 4.37 shows the uplink BER performance for a 1.25-Gbps G-PON OLT Rx tested at different PLM modes. The ONU launch-power adjustments were controlled by the OLT with the following three modes: Pavg = 0.5 dB (normal), Pavg = −2.5 dBm (Mode 1), and Pavg =−5.5 dBm (Mode 2). The insertion of 20-km fiber introduces a dispersion

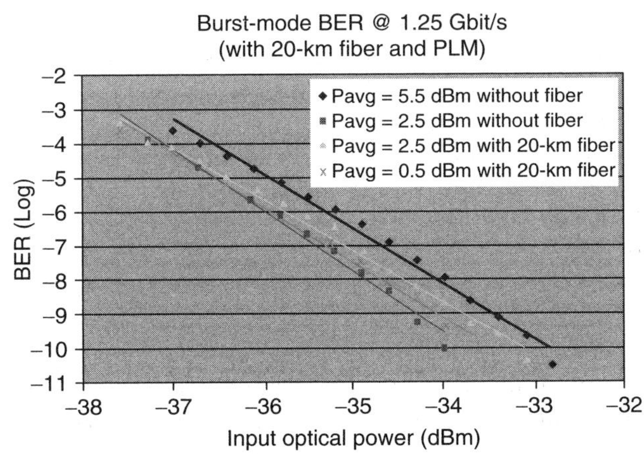

Figure 4.37 1.25-Gbps G-PON uplink performance with PLM.

penalty, causing the receiver sensitivity to degrade by 0.9 dB when the PLM mode was set to Mode 1.

Table 4.7 lists the state-of-the-art G-PON OLT transceiver performance parameters achievable from a DC-coupled APD with burst-mode TIA and limiting amplifier chipset for upstream link speed of 1.244 Gbps [85]. For comparison, the same table also lists the ITU-T G.984.2 specification requirements. Two ONUs were measured with an 8-bit reset signal pulse. The achieved OLT Rx sensitivity of −31.6 dBm and dynamic range of above 26.9 dB were much better than the G.984.2 specification of −28 dBm (with a 1310/1490 nm WDM) as shown in Table 4.7.

4.6.2 EPON Transceivers

The EPON specification allows AC-coupled burst-mode OLT transceivers for simplicity and economic reasons [25]. The burst-mode receiver measurement setup employing two ONUs shown in Fig. 4.34 is also commonly used to characterize EPON transceivers. The EPON burst-mode test patterns can be defined with a guard time of 512 ns, preamble sequence of <400 ns and continuous random test pattern (CRPAT) or continuous jitter test pattern (CJTP) payloads (8 b/10 B coded) of multiple 3,360 bits.

EPON modules are under extensive system evaluations in industry [22, 25, 86–91]. Figure 4.38 (a) shows the BER characteristic of an EPON burst-mode OLT receiver at room temperature [25]. The sensitivity of this receiver with PIN photodiode is changed little (less than 0.5 dB) as the temperature changes from 0 °C to 70 °C. The straight line represents the BER performance of continuous P2P mode. Receiver sensitivity at a BER of 10^{-12} is −26.0 dBm under 1.25-Gbps continuous-mode PRBS data streams from ONU#1 to OLT. The other traces in the graph show P2MP BER characteristics with preamble lengths of 512 bits and 960 bits, respectively. In both P2MP cases, the receiver sensitivities meet the IEEE 802.3ah specified value of −24.0 dBm. A 1.1-dB sensitivity difference between the two burst-mode measurement results was observed, owing to the preamble length difference. This graph also shows the burst-mode penalty of 1.6 dB between continuous-mode versus P2MP burst-mode. In the short preamble case, the OLT burst-mode receiver has a −1.5 dBm overload power, and a −24.4 dBm sensitivity at the BER of 10^{-12}, so the system dynamic range is 22.9 dB. It successfully meets the recommendations of IEEE802.3ah.

Figure 4.38 (b) clearly shows the burst-mode receiver time domain performance in case of preamble time of 512 bits. This burst-mode receiver can adapt the abrupt power-level change; in other words, it has 22.9 dB of system dynamic range during the guard time of 32 bits. In the case of strong power to weak power

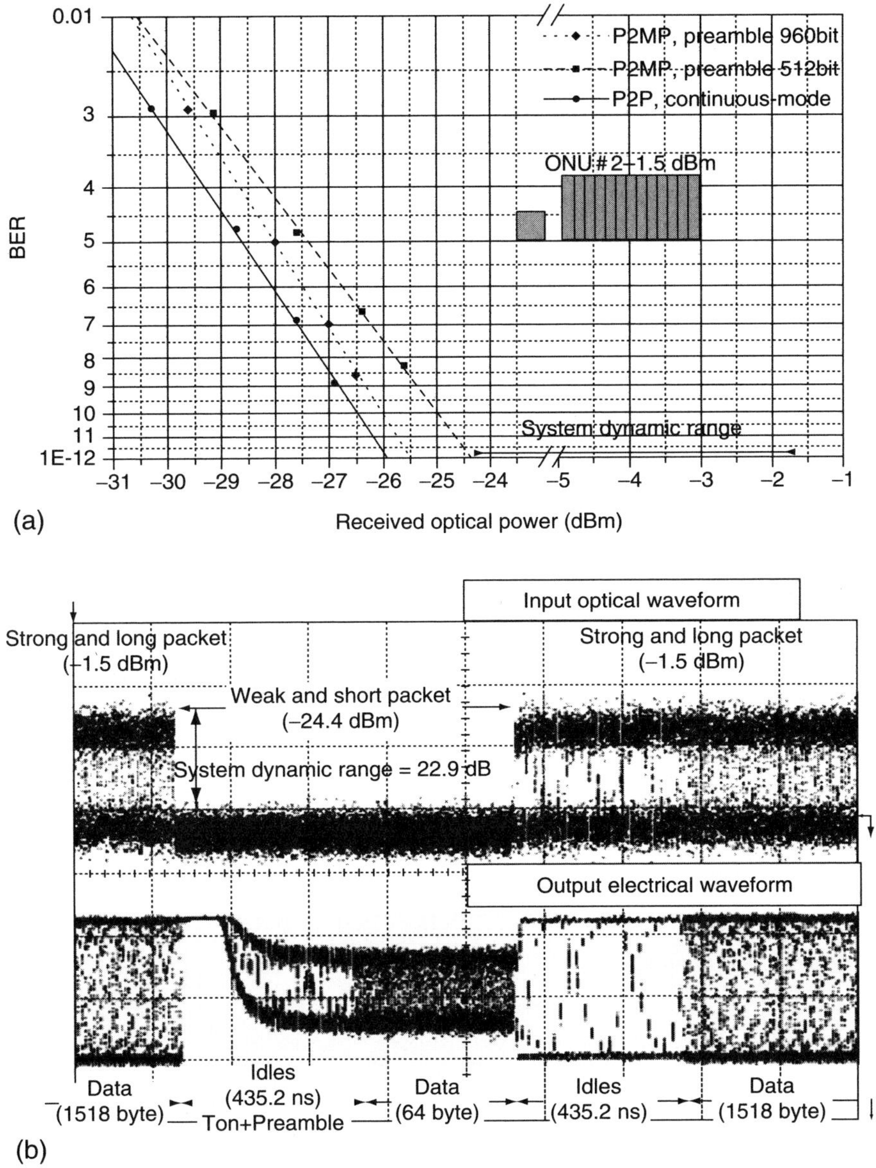

Figure 4.38 (a) Measured continuous-mode (P2P) versus burst-mode BER. (b) Input and output for burst-mode receiver.

transition, weak packet power level is more slowly recovered than the opposite case because of an AC-coupling effect.

Figure 4.39 shows the burst-mode BER curves of an EPON OLT receiver based on APD photodiode. The BER measurements are performed on the payload of the weak packet from BM-Tx as ONU#1, with varying optical power of ONU#2. Bit errors are only counted within the weak packet. It clearly shows a very good sensitivity of about −36 dBm, negligible penalty due to BM packet, and about 2-dB burst-mode penalty when ONU#2 has power as high as −6 dBm.

The summary of a state-of-the-art EPON transceiver performance [22] is given in Table 4.8. The system performance demonstrated here exceeds the specifications defined in the IEEE 802.3ah standard [9]. For the APD/TIA receiver, the measured overload power for this BM Rx is over −1.5 dBm, the highest ever for an APD detector.

It should be noted that, for optical Ethernet access, economical F-P laser diodes are recommended for distances of up to 20-km standard single-mode fiber (SMF-28) over extended temperature ranges. Historically, to operate at wavelengths close to the low dispersion window of SMF-28 has not been an issue. But extended temperature requirements and wavelength variation impose a stringent

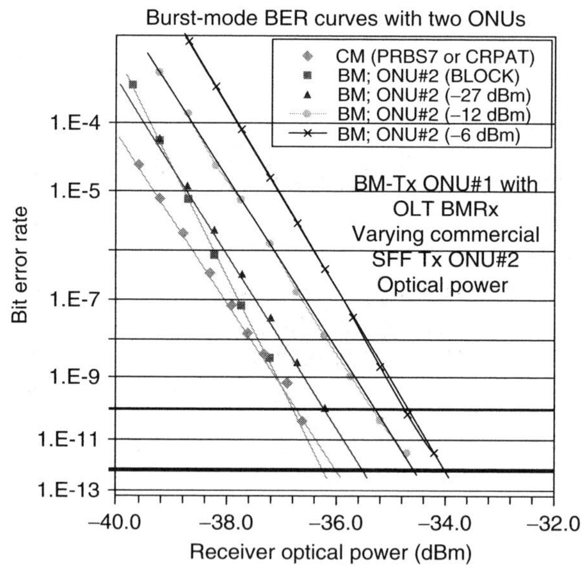

Figure 4.39 Burst-mode BER curves for the weak packet as a function of the power levels of the strong packet at 1.25 Gbps, indicating the degradation of OLT Rx sensitivity due to the influence of the strong packet.

Table 4.8

A state-of-the-art EPON transceiver uplink performance summary in comparison with the IEEE 802.3ah standard

Paramet Ers	Measured	IEEE 802.3ah	Unit
Bit rate	1.25	1.25	Gbps
BER	10–12	10–12	
Laser burst off	2	512	ns
Laser burst on	<8	512	ns
Rx sensitivity	−34.0 (a)	−27	dBm
Rx overload	>−1.5	−6	dBm
Settling-time	<100	400	ns
Dynamic range	>28 (b)	21	dB

Includes 2-dB BM penalty due to strong packet influence. Measured value is limited by the maximum 2nd ONU power level in setup.

dispersion budget [92, 93]. We did a series of experiments demonstrating significant system limitations due to various dispersion penalties [22]; the measured dispersion penalty could be as high as 3 dB under high temperatures. This limits EPON systems to work either in reduced temperature ranges (0–70 °C) or to count on more costly DFP lasers.

4.6.3 Impact of Analog CATV Overlay

It should be noted that PON transceivers may also support an RF video overlay using the additional 1550-nm wavelength in the downstream [10]. This puts an emphasis on the optical isolation parameters between the analog and digital downstream wavelengths at the ONU receiver, and determines the quality requirement of the ONT WDM filters [80, 94]. BER measurements of the digital signal and carrier-to-noise (CNR) of the overlaid video signal are feasible with BER testers and network analyzers.

It is required that the overlay of extra wavelength bands to not interfere with the data transmission. Figure 4.40 (a) shows the impact of the CATV cross-talk component on the BER degradation of the digital signal [80]. The measurement refers to the Rx sensitivity of −30 dBm. CATV noise component in the range of −50 dBm to −30 dBm is added and the receiver BER curves are plotted while setting the digital power level to its sensitivity level, sensitivity level −1 dB, and sensitivity level +1dB. It is clear that over 40-dB isolation is required to maintain negligible impact.

Given the fact that analog channels are the most sensitive to noise, the measurements on analog carriers are also performed. Figure 4.40 (b) shows the

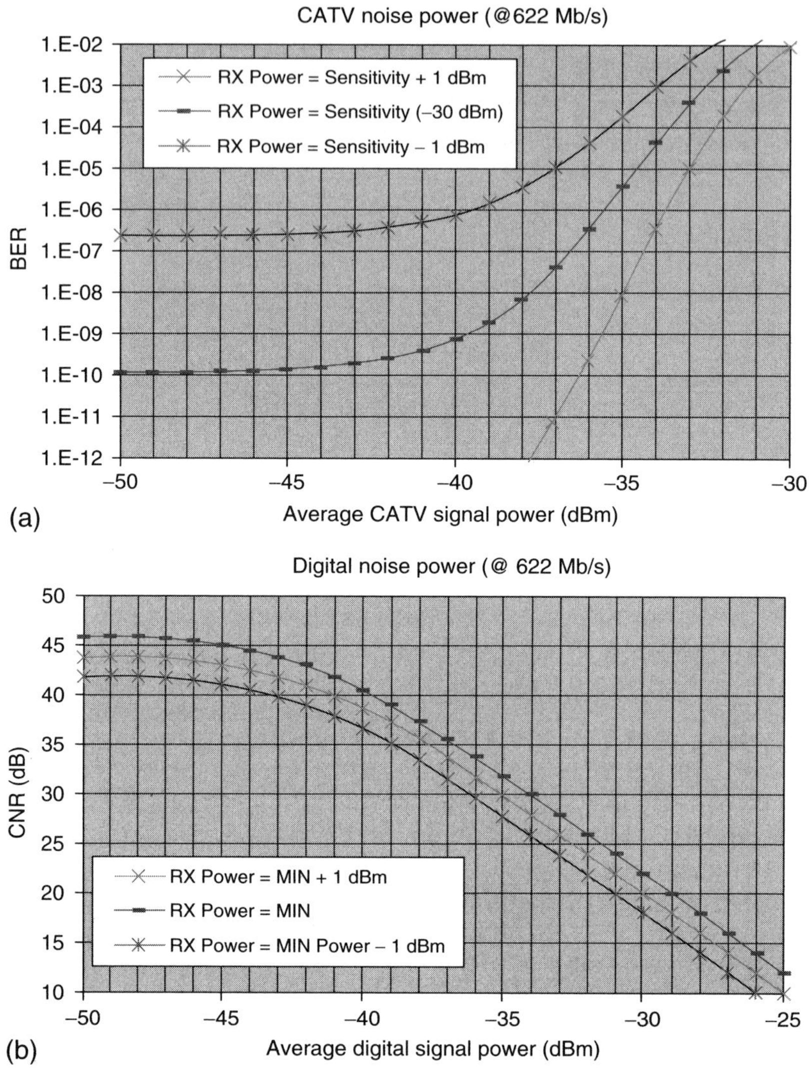

Figure 4.40 (a) Impact of CATV signal on BER of the digital receiver. (b) Impact of digital signal cross talk on the CNR of CATV receiver.

impact of the digital signal cross talk on the CNR degradation of the CATV receiver. Using a CATV receiver that has a CNR = 44 dB at −6 dBm, average digital signal power in the range of −50 dBm to −30 dBm is added, and the receiver

CNR curves are plotted while setting the CATV power level to its minimum sensitivity level (i.e. −6 dBm), sensitivity level −1 dB, and sensitivity level + 1 dB. CNR degradation is observed with increasing downstream data signal power, and the CNR limit of 41 dB is reached at −42-dBm digital cross talk.

4.7 SUMMARY AND OUTLOOK

This chapter has outlined the system requirements, enabling technologies, and system evaluation methodologies covering the various aspects of the current status of optical transceivers in the context of PON applications. PON transceivers are unique bidirectional devices that use different wavelengths to transmit and receive signals between the OLT and the ONUs over a single fiber. The different types of PONs, i.e. BPON, EPON, and G-PON, though all share the same wavelength plan, differ in the up- and downstream transmission bit-rates, reaches, split ratios, and burst-mode dynamics, which impacts the specifications of the respective transceivers. Their system requirements are dictated by the link power budget—a summation of channel losses, power penalties, and margin requirements. For this reason, advanced PON systems such as the PX20 EPON, class B and B+ G-PON, or even future WDM-PON and 10G PONs are still dominated by expensive optoelectronic components while high performance and added functionality are needed to meet the stringent system specifications. Besides, PON point-to-multipoint architectures raise the need for PON-specific test methods that differs from conventional optical test methods to characterize the physical-layer parameters of optical transceiver modules.

One of the critical issues for PON transceivers is the performance/cost ratio measured by the technical specifications and the unit cost of the optical transceivers. The mainstream of existing PON transceiver technologies still count on discrete optoelectronic devices, stand-alone TO-CANs, and coaxial packaged BOSAs that are assembled and tested manually and actively. Though commonly considered labor-intensive, and difficult to scale with drastically increasing volume, the continual improvement of this conventional PON transceiver technology in terms of board-level integrations makes it still the prevailing technology of choice [19]. Emerging waveguide-based PLC technologies, which promise to monolithically integrate the active and passive components onto the same substrate at the chip level, are under intensive development. It is expected to be the ultimate solution to achieve the required volume production at the desired cost; despite that it does not yet necessarily pose immediate challenges and impacts on the cost and performance of existing optical transceivers.

Obviously the huge business opportunity of economically deployable PONs compels intensive research and development efforts for innovative and disruptive technologies in many areas of the transceiver technologies, which promised to replace the current BOSAs made of discrete chips and parts with a highly integrated solutions based on hybrid or monolithical integration. These new technologies will eventually help reducing the number of discrete components, improving the manufacturing yield, increasing the production throughput, and reducing overall cost, and so on. One active area worth mentioning is the breakthrough in InP laser technology [95–97]. As it is well known that a long-reach PON covering 20-km transmission distance presents serious dispersion-related impairments for the downstream 1490-nm and 1550-nm signals [92, 93] so that costly narrow-linewidth DFBs either uncooled or cooled, have to be employed. Dispersion-induced power penalty is also an issue for the upstream signals around 1310 nm in G-PON running at 2.5 Gbps, which increases the overall cost of the ONU transceivers substantially. Conventional single-wavelength DFB lasers require optical isolators to reduce undesired noise feedback into the laser cavity for stable operation. They are thus expensive and contribute to significant portion of the cost in the BOSA module. The latest developments in low-cost, isolator-free laser sources in terms of discrete mode lasers [96] and gain-coupled DFB [97] fulfill this need and the market requirements. In particular, the recent demonstration by Vitesse and Eblana Photnonics of isolator-free lasers manufactured on a working IC fabrication lines using standard electronic IC process tools may lead to the electronic-type performance consistency in building laser diode devices [96].

The major technological challenges in PON transceivers are: (1) upstream burst-mode optical transmission technologies; (2) optical transmitter output power as well as optical receiver sensitivity to satisfy the increasing system link budget requirements; and (3) higher date rates. Burst-mode IC designs still represent one of the active areas of research, which entails the burst-mode laser driving circuit with feed-forward APC, and burst-mode optical receiving circuit by means of ATC and continuous and/or cell-AGC methods. A high sensitivity PIN or APD together with low-noise preamplifier IC and limiting amplifier IC [93, 98] were key elements to improve the optical sensitivity and its wide dynamic range. Furthermore, with increased photonic integration level and more compact optical front-end assembly in optical transceivers, the increase in electrical and optical cross talk [99] between the transmitting and receiving channels becomes another significant challenge. The cross-talk issue is particularly critical for the video channel in triplexer transceivers.

Although 1 Gbps PON systems such as EPON and G-PON have gained popularity, much faster PON systems are demanded in regions where many business users need large-capacity leased lines. In view of this trend, WDM-PON systems [101, 102] that multiplex high-speed services by means of WDM channels as well as 10-Gbps PON systems are currently under intensive study. Recently, IEEE 802.3 [103] has started the 802.3av task force to actively develop 10-Gbps EPON standards. The objectives of the 10-G EPON task force is to develop both symmetric and asymmetric 10-Gbps line rate operation where the symmetric operation will operate at 10 Gbps in both the downstream and upstream directions. This development is obviously posing even more stringent challenges in burst-mode IC developments [104, 105] and asking for more appealing transceiver designs. For cost-effective 10-Gbps transceivers, there is a possibility that the new small form–factor pluggable modules called SFP+ being specified by the SFF Committee (SFF 8431), may play a significant role in future PON-system developments [106].

REFERENCES

[1] J.R. Stern, C.E. Hoppitt, D.B. Payne, M.H. Reeve, and K.A. Oakley, "TPON-a passive optical networks for telephony," Fourteenth European Conference on Optical Communication (ECOC'88), vol.1, pp203–206, Brighton, UK, Sept., 1988.

[2] P.E. Green, "Fiber to the home: the next big broadband thing," *IEEE Comm. Mag.*, pp100–106, Sept., 2004.

[3] P.E. Green, *Fiber to the Home: The New Empowerment*; Wiley-Interscience, 2006.

[4] W. Diab and H. Frazier, "*Ethernet in the First Mile: Access for Everyone,*" Standard Information Network, IEEE press, 2006.

[5] M. Beck, *Ethernet in the First Mile*, McGraw-Hill, ISBN 007145506X, June, 2005.

[6] M. Aabrams, P.C. Becker, Y. Fujimoto, V.o'Byrne, and D. Piehler, "FTTP deployments in the United States and Japan—Equipment choices and service provider imperatives," *J. Lightwave Technol.*, vol.23, pp236–245, Jan., 2005.

[7] J.D. Angelopoulos, H. Leligou, T. Argyriou, S. Zontos, E. Ringoot, and T.V. Caenegem, "Efficient transport of packets with QoS in an FSAN-aligned G-PON," *IEEE Comm. Mag.*, vol.42, no.2, pp92–98, Feb., 2004.

[8] ITU-T Rec. G.983.1, Study Group 15, "Broadband optical access systems based on passive optical networks," Oct., 1998.

[9] *IEEE Standard for Information Technology*, IEEE 802.3ah Ethernet in the first Mile Task Force, D3.3, Apr. 19, 2004.

[10] ITU-T Rec. G.984.2. "Gigabit-capable passive optical networks (G-PON): physical media dependent (PMD) layer specification," Mar., 2003.

[11] M. Hajduczenia, H. da Silva, and P. Monteiro; "EPON versus APON and G-PON: a detailed performance comparison," vol.5, no.4, p298, *J. Optical Networking*, Apr., 2006.

[12] A. Girard, "*FTTx PON Technology and Testing,*" EXFO Electro-Optical Engineering Inc., 2005.

[13] G. Kramer, B. Mukherjee, and A. Maislos, "*Ethernet Passive Optical Networks,*" Chap. 8, McGraw-Hill Professional, ISBN: 0071445625, Publication date: March 2005, available from http://networks.cs.ucdavis.edu/~mukherje/links/gk_wiley_bc.pdf

[14] G. Kramer, "What is Next for Ethernet PON?" The 5th International Conference on Optical Internet (COIN 2006); MoB2–1.

[15] ITU-T Rec. G.984.2 Amendment, "New Appendix III—Industry best practice 2.488 Gbit/s downstream, 1.244 Gbit/s upstream G-PON," Feb., 2006.

[16] K. Kim, "On the evolution of PON-based FTTH solutions," *Information Sciences*, vol.149, no.1–2, pp21–30, Jan., 2003.

[17] T. Tatsuta†, Y. Yoshida, and Y. Maeda, "Standardization of G-PON (Gigabit Passive Optical Network) in ITU-T," *NTT Tech. Rev.*, vol.1, no.7, pp89–93, Oct., 2003.

[18] F. Effenberger, "Cost-Capacity-protocol Trade-offs in PON systems," NFOEC'2002, p1606.

[19] W. Huang, X. Li, C. Xu, X. Hong, C. Xu, and W. Liang, "Optical transceivers for fiber-to-the-premises applications: System requirements and enabling technologies," *J. Lightwave Technol.* vol.25, pp11–27, 2007.

[20] X.Z. Qiu, P. Ossieur, J. Bauwelinck, Y.C. Yi, D. Verhulst, J. Vandewege, B. De Vos, and P. Solina, "Development of G-PON upstream physical media dependent prototypes," *J. Lightwave Technol.*, vol.22, pp2498–2508, Nov., 2004.

[21] P. Vetter et al., "Study and demonstration of extensions to the standard FSAN BPON," 14th International Symposium on Services and Local access (ISSLS2002) in Seoul, Korea, on Apr. 14–18, 2002.

[22] Y. Chang and G. Noh, "1.25 Gb/s uplink burst–mode transmissions: System requirements and optical diagnostic challenges of EPON physical-layer chipset for enabling broadband optical Ethernet access networks," OFC/NFOEC'06 Paper JThB84.

[23] M. Fuller, "G-PON burst-mode receiver electronics *prove challenging,*" *Lightwave, PennWell*, vol.23, no.5, pp14–17, May, 2006.

[24] J. Redd and C. Lyon, "Challenges mark G-PON FEC receiver designs," *Lightwave, PennWell*, vol.23, no.5, pp11–14, May, 2006.

[25] J. Kwon, J. Lee, J. Baek, J. Cho, J. Seo, S. Park, J. Lee, Y. Oh, and D. Jang, "AC-coupled burst-mode OLT SFP transceiver for gigabit Ethernet PON systems," *IEEE Phot. Tech. Lett.*, vol.17, no.7, p1519, July, 2005.

[26] H. Oohashi, "Overview over InP technologies for optical access network modules and their reliability," Eleventh International Conference on Indium Phosphide and Related Materials, 1999. IPRM, vol.1999, pp375–380 (NTT Photonics Labs).

[27] S. Kaneko and H. Haneda, "Optical devices for optical access network systems," Mitsubishi Electric ADVANCE, vol.114, pp17–19, June, 2006.

[28] A. Behfar, M. Green, A. Morrow, and C. Stagarescu, "Monolithically integrated diplexer chip for PON applications," OFC 2005 paper OTuM5.

[29] M. Iwase, Y. Ishikawa, T. Komatsu, J. Kasahara, N. Hattori, M. Miura, N. Nakamura, and K. Odaka "Optical transceiver modules for gigabit Ethernet PON FTTH systems," *Furukawa Review*, no.28, p8, 2005.

[30] D. Shimura, M. Uekawa, R. Sekikawa, K. Kotani, Y. Maeno, H. Sasaki, and T. Takamori, "Ultra compact optical subassembly using integrated laser diode and silicon microlens for low-cost optical component," presented at the Electronic Components Technology Conf., Las Vegas, NV, 2004, Paper s05P7.

[31] D. Shimura, R. Sekikawa, K. Kotani, M. Uekawa, Y. Maeno, K. Aoyama, H. Sasaki, T. Takamori, K. Masuko, and S. Nakaya, "Bidirectional optical subassembly with prealigned silicon microlens and laser diode," *IEEE Phot. Tech. Lett.*, vol.18, no.16, pp1738–1740, Aug., 2006.

[32] K. Masuko, T. Ori, T. Tanaka, M. Inoue, H. Sasaki, M. Uekawa, Y. Maeno, K. Kotani, D. Shimura, R. Sekikawa, and T. Takamori, "A low cost PON transceiver using single TO-CAN type micro-BOSA," presented at the Electronic Components Technology Conf., pp1082–1086, June, 2006.

[33] H. Sasaki, M. Uekawa, Y. Maeno, K. Kotani, D. Shimura, R. Sekikawa, T. Takamori, T. Ori, K. Masuko, and Y. Katsuki; "A low-cost micro-BOSA using Si microlens integrated on Si optical bench for PON application," OFC 2006, Paper OWL6.

[34] M. Pearson, "PLCs poised to displace bulk-optics in FTTH," *Lightwave*, *PennWell*, vol.23, no.2, p15, Feb., 2006.

[35] M. Pearson, S. Bidnyk, A. Balakrishnan, and M. Gao, "PLC platform for low-cost optical access components," The 18th Annual Meeting of the IEEE Lasers and Electro-Optics Society, 2005, pp660–661, Paper WX1, Oct. 22–18, 2005.

[36] T. Kurosaki et al., "1.3/1.55-μm full-duplex WDM optical transceiver modules for ATM-PON (PDS) systems using PLC-hybrid-integration and CMOS-IC technologies," IEICE TRANSACTIONS on Electronics, vol.E82-C, no.8, pp1465–1474.
[37] T. Kurosaki et al., "Full duplex 1300/1550-nm-WDM optical transceiver modules for ATM-PON systems using PLC-hybrid-integration and CMOS IC technologies," ECOC'98, Technical Digests, vol.1, pp631–632, 1998.
[38] C. Xu, X. Hong, and W. Huang; "Design optimization of integrated BiDi triplexer optical filter based on planar lightwave circuit," *Opt. Exp.*, vol.14, no.11 p4675, May, 2006.
[39] Y. Inoue, M. Ishii, Y. Hida, M. Yanagisawa, and Y. Enomoto, "PLC components used in FTTH access networks," *NTT Technol. Rev.*, vol.3, no.7, pp22–26, July, 2005.
[40] K. A. McGreer, H. Xu, C. Ho, N. Kheraj, Q. Zhu, M. Stiller, and J. Lam; "Planar lightwave circuits for PON applications," OFC/NFOEC 2006, Paper NWD4.
[41] H. Blauvelt, A. Benzoni, J. Byrd, M. Downie, C. Grosjean, S. Hutchinson, R. Monzon, M. Newkirk, J. Paslaski, P. Sercel, D. Vernooy, and R. Wyss, "High performance planar lightwave circuit triplexer with passive optical assembly," OFC/NFOEC 2005 OThU7.
[42] T. Haslumoto, A. Kanda, R. Kasalnml, I. Ogawa, Y. Shuto, M. Yanagisawa, A. Ohki, S. Mino, M. Ishui, Y. Suzuki, R. Nagasc, and T. Kitagawa, "A bidirectional single fiber 1.25 Gb/s optical transceiver module with SFP package using PLC," in *Proc. Electron. Compon. and Technol. Conf.*, pp279–283, 2003.
[43] J.H. Song et al., "Bragg grating-assisted WDM filter for integrated optical triplexer transceivers," *IEEE Phot. Technol. Lett.*, vol.17, no.12, pp2607–2609, Dec., 2005.
[44] Small Form Factor Pluggable (SFP) Transceiver MultiSource Agreement (MSA), Sept. 14, 2000.
[45] Gigabit Interface Converter (GBIC). SFF document number: SFF-0053, rev. 5.5, Sept. 27, 2000.
[46] Luminent Inc. Triplexer transceiver datasheet, available from http://www.luminentoic.com/downloads/OFC_FTTx.pdf
[47] C. Su, L. Chen, K. Cheung; "Theory of burst-mode receiver and its applications in optical multiaccess networks," *J. Lightwave Technol.*, vol.15, no.4, pp590–606, Apr., 1997.
[48] J. Bauwelinck, Y. Martens, P. Ossiieur, K. Noldus, X.Z. Qiu, J. Vandewge, E. Gilon, and A. Ingrassia "Generic and Intelligent CMOS 155 Mb/s burst mode laser driver chip design and performance," ESSCIRC 2002, 28th European Solid-State Circuits Conference, Florence, Italy, pp495–498, Sept. 24–26, 2002.
[49] E. Sackinger, Y. Ota, T. J. Gabara, and W. C. Fischer, "A 15-mW, 155-Mb/s CMOS burst-mode laser driver with automatic power control and end-of-life detection," *IEEE J. Solid-State Circuits*, vol.35, no.2, pp269–275, Feb., 2000.
[50] Dieter Verhulst, Yves Martens, Johan Bauwelinck, Xing-Zhi Qiu, and Jan Vandewege, "Fast consecutive zero and one bits detection circuits for a 1.244 Gbps burst mode laser driver," IEICE Trans. on Electronics, vol.E87-B, no.8, pp2377–2379, Aug., 2004.
[51] J. Bauwelinek, D. Verhulst, P. Ossieur, and J. Bauwelinek, "DC-coupled burst-mode transmitter for 1.25 Gbit/s upstream PON." *Elect. Lett.*, vol.40, no.8, pp501–502, 2004.
[52] X. Qiu, J. Vandewege, Y. Martens, J. Bauwelinck, P. Ossieur, E. Gilon, B. Stubbe, "A burst-mode laser transmitter with fast digital power control for a 155 Mb/s upstream PON," IEICE Transactions on Communications, special issue on "Recent Progress in Optoelectronics and Communications," vol.E86-B, no.5, pp1567–1574, May, 2003.
[53] Y. Oh, L. Quan, G. Lee et al., "Burst-mode transmitter for 1.25Gb/s Ethernet PON applications," in ESSCIRC, 2004, Tech. Papers, pp283–286, Sept., 2004.
[54] X.Z. Qiu, J. Vandewege, F. Fredricx, and P. Vetter, "Burst mode transmission in PON access systems," NOC'2002, 7th European Conference on Networks & Optical Comm., Darmstadt, Germany, pp127–132, June 18–21, 2002.

[55] Y. Ota and R. G. Swartz, "Burst mode compatible optical receiver with large dynamic range," *J. Lightwave Technol.*, vol.8, pp1897–1903, Dec., 1990.
[56] C.A. Eldering, "Theoretical determination of sensitivity penalty of burst-mode fiber optic receiver," *J. Lightwave Technol.*, vol.11, pp2145–2149, Dec., 1993.
[57] S. Ide, H. Nobuhara, N. Nagase, A. Hayakawa, K. Mori, and M. Kawai, " + 3.3V PON receiver IC with a high-speed ATC circuit," in Proc. IEEE Int. Solid-State Circuit Conf. (ISSCC), *(ISSCC)*, pp244–245, Feb., 1997.
[58] P. Ossieur, Y.C. Yi, J. Bauwelinck, X.-Z. Qiu, J. Vandewege, and E. Gilon, "DC-coupled 1.25 Gbit/s burst-mode receiver with automatic offset compensation," *Electron. Lett.*, vol.40 (7), pp447–448, Apr., 2004.
[59] P. Ossieur, D. Verhulst, Y. Martens, W. Chen, J. Baauwlinck, X. Qiu, and J. Vandewege, "A 1.25-Gb/s burst-mode receiver for G-PON applications," *IEEE J. Solid-State Circuits*, vol.40, pp1180–1189, May, 2005.
[60] S. Yamashita et al., "Novel cell—AGC technical for burst-mode CMOS preamplifier with wide dynamic range and high sensitivity for ATM-PON systems," *IEEE J. Solid-State Circuits*, vol.37, pp881–886, July, 2002.
[61] J.M. Baek et al., "Low-cost and high-performance APD burst-mode receiver employing commercial TIA for 1.25-Gb/s EPON," *IEEE Phot. Technol. Lett.*, vol.17, no.10, p2170, Oct., 2005.
[62] Q. Le, S. Lee, Y. Oh, H. Kang, and T. Yoo, "A bursts-mode receiver for 1.25Gb/s Ethernet PON with AGC and internally created reset signal," *IEEE J. Solid-State Circuits*, vol.39, pp2379–2388, Dec., 2004.
[63] S.R. Cho, S.K. Yang, J.S. Ma, S.D. Lee, J.S. Yu, A.G. Choo, T.I. Kim, and J. Burm, "Suppression of avalanche multiplication at the periphery of diffused junction by floating guard rings in a planar In-GaAs-InP avalanche photodiode," *IEEE Phot. Technol. Lett.*, vol.12, no.5, pp534–536, May, 2000.
[64] Le Q., Y. Oh, and S. Lee "Integrated differential preamplifier for 155Mb/S ATM-PON system with fast response, high sensitivity and wide dynamic range," *in Proc. Asia-Pacific Microwave conf. (AMPC)*, vol.2, pp478–481, Nov., 2002.
[65] H. Ishikawa, M. Soda, T. Takeuchi, H. Kaneko, and T. Suzaki, "2.4-Gbit/s compact optical receiver module with one-chip 3R IC," in *Optical Fiber Communication Conference*, vol.2 of 1996 OSA Technical Digest Series (Optical Society of America, 1996), Paper FC3.
[66] S. Han and M. Lee, "Burst-mode penalty of AC-coupled optical receivers optimized for 8B/10B line code," *IEEE Phot. Technol. Lett.*, vol.16, no.7, pp1724–1726, July, 2004.
[67] SFF-8472 Specification for Diagnostic Monitoring Interface for Optical Xcvrs, Revision 9.1, Mar., 2002.
[68] Finisar Appl. Notes; "Digital Diagnostic Monitoring Interface for SFP Optical Transceivers," available from http://protocoltransport.com/finisar/optics/wdmmain/pdf/AN-2030RevDDKFinal9_26_02.pdf.
[69] Avago Appl. Note 5016 "Digital Diagnostic Monitoring Interface (DMI) on Avago Enterprise and Storage Fiber-Optic Transceivers: Applications and Implementation," AN5016; June, 2006.
[70] Z. Gong, "Digital-diagnostic-monitoring functions for optoelectronic transceivers," EDN (Vitesse) July, 2007.
[71] C.-H. Yeh and S. Chi, "Optical fiber-fault surveillance for passive optical networks in S-band operation window," *Opt. Exp.*, vol.13, pp5494–5498, 2005.
[72] I. Sankawa, "Fault location technique for in-service branched optical fiber networks," *IEEE Phot. Technol. Lett.*, vol.2, pp766–768, 1990.

[73] C.K. Chen, F. Tong, L.K. Chen, J. Song, and D. Lam, "A practical passive surveillance scheme for optical amplified passive branched optical networks," *IEEE Phot. Technol. Lett.*, vol.9, pp526–528, 1997.

[74] J. Park, J. Baik, and C. Lee, "Fault-detection technique in a WDM-PON," *Opt. Exp.*, vol.15, pp1461–1466, 2007.

[75] Y. Koyamada, T.K. Horiguchi, and S. Furukawa, "Recent progress in OTDR technologies for maintaining optical fiber networks," in Tech. Dig., IOOC'95, FA1–4, Hong Kong, 1995.

[76] W. Chen, B.D. Mulder, J. Vandewege, X.Z. Qiu, J. Bauwelinck, and B. Baekelandt, "A novel technique for low-cost embedded non-intrusive fiber monitoring of P2MP optical access networks," OFC'07, Paper OThE4.

[77] B. De Mulder, W. Chen, J. Bauwelinck, J. Vandewege, and X.-Z. Qiu, "Nonintrusive fiber monitoring of TDM optical networks," *J. Lightwave Technol.*, vol.25, pp305–317, 2007.

[78] H. Nakanishi, T. Okada, Y. Yamaguchi, K. Hirayama,Y. Iguchi, A. Yamaguchi, N. Yamabayashi, and Y. Kuhara, "Development of bidirectional 1.3/1.55-μm optical transceiver module compliant with Ethernet standard," *SEI Tech. Rev.*, no.54, p99, June, 2002.

[79] H. Schmuck, J. Hehmann, M. Straub, and Th. Pfeiffer, "Embedded OTDR techniques for cost-efficient fibre monitoring in optical access networks," in Proc.of the ECOC'06, Sept., 2006.

[80] Broadlight Inc. White paper, "Optical diagnostic challenges in BPON networks."

[81] X.Z. Qiu, P. Ossieur, J. Bauwelinck, Y.C. Yi, D. Verhulst, S. Verschuere, Z. Lou, W. Chen, Y. Martens, X. Yin, and J. Vandewege, "FSAN G-PON Upstream Burst-Mode Transmission Experiments," in Proc. 30th European Conference on Optical Communication (ECOC 2004), pp398–399, Sept., 2004.

[82] Lou, Z., S. verschuere, T. Yi, D. Verhulst, X. Qiu, and J. Vandewege, "Lab test bed development for evaluation of the GigaPON uplink performance," Symposium IEEE/LEOS Benelux chapter, 2003.

[83] C. Lee, H Kim, S. Kang, and M. Lee, "A systematic evaluation on burst-mode data transmission in passive optical access network," 1996 International Conference on Communication Technology Proceedings, vol.1, pp578–581, May 5–7, 1996.

[84] P. Ossieur, X. Qiu, J. Bauwelinck, and J. Vandewege, "Sensitivity penalty calculation for burst-mode receivers using avalanche photodiodes," *IEEE J. Lightwave Technol.*, vol.21, no.11, pp2565–2575, Nov., 2003.

[85] Vitesse Semiconductor Corp., Internal communication.

[86] J. Nakagawa, M. Nogami, M. Noda, Y. Kato, T. Uo, and K. Motoshima, "Newly developed OLT optical transceiver for GE-PON systems compliant with IEEE 802.3ah," in Proc. ECOC, vol.3, pp746–747, 2004.

[87] K. Nishimura, H. Kimura, M. Watanabe, T. Nagai, K. Nojima, K. Gomyo, M. Takata, M. Iwamoto, and H. Asano, "A 1.25Gbit/s CMOS burst-mode optical transceiver for Ethernet PON system," Matsushita Electr. Ind. Co. Ltd., Osaka, Japan in: VLSI Circuits, 2004. Digest of Technical Papers, pp414–417, 2004 Symposium Date: June 17–19, 2004.

[88] J. Seo, S. Han, S. Lee, M. Lee, and T. Yoo, "A 1.25 Gb/s high sensitive peak detector in optical burst-mode receiver using a 0.18um CMOS technology," Communication Technology Proceedings, 2003. ICCT, vol.1, pp644–646, Apr., 2003.

[89] H. Ishizaki, H. Okada, T. Tanaka, and Y. Arai, "Development of optical transmission module for access networks," *OKI Technical Review*, vol.69, no.2, p76, Apr., 2002.

[90] T. Tanaka, Y. Yoshida, and Y. Maeda, "Development of a Gigabit Ethernet Passive Optical Network (GE-PON) System," *NTT Tech. Rev.*, vol.1, no.7, p89, Oct., 2003.

[91] M. Kao, Y. Cheng, S. Cheng and C. Shaw, "Small-form-pluggable PON burst-mode transceiver," CLEO, Paper CThT14, May, 2004.

[92] G.P. Agrawal, P.J. Anthony, and T.M. Shen, "Dispersion penalty for 1.3-μm lightwave systems with multimode semiconductor lasers," *J. Lightwave Technol.*, vol.6, pp620–625, May, 1988.

[93] G.P. Agrawal, *Photodetectors* from *Lightwave Technology: Components and Devices*, Chap. 7, pp253, Wiley-Interscience, 2004.

[94] R. Schoop, F. Fredricx, T. Koonen, and C. Hardalov, "WDM Isolation Requirements for CATV in BPON," 28th European Conference on Optical Communication, ECOC 2002, vol.4, pp1–2, 2002.

[95] K. Terada et al., "Isolator-free DFB-LD module with TEC control using silicon wafer board," IEEE 5th Topical Meeting on Electrical Performance of Electronic Packaging, 1996, pp71–73, Oct. 28–30, 1996.

[96] Eblana Photonics, White paper, "Cost-effective optical products for G-PON physical layer requirements," May, 2005.

[97] K. Nakamura et al., "Optical feedback-tolerant 1.3 μm gain-coupled DFB lasers for isolator-free micro-BOSA modules," OFC'2006, Paper OMK5.

[98] B. Razavi, *Design of Integrated Circuits for Optical Communications*, pp87–91, New York: McGraw-Hill, 2003.

[99] S. Kim, S. Park, J. Moon, and H. Lee, "A low-crosstalk design of 1.25 Gbps optical triplexer module for FTTH systems," *ETRI Journal*, vol.28, no.1, pp9–16, Feb., 2006.

[100] A. Banerjee, Y. Park, F. Clarke, H. Song, S. Yang, G. Kramer, K. Kim, B. Mukherjee, "Wavelength-division-multiplexed passive optical network (WDM-PON) technologies for broadband access: a review," *J. Optical Networking*, vol.4, no.11, p737, Nov., 2005.

[101] M. Choudhary and B. Kumar, "Analysis of next generation PON architecture for optical broadband access networks," available from http://hosteddocs.ittoolbox.com/MC120606.pdf

[102] D. Shin et al., "Low-cost WDM-PON with colorless bidirectional transceivers," *J. Lightwave Technol.*, vol.24, no.1, p158–165, Jan., 2006.

[103] IEEE Standard 802.3av, 10G EPON task force, available from http://www.ieee802.org/3/av/index.html.

[104] M. Nogawa, K. Nishimura, S. Kimura, T. Yoshida, T. Kawamura, M. Togashi, K. Kumozaki, and Y. Ohtomo, "A 10 Gb/s burst-mode CDR IC in 0.13 /spl mu/m CMOS," ISSCC Dig. Tech. Papers, pp228–229, Feb. 2005.

[105] IEEE International Solid-State Circuits Conference, 2005. Digest of Technical Papers ISSCC, 2005, vol.1, pp228–595, Feb., 2005.

[106] S. Nishihara et al., "A 10.3125-Gbit/s SiGe BiCMOS burst-mode 3R receiver for 10G-EPON systems," OFC'07, Paper PDP8, Mar., 2007.

[107] S. Bhoja, A. Ghiasi, Y. Chang, B. Mayampurath, M. Dudek, S. Inano, and E. Tsumura, "Next-generation 10 GBaud module based on emerging SFP + with host-based EDC," *IEEE Comm. Mag.*, vol.45, no.3, ppS32–S38, Mar., 2007.

Chapter 5

Ranging and Dynamic Bandwidth Allocation

Noriki Miki and Kiyomi Kumozaki

NTT Access Network Service Systems Laboratories

5.1 RANGING

5.1.1 Purpose of Ranging

Since several optical network units (ONUs) are connected to one optical line termination (OLT) interface in a passive optical network (PON) system, a method to multiplex signals from each ONU is required. One of the methods is time division multiple access (TDMA) which separates the signals into time slots so that they do not collide. Other methods besides TDMA include wavelength division multiple access (WDMA) which separates the signals with optical wavelengths and code division multiple access (CDMA) which is widely used in cellular phone systems. Due to the fact that the TDMA method is the least expensive now, PON systems using the TDMA method are the only ones which have been standardized and commercially introduced into optical access systems[1–3].

Now, let us explain more specifically about the TDMA method which involves Ranging. Firstly, the OLT measures the round-trip delay (RTD) to the ONU. As shown in Fig. 5.1, the signal transmission timing of each ONU is adjusted according to this RTD so that the signals from individual ONUs arrive at different times and do not overlap. The TDMA method multiplexes the signals from different ONUs in nonoverlapping time slots. The carrier sense multiple access with collision detection (CSMA/CD) protocol used in early generation Ethernets is a well-known method which also separates user signals by time. The CSMA/CD method can also be categorized as a TDMA protocol in a broader sense.

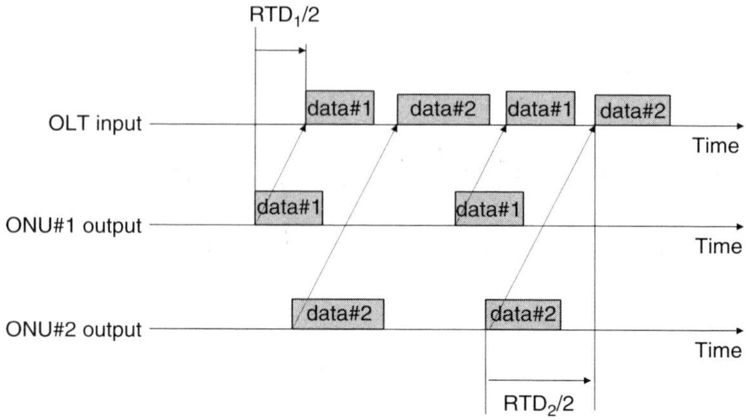

Figure 5.1 Time division multiple access.

With the CSMA/CD method, if other terminals are transmitting signals, a terminal must wait for the transmission to complete. It can transmit its own data immediately if other terminals are not transmitting signals. If the data unfortunately collides with the signals from other terminals, the transmitter waits for a random length of time and retransmits the data. The reason why random delay is used is to minimize the probability of another collision. When random delay and retransmission are repeated, the probability of continued collision becomes minimized exponentially. This method requires no measurement of the RTD and if the distance is within several hundred meters, it has very low delay and high efficiency. However, if the distance reaches several tens of kilometers like in fiber-access networks, it would take much time to detect if other terminals are transmitting signals or if there has been a collision. Therefore, there will be a tendency of frequent collision and retransmission, and adversely this will cause a significant drop in efficiency and a drastic increase in delay.

Ranging is executed to eliminate the necessity for data retransmission, hence using the bandwidth efficiently and keeping the maximum delay time to the minimum by preventing the signals from multiple ONUs from colliding.

5.1.2 Ranging Procedures Overview

The distance between the OLT and ONU normally differs from ONU to ONU. Unless each RTD is accurately determined, the transmission timing cannot be established. With that in mind, if we newly connect an ONU, we must first measure the RTD. By the command of operation system (OPS), the

OLT automatically and regularly makes available the ranging window for delay measurement (as shown in Fig. 5.2) and specifies an ONU to transmit signals for delay measurement. The length of the ranging window is set up according to the distance between the OLT and ONU. If the ONU is located in the range of 5–10 km from the OLT, the length of the ranging window would need to be at least (10–5 km) × 2/(300,000 km/sec/1.5) = 50 msec, according to the time it takes for light to travel back and forth between the OLT and ONU. In this expression, 1.5 is the refractive index of the optic fiber.

There are two ways of specifying the ONU for Ranging. One method specifies only one already-registered ONU and the other method specifies all unregistered ONUs. In the first method, an ONU with a unique ID number is specified in advance from the OPS. In the second method, the OLT does not know the unique ID number of each ONU. In this case, we must also take into account that several ONUs may transmit signals for delay measurement simultaneously and that the RTDs are close enough for the signals to overlap. One of the methods to reduce collision during ranging is to transmit signals for delay measurement with a random delay, similar to the method used in the Ethernet. This method is applied to both G-PON [2] and EPON [3]. Even if a collision unfortunately occurs in the first place, it is possible to execute a delay measurement without collision by repeating the transmission twice or three times, etc. Since user data transmission is not yet started before ranging is finished, the latency is not increased. Only the time from connecting to the ONU to actually starting communication is slightly prolonged. Of course, the random delay used to avoid contention must not be included in the measured RTD.

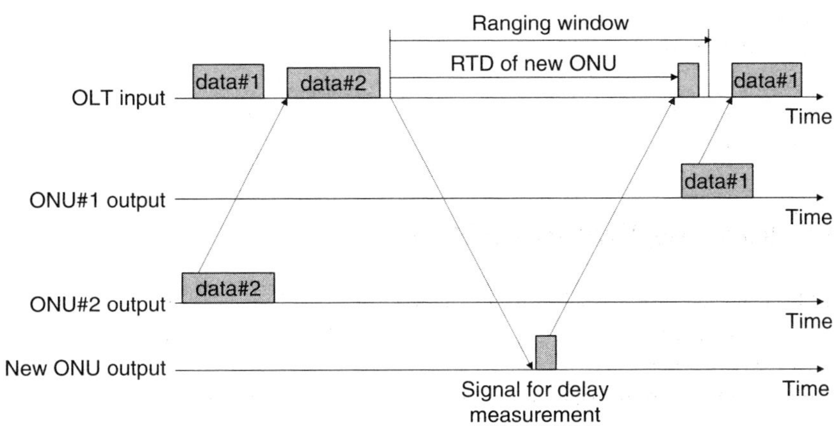

Figure 5.2 Ranging window.

After the RTD is measured, the transmission timing of the ONU has to be adjusted and there is one more point to keep in mind. That is the time interval (or guard time) between the signals transmitted by a certain ONU and those transmitted by other ONUs. There must be sufficient guard time so that the signals from different ONUs do not collide. This guard time interval is called burst overhead (BOH). The following factors influence the size of burst overhead:

1. Error in RTD measurement: RTD measuring comes with errors and we must anticipate errors caused by clock jitters and the quantization of RTD values.
2. Change in RTD: The change in RTD is about 36 ps/km/°C, owing to the temperature dependence of the refractive index and thermal expansion of the optical fiber. In addition to the change in the signal propagation time, we must also accurately determine the fluctuation of processing time in the device.
3. ON/OFF time of laser: The time for transmitted power to stabilize (laser on time) when the ONU laser is switched ON and the time for optical power to drop to a level that does not hinder the next ONU signal reception when the laser is switched OFF (laser off time) must be made available.
4. Receiver setup time: We must take into account the standard time it takes to enable the extraction of the clock and data from the received signals by the OLT. This includes the time required to adjust the reception level such as automatic gain control (AGC) settling time, bit synchronization time of received signals, and code synchronization time.

When the BOH is lengthened, it leads to the deterioration of efficiency and when it is shortened on the other hand, the cost of components increases. Implementation requires a design which takes into account this trade-off. Those who are interested in the specific amount of BOH should refer to the paper in [4].

In the following section we will introduce more specific ranging steps concerning G-PON and EPON.

5.1.3 Ranging Protocol of G-PON

ITU-T Rec. G.983.2 B-PON and its successor ITU-T Rec. G.984.3 G-PON use a similar Ranging procedure. Here, we will introduce only the Ranging procedure of G-PON.

The G-PON ranging procedure is characteristically divided into two phases. In the first phase, registration of the serial number for unregistered ONU and allocation of ONU-ID for the ONU are executed.

The serial number is an ONU-specific ID and must be universally unique. The ONU-ID on the other hand is used for controlling, monitoring, and testing the ONU. So it can be unique only within its own PON.

The steps taken in the first phase are as follows (Fig. 5.3):

1. The OLT specifies all ONU presently communicating to halt upstream transmissions (ONU halt).
2. The OLT specifies ONU without an ONU-ID to transmit the serial number (serial_number request).
3. Upon receiving the serial number transmission request, an ONU without an ONU-ID transmits the serial number (SN-transmission) after waiting for a random time (max 50 msec).
4. The OLT assigns an ONU-ID to an unregistered ONU whose serial number was received normally (assign ONU-ID message).

In the next phase, the RTD is measured for each newly registered ONU. Moreover, this phase is also applied to ONUs which lost their signals during communication. The steps in detail are as follows (Fig. 5.4):

5. The OLT specifies all ONUs presently communicating to halt upstream transmission (ONU halt).
6. Using the registered serial numbers, the OLT specifies a specific ONU only to transmit signals for delay measurement (Ranging request).
7. The ONU which has a serial number that matches the specified serial number transmits the signal for delay measurement (Ranging transmission), which includes the ONU-ID assigned in the first phase.

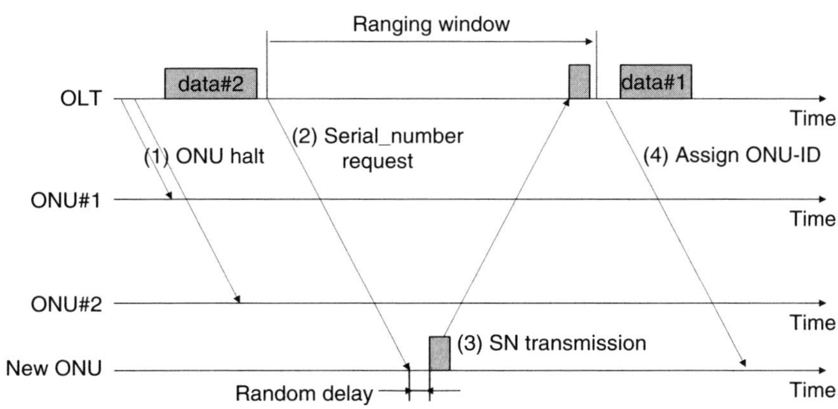

Figure 5.3 G-PON ranging phase 1: serial number process.

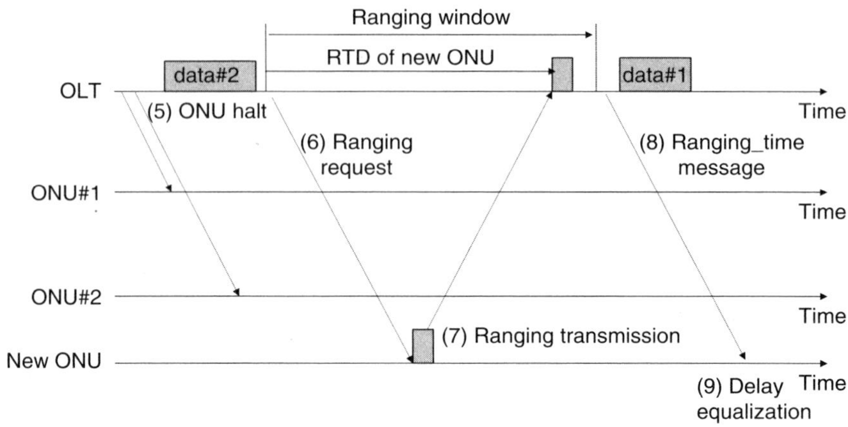

Figure 5.4 GPON ranging phase 2: delay measurements.

8. The OLT measures the RTD according to the time that signals for delay measurement were received. Furthermore, after confirming that the combination of the serial number and ONU-ID is correct, it notifies the *Equalization Delay* (= *Teqd* − RTD) to the ONU (Ranging_time message). Here, Teqd is the constant number and the maximum RTD value setting in the PON. For instance, in the case of 20 km maximum RTD, *Teqd* is 200 msec.
9. The ONU memorizes the *Equalization Delay* it receives and delays the subsequent upstream transmission timing by this value.

Through this process, all ranged ONU become equivalent with an RTD of *Teqd* (= original RTD + *Equalization Delay*).

5.1.4 Ranging Protocol of EPON

In the case of EPON standardized by IEEE 802.3, all control messages for PON (MPCPDU) have a *Timestamp* and the method is characterized by the Ranging and clock synchronization using this Timestamp.

Moreover, EPON is different from G-PON in that the Ethernet MAC address is used in place of the serial number, logical link ID (LLID) is used in place of the ONU-ID, and LLID is assigned and RTD is measured in a single phase.

The Ranging procedure in EPON is as follows:

1. The OLT halts the allocation of upstream bandwidth for all ONUs presently communicating to create a vacant time zone (Ranging window). The starting time of this Ranging window is called the *StartTime*.

2. The OLT transmits a Discovery GATE message with *StartTime*. GATE is a message to the ONU which notifies the time range in which upstream data can be transmitted. The GATE used for ranging is called the Discovery GATE.
3. An ONU which has no LLID assigned first set its local time to the value of Timestamp of the Discovery GATE received. After waiting for a random length of time *Twait* from the *StartTime* specified in the Discovery GATE, it transmits a Register Request message. The maximum value of random wait is also specified in the Discovery GATE. The ONU local time (= *StartTime* + random delay) is written in the *Timestamp of Register Request*.
4. For a Register Request which was received normally at the OLT, the OLT calculates the *RTD* = OLT local time − *Timestamp of Register Request* = (*StartTime* + Tdownstreatm + *Twait* +Tupstream) − (*StartTime* + *Twait*) = Tdownstream + Tupstream. Tdownstream and Tupstream represents the downstream and upstream transmission delay respectively. The OLT also allocates a new LLID to the ONU, and memorizes the LLID and RTD associated with the MAC address of that ONU. It then transmits a Register message to that ONU.
5. The ONU which received the Register message, memorizes the LLID and returns a Register Ack message to the OLT to finish Ranging.

Subsequently, the OLT transmits the StartTime corrected by subtracting the RTD from the target's StartTime and transmits it to the ONU every time it allocates the bandwidth by the GATE.

Contrary to the G-PON making corrections using RTD on the ONU side, EPON makes corrections on the OLT side.

5.2 DYNAMIC BANDWIDTH ALLOCATION (DBA)

5.2.1 DBA Overview

In Sect. 5.1, we described the method of preventing overlap of allocated time slots. In this section, we will describe how to allocate the bandwidth.

Fixed bandwidth allocation (FBA) is the method which allocates a deterministic transmission window in a deterministic cycle for individual ONUs. With FBA, while the bandwidth and delay are steady, it is not efficient since bandwidth is consumed even when there is no upstream traffic. In contrast, DBA allocates bandwidth to each ONU according to upstream traffic demand and requirements.

Although it is intended for better efficiency, if the DBA is not appropriately designed to meet the service specifications, it can lead to increased delay or even deteriorated bandwidth usage efficiency.

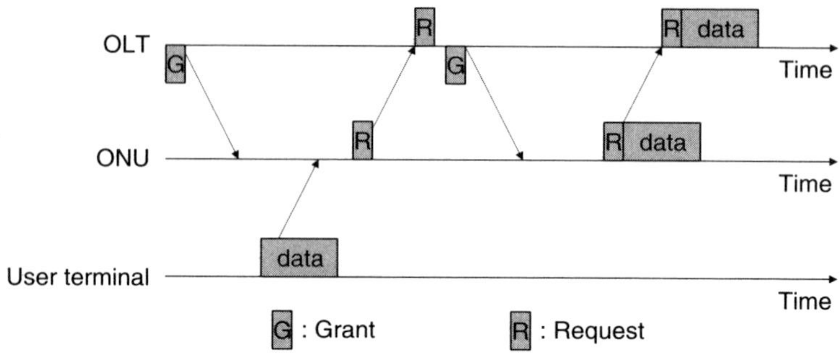

Figure 5.5 Dynamic bandwidth allocation.

Let us first give a simple explanation on how DBA operates (Fig. 5.5).

1. The ONU stores the upstream traffic received from the user in the buffer.
2. Next, the data volume stored in the buffer is notified to the OLT as a request at a time prescribed by the OLT.
3. The OLT specifies the transmission start time and available duration (=transmission window) to ONU as a grant, taking into account the notified volume and service specifications.
4. The ONU waits for the granted time and then transmits the OLT-specified data volume.

In point 3, the service specifications mentioned here refer to guaranteed minimum bandwidth, maximum bandwidth, maximum delay time, etc. In this way, the DBA links directly to the service specifications. To put it in another way, a completely different design is required if the services are different. Here, we hypothesize the target service and show examples of DBA designs based on the hypothesis.

5.2.2 Target Service

The system must satisfy the functions and performance required for providing the target service and must also be designed at a minimum cost. This is because if the system is for general users and the number of users exceeds 10,000,000, the difference of several dollars per user will show up as a difference of tens of millions of dollars of investments. DBA depends largely on service specifications and if the services change, the DBA has no choice but to change drastically. Therefore, the DBA must be designed with clear definitions of the target services.

Today, the greatest demand for FTTH is the Internet access service. The general characteristics of the Internet access service is summarized as follows:

1. Transports variable-length packets of Internet Protocol (IP) between the user and the service provider.
2. The user network interface (UNI) has high affinity with IP and Ethernet (100Base-TX or 1000Base-T) is the most popular in user premises.
3. The bandwidth must be distributed fairly among the users unlike a LAN system which distributes bandwidth on a first-come-first-served basis.
4. Furthermore, recently, there has been a need to provide additional services with assured bandwidth and delay (QoS), such as voice-over IP (VoIP).

Now let us take a deeper look at the characteristics of Internet access traffic.

It is well known that the Internet access traffic is highly bursty. We can understand this more when we think about Web access. A user downloads data from a website, displays it on his/her PC screen and views it. The time that the user views the screen does not shorten even if the speed of Internet access increases and the time of data download shortens. The higher the speed of Internet access links, proportionately the less the percentage of the time to download data, and the more bursty is the traffic. Moreover, even the traffic from P2P (peer-to-peer) type file-sharing applications, which makes up a majority of the Internet traffic nowadays, is bursty, since the data is not constantly transmitted but transmitted only when requested. Meanwhile, streamed contents for moving picture have increased recently; but these are constant-rate traffic of only several Mbps which continue for a few hours and there are not many people who spend all day viewing streamed moving picture.

Specifically, how much data are users transmitting in a day? We can refer to the November 2005 report on the findings about Internet user traffic in Japan [5]. The crucial points quoted in this Japanese article are: "When we define heavy users as someone who creates 2.5 GB of traffic in a day (average 230 kbps), the percentage of heavy users is four percent of the whole, and equivalent to, two percent of ADSL users, ten percent of FTTH users respectively. The traffic used by this four percent heavy users account for approximately seventy five percent of the whole traffic (upstream)."

The upstream traffic of heavy users defined in the article is only 230 kbps on average per day. Even if we gather 32 heavy users, the total traffic only reaches 7.4 Mbps. By taking into account the very bursty nature of traffic in reality, let us calculate a day's data transmission time. When we postulate that the upstream throughput of the FTTH service is between 10 Mbps and 100 Mbps, it would be 2.5 Gbytes/10 Mbps –100 Mbps = 2000 –200 [sec]. This means that even a heavy user is only transmitting data of about several minutes to several dozen minutes per day. If the throughput is shared by 32 users including nonheavy users, we can see that there is a considerable proportion of time within which a single person can occupy the full upstream bandwidth.

What role does DBA play for such highly bursty traffic? DBA improves the experience of access by shortening the wait time of user data transmission through increasing the peak rate. If the peak rate is ten times, the transmission time of a file can be just one-tenth. A service which can transmit a file in a third of the time (or 1/30) with at least 30 Mbps and if available, 100 Mbps (or 1 Gbps) is easier to use than one which can only support 30 Mbps all the time. On the other hand, it is possible for a carrier to reduce the cost by having the DBA emulate an expensive link aggregation switch which supports QoS, in addition to reducing the expensive optical fibers that a PON achieves.

Since the Internet made always-on connection possible at a low cost by having many users sharing the essentially expensive leased line, it continues to flourish as more and more users are added. With this in mind, PON systems using DBA described here are meant for providing an easy and economical Internet access service with a good peak rate by sharing optical fibers with several dozen users.

5.2.3 Requirements of DBA

This section illustrates the requirements of DBA with an example network configuration which offers a target service in the form of Layer-2 link provider for ISPs as shown in Fig. 5.6. Here, we must be aware that the QoS policy may be

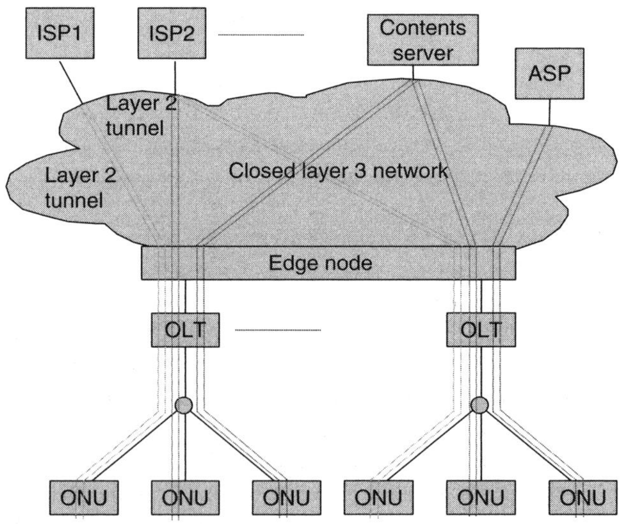

Figure 5.6 An example network configuration which offers Layer-2 link services to ISPs.

different for each ISP. In addition, if the number of Layer-2 links in traffic control increases, it would carry a cost, and the efficiency of the bandwidth would decrease since the burst overhead described in Sect. 5.1.2 would increase. When we consider these factors, it would appear that dividing the traffic control into two layers would be appropriate. This is premised on making it possible to share the bandwidth fairly among the users by inter-ONU scheduling and carrying out the traffic control of class base/flow base independently by each ISP using intra-ONU scheduling. References [6] and [7] discuss intra-ONU schedulings for PON systems with DBA in detail.

If it is possible to share the bandwidth among the users fairly and efficiently by limiting the number of users accommodated on one link, it would be possible to ensure the minimum bandwidth for the users and the delay for traffic less than the guaranteed minimum bandwidth. Thus the fairness, efficiency, and the number of users sharing the bandwidth are factors that determine the minimum guaranteed bandwidth per user. Ensuring minimum bandwidth and limiting the maximum latency are essential for providing services with guaranteed throughput and delay (QoS) such as VoIP. In ITU-T, the conditions for different quality of IP services are standardized and the most stringent end-to-end delay condition is a maximum of 100 msec [9] for real-time, jitter-sensitive, and highly interactive applications such as VoIP. When designing a system, the DBA must be designed based on the established maximum delay time allocated to the PON access portion of the network, which is part of the 100-msec end-to-end delay mentioned before. Moreover, for the TCP protocol, which is widely used in the Internet, the average delay time must be reduced as much as possible to achieve a high throughput [8].

We summarize requirements for a DBA algorithm as the following, based on the use of variable-length Ethernet packets:

A. Fairness: Allocates the bandwidth between the users fairly.
B. Low delay: Can achieve the maximum delay time below the designated delay value and minimize the latency as much as possible.
C. High efficiency: Can increase the efficiency of the bandwidth and increase the peak rate as much as possible.

Since our focus is on the Internet access service which uses the Ethernet as the target interface, from the next section onwards, we will study specific DBA methods based on EPON. In EPON, Ethernet packets are not fragmented to keep it simple and cost-effective. This issue influences the fairness and the efficiency of bandwidth sharing. We will clarify this issue in Sect. 5.2.5, and introduce a solution for it in Sect. 5.2.6. DBA itself can be applied to B-PON and G-PON. However, with B-PON, we must be aware that its speed is low by nature and that the efficiency reduces due to the large overhead when encapsulating the variable length Ethernet packet in fixed length ATM cells [4].

On the other hand, in G-PON, Ethernet packets can be fragmented by using GEM encapsulation [2]. It makes DBA design easier if packet assemblers/disassemblers are implemented.

5.2.4 Traffic Control Fundamentals—Fair Queuing

In this section, we will describe Fair Queuing which is the basic technique of traffic control [10]. Fair Queuing makes fair bandwidth allocation possible. The principle of Fair Queuing is described in Fig. 5.7. The input packets are classified by the classifier and input into separate queues. The classifier separates the packets to groups which are to be allocated the output bandwidth fairly. As an example, when packets are classified based on terminal addresses, the output bandwidth is fairly allocated to each terminal with a different address. In this case, the number of queues and the number of terminals must match. In the case of DBA for inter-ONU control, in terms of the requirements mentioned in Sect. 5.2.3 the input packets are classified by users and distributed to the corresponding queue. Naturally, the number of queues and the number of users must match. On the exit side of the queue, the scheduler reads the queues in a round-robin manner and outputs the packet from each queue so as to achieve fairness. Even if a user transmits a large amount of traffic compared to other users, his/her traffic is directed into a separate queue by the classifier at the input. This makes it possible to design the maximum waiting time for the output. In a nutshell, the maximum delay time in the worst condition happens when maximum-length packets are stored in all the queues.

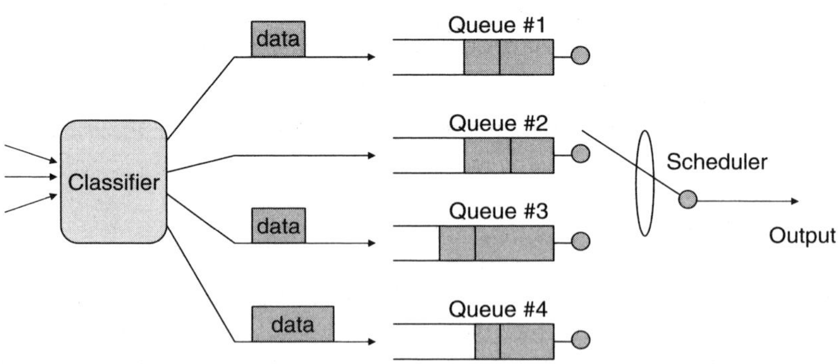

Figure 5.7 The principle of Fair Queuing.

In this exercise, it is the scheduler's operation which holds the key to fairness. The following is the overview of the typical weighted fair queuing (also called general processor sharing) scheduler's operation:

1. Sets up a weight w_i [bit] in each queue where "i" represents the queue number.
2. Executes the round-robin scheduling which goes around each queue sequentially. Each queue is selected to receive w_i bits virtually in one iteration of the round-robin.
3. When a queue's accumulated number of bits to transmit S_i exceeds the queue's head-of-line packet length d_i, i.e. when a full packet has been stored in the queue, the queue empties that packet and reduces its packet length from S_i.
4. Skips the round-robin emptying for queues which have not yet completely stored a packet.

Here is a simple example showing how this scheduler operates. First, we assume that w_1 is 2 and w_2 is 1, and the first queue is receiving 3000-bit packets and the second queue 1000-bit packets. Here, we ignore the Ethernet's preamble and interframe gaps to simplify the explanation.

1. Initially, $S_1 = 2$ and $S_2 = 1$.
2. The virtual round-robin is iterated 1000 times, resulting in $S_1 = 2002$ and $S_2 = 1001$. The first 1000-bit packet is then transmitted from the second queue. This reduces S_2 to 1.
3. The virtual round-robin is iterated 500 more times, resulting in $S_1 = 3002$ and $S_2 = 501$. The first 3000-bit packet is then transmitted from the first queue. This reduces S_1 to 2.
4. The virtual round-robin is iterated 500 more times, resulting in $S_1 = 1002$ and $S_2 = 1001$. The second 1000-bit packet is transmitted from the second queue. This reduces S_2 back to 1.
5. The virtual round-robin is iterated 1000 more times, resulting in $S_1 = 3002$ and $S_2 = 1001$. The second 3000-bit packet is transmitted from the first queue. This reduces S_1 back to 2. In the same iteration, the next 1000-bit packet is transmitted from the second queue and S_2 is reduced to 1.

The operations from (2) to (5) are repeated until the input traffic ceases.

In one repetition from (2) to (5), the output is 6000 bits from the first queue and 3000 bits from the second queue, indicating that the bandwidths are allocated according to the proportion of w_i. As we can clearly see from this example, it is possible to allocate the bandwidth fairly regardless of packet lengths.

In other words, the guaranteed minimum bandwidth for the jth queue can be determined by

$$\text{Output transmission rate} \times \frac{w_j}{\sum w_i}.$$

In addition, we can see from the example that the delay incurred on shorter packets from the second queue is also reduced.

The problem with this WFQ scheduler is the expense in implementing the virtual round-robin algorithm. Because of this, various schedulers have been suggested which can be implemented at a lower cost [10–13]. More details can be found from these references.

Now, when we apply the Fair Queuing technique to the upstream transmission of a PON system, the queues correspond to individual ONUs and the scheduler corresponds to the DBA at the OLT. However, in this case, it would be difficult to install WFQ as DBA without modification due to the following constraints:

1. Only limited information transmitted from the ONU can be identified by the DBA.
2. Furthermore, due to the separation between ONU and OLT, that information is not real time and is slightly delayed.
3. Because of the presence of burst overhead (BOH) mentioned in Sect. 5.1.2, the efficiency deteriorates and the throughput would drop unless several short packets are transmitted in a concatenated fashion.

In Sect. 5.2.5, we will study the DBA which can be implemented at a low cost with these restrictions.

5.2.5 IPACT and its Variants

Interleaved polling with adaptive cycle time (IPACT) is a well-known DBA which can be implemented at a low cost due to its simple processing [14]. With IPACT, the ONU transmits the volume of all packets stored in the upstream receiving buffer as a request to the OLT, and the OLT allocates the transmission window, in the arrival order of the requests. The following five different methods may be used by the OLT to determine the granted window size for each ONU:

1. Fixed service: OLT allocates a certain window size regardless of the amount of ONU requests.
2. Limited service: If the amount of ONU requests exceeds the maximum transmission window (MTW), MTW is allocated and if not, allocate window size in the amount of ONU requests.
3. Constant Credit service: Allocate the window size with a certain amount added to the ONU's requests.
4. Linear Credit service: Allocate the window size with additional window size which is proportionate to the amount of requests to the ONU's requests.

5. Elastic service: Allocate the window size according to the amount of ONU's requests with the restriction that the accumulation of last N grants (including the one being granted) does not exceed $N \times$ MTW. Here, N stands for the number of ONUs. If the accumulation of last $N - 1$ grants and the next ONU's requests exceeds $N \times$ MTW, allocate $N \times$ MTW – (the accumulation of last $N - 1$ grants).

Since (1), (3), and (4) allocate a window size which exceeds the fixed bandwidth or the volume of requests, the efficiency deteriorates if there is no traffic. This would certainly not meet the requirements described in Sect. 5.2.3. Meanwhile, (5) is fine in that it can allocate a maximum window size of $N \times$ MTW if there is only one ONU with input traffic, but when there are several ONUs with input traffic, it allocates unfairly on a first-come-first-served basis. In other words, it does not meet 5.2.3 A. Therefore, in this chapter, we will study the characteristics and specific operations concerning method (2) above.

The limited service DBA (2) operates in the following manner and can be regarded as a DBA method which is derived by thoroughly simplifying the Fair Queuing approach in Sect. 5.2.4:

(i) The OLT memorizes the maximum transmission window size (MTW_i) of each ONU. This MTW_i is equivalent to the weight of WFQ.
(ii) First, the OLT allocates the transmission window in the arrival order of the requests from each ONU.
(iii) Each ONU transmits the volume of the data stored in the upstream receive buffer as a request.
(iv) Let us define Wreq as the window size required for an ONU to send their next bandwidth request to the OLT in a request report. In the order of the received requests and while also considering RTD, the OLT allocates MTW_i if the sum of the ith ONU's requested volume + the window size for the next request report (Wreq) exceeds MTW_i. Otherwise, the OLT allocates the window size for the requested volume + Wreq. If the requested volume is zero, only the window size for the next request report (i.e. Wreq) is allocated.
(v) Each ONU transmits the packets stored in the upstream receiving buffer according to the allocated window size, followed by a request report containing the volume of data remaining in the upstream receiving buffer. The process returns to (iv).

Examples of the operation are shown in Fig. 5.8 [14]. Figure 5.8 (a) shows an example where there is only one ONU with input traffic and Fig. 5.8 (b) shows an example where there is input traffic at all ONUs.

The delay time caused by DBA is the waiting time from request transmission to the start of transmission of the requested data. This is equivalent to the time

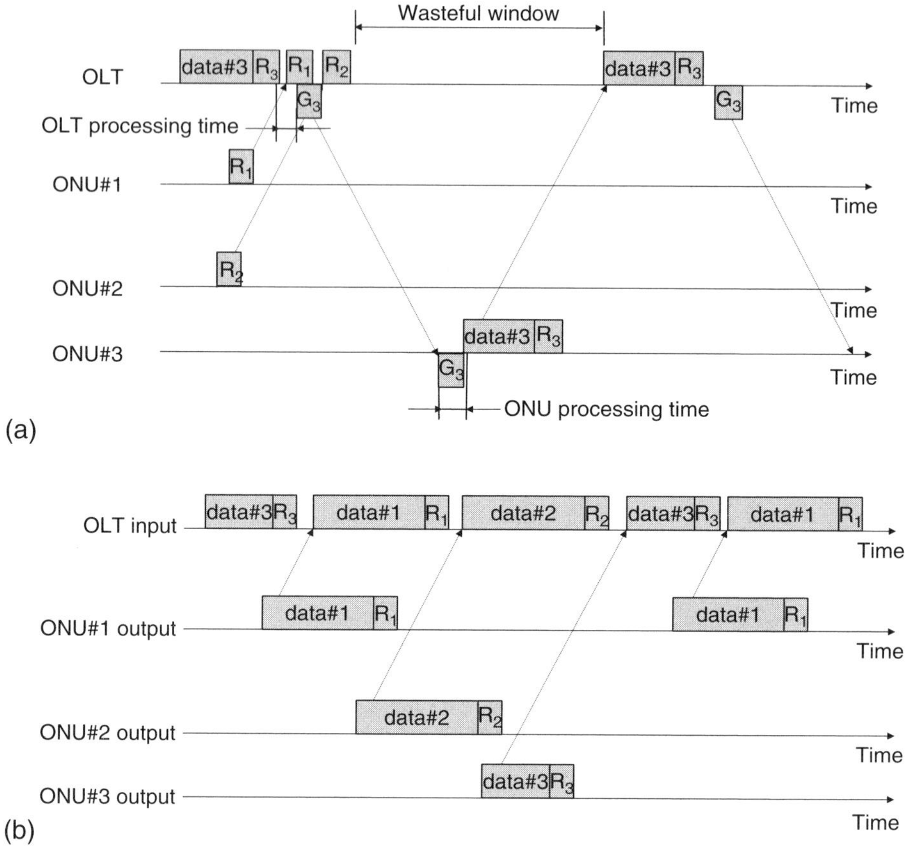

Figure 5.8 Operation of limited service IPACT: (a) when only ONU#3 has upstream traffic; and (b) when all ONUs have upstream traffic. Here, R1, R2, and R3 represent request messages from each ONU respectively; data#3 represents user data transmitted from ONU#3, G3 represents grant for ONU#3. In (a), the values of R1 and R2 are equal to 0 and the value of R3 indicates amount of upstream user data stored in ONU#3.

for the round-robin makes one cycle. From Fig. 5.8, when we set the number of ONUs as N, the maximum duration Rmax it took for the round-robin to make a cycle would be:

$$\mathrm{Rmax} = \sum \mathrm{MTW_i} + N \times (\mathrm{BOH} + \mathrm{Wreq}) \tag{5.1}$$

Moreover, if the only ONU which transmits upstream traffic is the ith ONU, the efficiency Emax (or peak rate) will be given by the following formula:

$$\text{Emax} = (\text{MTW}_i - \text{Wreq})/\{\text{Max}[(N-1)\times(\text{BOH}+\text{Wreq}), \text{transmission rate}\times (\text{RTD}+\text{OLT processing time}+\text{ONU processing time})] + \text{MTW}_i + \text{BOH} \quad (5.2)$$

If MTW_i is set to a small value, the limited service IPACT will meet the requirements of minimizing the delay (Sect. 5.2.3 B). However, we still have to satisfy requirements A and C of Sect. 5.2.3. Due to the fact that Ethernet packets cannot be fragmented in EPON, a small MTW_i setting has the following problems regarding these requirements.

Requirement 5.2.3 A: The fairness diminishes when the packet sizes used between the users are different. For instance, if MTW_i is 4000 bytes, an unfair allocation with a difference in bandwidth of up to 25% (=(4000−3000) × 100/4000) will occur between a user who transmits 1500-byte packets and a user who transmits 500-byte packets.

Requirement 5.2.3 C: In a variable-size packet scenario as EPON, if the volume of data packets stored in the upstream receiving buffer is large and a fixed MTW_i is allocated, a wasteful gap occurs in which data cannot be transmitted. In the worst case, the maximum Ethernet frame length (Fmax) minus one byte will be wasted. If the MTW_i is small, this waste ratio $(\text{Fmax} - 1)/\text{MTW}_i$ becomes comparatively large.

Furthermore, if MTW_i is far smaller than RTD and there is only one ONU with input traffic, the wasteful window which cannot be allocated for data traffic becomes relatively large and the peak rate deteriorates further (Fig. 5.8 (a)).

5.2.6 Improved DBA

5.2.6.1 Deficit Round-Robin Scheduling

Here, we will study a DBA which uses deficit round-robin (DRR) scheduling [11] as a method to improve problems A and C in Sect. 5.2.5. DRR is a method which simplifies WFQ and is also very compatible with burst transmission. In the DBA algorithm described in [15], a circular buffer is used as the ONU's upstream receiving buffer to implement a process equivalent to a deficit counter. The DBA method with DRR applied is reviewed in the following:

1. The ith ONU memorizes the average burst length w_i which can be transmitted with each round-robin. The average burst length is equivalent to the weight of WFQ. Each ONU has one deficit counter (S_i) whose initial value is set to the average burst length w_i.
2. The ONU requests the maximum data volume less than the value of its deficit counter and can be transmitted without dividing any packet in its upstream

receive buffer. It then transmits the packets from the transmit buffer. If the requested volume is the entire volume of data stored in the receive buffer, it resets the value of the deficit counter to the average burst length. If packets other than those requested remain, (w_i – requested volume) is added to the deficit counter.

3. Round-robin scheduling: While considering RTD, the OLT allocates the transmission windows in the arrival order of the requests it receives from ONUs. The allocated window size includes the window size required to transmit the next request (Wreq). If the request is zero, a window size only for the next request (Wreq) is allocated.
4. At the ONU, packets for the window size allocated are moved to the transmit buffer and the process returns to (2).

In step (2) above, the difference between the preassigned average burst length and the present request volume is saved so that slightly more requests can be made by carrying it forward to the next iteration of the algorithm. In this way, a fair allocation which absorbs the differences in packet lengths is achieved.

Let us show a simple example of this operation. First, we set the average burst length for the first ONU as $w_1 = 4000$ bytes and that for the second ONU as $w_2 = 2000$ bytes. We assume continuous input of 1500-byte packets and 500-byte of packets to the first and second ONU respectively, and that a data volume which always exceeds the average burst length is stored. Here, we will ignore Ethernet preambles and interframe gaps to simplify the explanation.

1. The initial condition is set as $S_1 = 4000$, $S_2 = 2000$.
2. The first ONU transmits a 3000-byte request, which is enough to fit two 1500-byte packets in the receive buffer. S_1 is updated as $S_1 = 4000 + (4000 - 3000) = 5000$.
3. The second ONU transmits a 2000-byte request, enough for four 500-byte packets. S_2 is updated as $S_2 = 2000 + (2000 - 2000) = 2000$.
4. The OLT allocates the window size of (Request data volume + Wreq) to each ONU.
5. The first ONU transmits a 4500-byte request and 3000 byte of data, and update S_1 as $S_1 = 5000 + (4000 - 4500) = 4500$.
6. The second ONU transmits a 2000-byte request and 2000 bytes of data, and update S_2 as $S_2 = 2000 + (2000 - 2000) = 2000$.
7. The OLT allocates the window size of (Request data volume + Wreq) to each ONU.
8. The first ONU transmits a 4500-byte request (enough for three 1500-byte packets) and 4500 bytes of data. S_1 is updated to $S_1 = 4500 + (4000 - 4500) = 4000$.

9. The second ONU transmits a 2000-byte request (enough for four 500-byte packets) and 2000 bytes of data. S_2 is updated to $S_2 = 2000 + (2000 - 2000) = 2000$.
10. The OLT allocates the window size of (Request data volume + Wreq) to each ONU.
11. The first ONU transmits a 3000-byte request and 4500 bytes of data. S_1 is updated to $S_1 = 4000 + (4000 - 3000) = 5000$.
12. The second ONU transmits a 2000-byte request and 2000 bytes of data. S_2 is updated to $S_2 = 2000 + (2000 - 2000) = 2000$.

After this, the process returns to (3) and is repeated.

In the above sequence from (3) to (8), the round-robin makes three cycles, with the first ONU transmitting 12,000 bytes and the second ONU 6000 bytes of data. The average volume of data transmitted per cycle is 4000 bytes from the first ONU and 2000 bytes from the second ONU, which are consistent with the preassigned average burst length. From this example, we can see that the bandwidth is fairly allocated in accordance with the average burst length regardless of the packet length.

In step (2) above, the deficit counter is reset to the average burst length when there is little data stored. Without this reset function, the behavior of the deficit counter will be abnormal if there is little input traffic and the stored data volume of the receive buffer is small. Due to a continued condition of little requested volume, the value of the deficit counter would increase by (average burst length – requested volume) toward infinity in every cycle. The deficit counter was originally introduced to carry over the shortfall of allocation caused by the difference in packet lengths to the next allocation and it is not for carrying over the shortfall of input traffic. Therefore, the deficit counter is reset to the average burst length w_i when there is little data stored. As a result, a carryover addition occurs when the amount of data stored in the receive buffer is greater than the value of the Deficit Counter. In this case, the condition:

Deficit counter value – Requested volume < Ethernet's maximum frame length (Fmax) is satisfied. When we alter the formula for the carryover addition process and substitute this condition, the result will be:

$$\begin{aligned}
&\text{Next deficit counter value} \\
&\quad = \text{Present deficit counter value} + (w_i - \text{requested volume}) \\
&\quad = w_i + (\text{present deficit counter value} - \text{request volume}) \\
&\quad < w_i + \text{Fmax}
\end{aligned} \tag{5.3}$$

Therefore, the maximum value of the deficit counter will be $w_i + \text{Fmax} - 1$.

When we set the number of ONUs as N, the maximum duration Rmax spent by one cycle of the round-robin, i.e. the maximum delay from requesting the transmission to the start of transmitting the requested data will be:

$$\text{Rmax} = \sum w_i + N \times (\text{Fmax} - 1 + \text{BOH} + \text{Wreq}) \quad (5.4)$$

Moreover, the efficiency Emax (or peak rate) if only the ith ONU is transmitting upstream traffic is given by the following formula:

$$\begin{aligned}\text{Emax} = w_i/\text{Max}\{&N \times (\text{BOH} + \text{Wreq}) + w_i,\\ &\text{RTD} \times \text{transmission rate} + \text{OLT processing time}\\ &+ \text{ONU processing time}\}\end{aligned} \quad (5.5)$$

The difference between the DRR approach and the IPACT described in the previous section is that here the request is sent in front of the burst. In this way, when there is only one ONU transmitting upstream traffic, a waste in bandwidth can be eliminated by setting w_i long enough to improve the efficiency (Fig. 5.9). This difference is clear when we make a comparison with IPACT shown in Fig. 5.8 (a).

Nonetheless, when the average burst length w_i is lengthened, the maximum delay time is increased accordingly, so the maximum delay time of requirement 5.2.3 (B) and the efficiency requirement 5.2.3 (C) are in a trade-off relationship. In Sect. 5.2.6.2, we will describe a method which alleviates this trade-off.

5.2.6.2 DBA using Multiple Queue Report Set

In order to alleviate the trade-off between the maximum delay time and efficiency mentioned in the previous section, a method which transmits multiple requests with different thresholds from an ONU at the same time was proposed

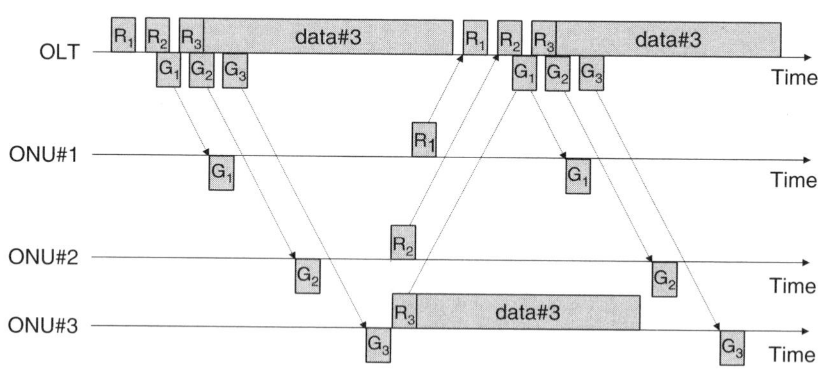

Figure 5.9 Deficit round-robin (DRR) scheduling.

in [16]. The ONU transmits two request volumes, #1 which may be longer than the RTD and #2 which is shorter than the RTD. The OLT selects and allocates the appropriate value. In this way, a transmission window which is longer than the RTD can be allocated if communication is made with only one ONU while maintaining the round-robin cycle short (Fig. 5.10).

The scheduler proposed in [16] has the property that even with significant differences created in weights w_i's among the ONUs, it is of low-delay and maintains high-precision bandwidth control. However, this scheduler is quite complex. In this chapter, we will offer a simpler explanation of how it meets the requirements in Sect. 5.2.3 using several request examples.

First, as shown in Fig. 5.11, we postulate that two requests, request volume #1 which contains the deficit counter and request volume #2 which contains a small threshold, are transmitted together. By nature, in an EPON conforming to IEEE

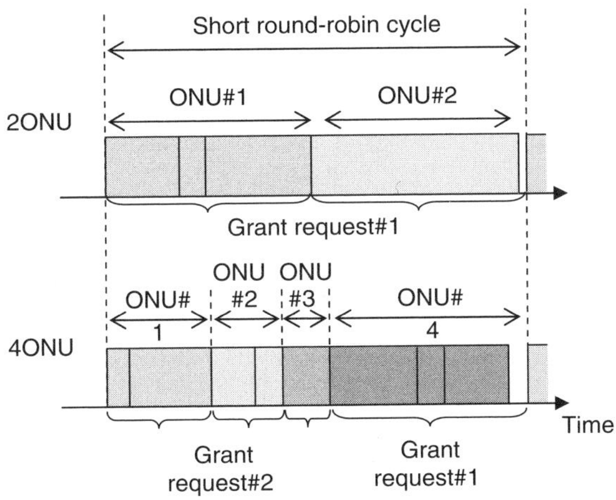

Figure 5.10 Maintaining short round-robin cycle using multiple queue request.

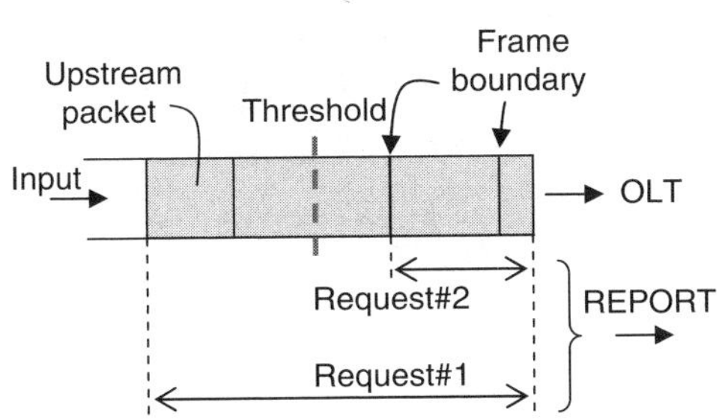

Figure 5.11 An example with two types of requests.

802.3ah, one REPORT frame can contain several requests. For instance, as shown in Fig. 5.12, an ONU can transmit request volume #1 in the 1st queue set and request volume #2 in the 2nd queue set.

Since the algorithm described in Sect. 5.2.6.1 requires that the requested volume be always allocated in the next allocation, the reset condition for the deficit counter is quite simple. However, when several requests are transmitted, it cannot be determined if it is fine to reset the deficit counter until an allocation is actually made. For this reason, the reset condition of the deficit counter is slightly more complex than the algorithm described in Sect. 5.2.6.1. The following explanation illustrates the specific operation:

1. The maximum allowed transmission burst length w_i for one allocation is set for the ith ONU.[1] The maximum burst length w_i must be longer than the RTD. The maximum burst length is equivalent to the weight of WFQ.
2. Each ONU has one deficit counter for determining request #1 and its initial value is set as the maximum burst length w_i. Each ONU also has a threshold

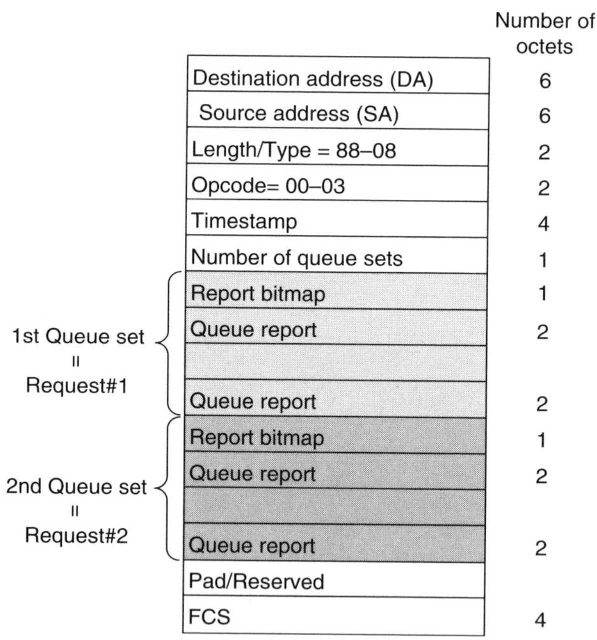

Figure 5.12 An IEEE802.3ah REPORT frame with multiple requests.

[1] As explained previously, the maximum length of burst that can actually be transmitted is (w_i + Fmax – 1) by the function of deficit counter.

besides the deficit counter to determine request #2. The threshold is set at the maximum packet length used in services which require low-delay or at a small value about Ethernet's maximum frame length (Fmax).

3. Furthermore, each ONU has a *reset_flag* as the flag for memorizing the reset conditions of the deficit counter. The initial value of the *reset_flag* is False.
4. Each ONU subtracts the whole volume of data packets to be transmitted in the transmission window allocated in this cycle from the deficit counter. If the allocated window size in this cycle is above "previous request #1 + Wreq" and the previous request #1 differs from previous request #2, the *i*th ONU adds w_i to the deficit counter if the *reset_flag* is True; and it resets the deficit counter value to w_i if *reset_flag* is False.
5. Each ONU transmits the maximum volume of data below the deficit counter value in request #1, without dividing any packet, and the maximum volume data below the threshold in request #2 without dividing any packet in the allocated windows. Then, it transmits the packets stored in the transmit buffer.
6. If the value of request #1 is the whole volume of the stored data packets, the *reset_flag* is set to False. If there are packets left from the volume of request #1, the *reset_flag* is set to True.
7. New round-robin scheduling: While considering RTD, the OLT allocates the transmission window with length "request #2 + Wreq" in the arrival order of the requests from ONUs. However, this allocation is not executed on the first ONU with request #1 which is larger than request #2 in one cycle. It waits until the cycle is completed by the round-robin to allocate the transmission window with length "request #1 + Wreq." In other words, instead of skipping the transmission window allocation to the end of its round, the OLT allocates a window long enough for request #1 (Fig. 5.13) at the end of the cycle. The ONU which is allocated a request #1 is also shifted to the end of the ONU list in the next cycle of the round-robin.
8. At the ONU, packets for the allocated transmission window are moved to the transmit buffer and the process returns to (3).

Let us present a simple example of this operation. First, we assume that the threshold of ONU#1 is $Th_1 = 1500$ bytes and its maximum burst length is

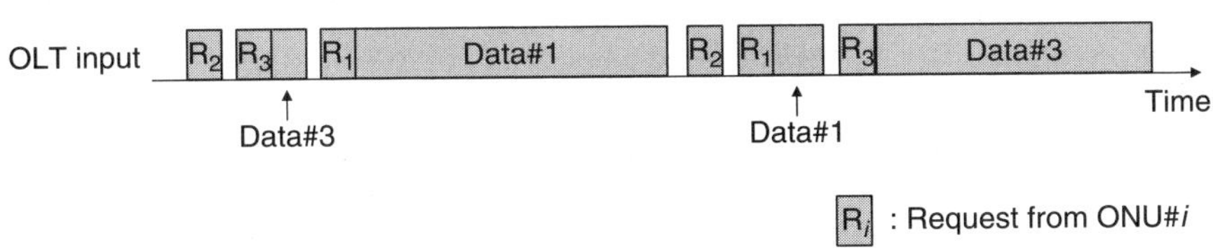

Figure 5.13 Burst allocation example of DRR using multiple queue report set.

$w_1 = 30{,}000$ bytes; the threshold of ONU#2 is $\mathrm{Th}_2 = 500$ bytes and its maximum burst length is $w_2 = 20{,}000$ bytes. 1500-byte and the 300-byte packets are continuously input to the first ONU and second ONU respectively, with the stored data volume always exceeding the maximum burst length. Again, we will ignore Ethernet's preamble and interframe gaps to simplify the explanation.

1. Initial condition: $S_1 = 30{,}000$, reset_flag 1 = False, $S_2 = 20{,}000$, reset_flag 2 = False.
2. ONU#1 transmits 30,000-byte request #1 and 1500-byte request #2. ONU#1 updates S_1 and reset_flag 1 to $S_1 = 30{,}000$, reset_flag 1 = True.
3. ONU#2 transmits 19,800-byte request #1 and 300-byte of request #2. ONU#2 updates S_2 and reset_flag 2 to $S_2 = 20,000$, reset_flag 2 = True.
4. The OLT skips the allocation for ONU#1. ONU#2 is allocated a transmission window with "300 bytes+ Wreq."
5. The skipped ONU#1 is allocated a transmission window with "30,000 bytes + Wreq."
6. ONU#2 transmits 19,500-byte request #1, 300-byte request #2, and 300 bytes of data.
7. ONU#2 updates S_2 and reset_flag 2 to $S_2 = 19{,}700$, reset_flag 2 = True.
8. ONU#1 transmits 30,000-byte request #1, 1500-byte of request #2, and 30,000 bytes of data.
9. ONU#1 updates S_1 and reset_flag 1 to $S_1 = 30{,}000$, reset_flag 1=True.
10. The OLT skips the allocation for ONU#2. ONU#1 is allocated a transmission window with (1500 bytes +Wreq).
11. The skipped ONU#2 is allocated a transmission window with "19,500 bytes + Wreq."
12. ONU#1 transmits 28,500-byte request #1, 1500-byte request #2, and 1500 bytes of data. ONU#1 updates S_1 and reset_flag 1 to $S_1 = 28{,}500$, reset_flag 1 = True.
13. ONU#2 transmits 20,100-byte of request #1, 300-byte of request #2, and 19,500 bytes of data. ONU#2 updates S_2 and reset_flag 2 to $S_2 = 20,200$, reset_flag 2=True.
14. The OLT skips the allocation for ONU#1. ONU#2 is allocated a transmission window with (300 bytes + Wreq).
15. The skipped ONU#1 is allocated a transmission window with (28,500 bytes + Wreq).
16. ONU#2 transmits 19,800-byte request #1, 300-byte request #2, and 300 bytes of data. ONU#2 updates S_2 and reset_flag 2 to $S_2 = 19{,}900$, reset_flag 2 = True.
17. ONU#1 transmits 30,000-byte request #1, 1500-byte request #2, and 28,500 bytes of data. ONU#1 updates S_1 and reset_flag 1 to $S_1 = 30{,}000$, reset_flag 1 = True.

18. The OLT skips the allocation for ONU#2. ONU#1 is allocated a transmission window with (1500 bytes + Wreq).
19. The skipped ONU#2 is allocated a transmission window with "19,800 bytes + Wreq."
20. ONU#1 transmits 28,500-byte request #1, 1500-byte request #2, and 1500 bytes of data. ONU#1 updates S_1 and reset_flag 1 to $S_1 = 28{,}500$, reset_flag 1 = True.
21. ONU#2 transmits 20,100-byte request #1, 300-byte request #2, and 19,800 bytes of data. ONU#2 updates S_2 and reset_flag 2 to $S_2 = 20{,}100$, reset_flag 2 = True.
22. The OLT skips the allocation for ONU#1. ONU#2 is allocated a transmission window with (300 bytes + Wreq).
23. The skipped ONU#1 is allocated a transmission window with (28,500 bytes + Wreq).
24. ONU#2 transmits 19,800-byte request #1, 300-byte request #2, and 300 bytes of data. ONU#2 updates S_2 and reset_flag 2 to $S_2 = 19{,}800$, reset_flag 2 = True.
25. ONU#1 transmits 30,000-byte request #1, 1500-byte request #2, and 28,500 bytes of data. ONU#1 updates S_1 and reset_flag 1 to $S_1 = 30{,}000$, reset_flag 1=True.
26. The OLT skips the allocation for ONU#2. ONU#1 is allocated a transmission window with (1500 bytes + Wreq).
27. The skipped ONU#2 is allocated a transmission window with (19,800 bytes + Wreq).
28. ONU#1 transmits 28,500-byte request #1, 1500-byte request #2, and 1500 bytes of data. ONU#1 updates S_1 and reset_flag 1 to $S_1 = 28{,}500$, reset_flag 1=True.
29. ONU#2 transmits 19,800-byte request #1, 300-byte request #2, and 19,800 bytes of data. ONU#2 updates S_2 and reset_flag 2 to $S_2 = 20{,}000$, reset_-flag2 = True.
30. The OLT skips the allocation for ONU#1. ONU#2 is allocated a transmission window with (300 bytes + Wreq).
31. The skipped ONU#1 is allocated a transmission window with (28,500 bytes + Wreq).
32. ONU#2 transmits 19,500-byte request #1, 300-byte request #2, and 300 bytes of data. ONU#2 updates S_2 and reset_flag 2 to $S_2 = 19{,}700$, reset_flag 2 = True.
33. ONU#1 transmits 30,000-byte request #1, 1500-byte request #2, and 28,500 bytes of data. ONU#1 updates S_1 and reset_flag 1 to $S_1 = 30{,}000$, reset_flag 1 = True.

The process returns to (5).

In one sequence from (5) and (16), the round-robin makes six cycles. ONU#1 transmitted $3 \times (1500 + 28{,}500) = 90{,}000$ bytes of data, and ONU#2 transmitted

19,500 + 300 + 19,800 + 300 + 19,800 + 300 = 60,000 bytes of data. On average, the amount of data transmitted in two cycles of round-robin is 30,000 bytes from the first ONU and 20,000 bytes from the second ONU, which are consistent with the maximum burst lengths selected. From this example, we can see that regardless of the packet lengths, the bandwidth is fairly allocated according to the selected maximum burst lengths.

In the OLT process described in step (4) above, a long window size equivalent to the value of the deficit counter is allocated to only one ONU per round-robin cycle. In this way, even when there is only one ONU in transmission, there is no unused gap and the peak rate can be increased by boosting the efficiency. By skipping the allocation of a long window size to the end of the round-robin cycle, the ONU to receive the next long window size is allocated at last. This makes it possible for the order and frequency of allocating a long window size fair among the ONUs. The skipping increases the delay time slightly but by keeping the value of each threshold at the requisite minimum, the increase in delay time can be kept to minimum as well.

Besides, if there are many ONUs with upstream traffic, the number of times request #2 is allocated will increase and the value of each deficit counter is decreased accordingly by the process in step (3) above. Request #1 limited by the value of each deficit counter also decreases when request #1 is allocated. For above examples 1–16, if 16 ONUs have upstream traffic, request #2 is allocated 15 times of 16 rounds and the value of the deficit counter S_1 becomes $30{,}000 - 15 \times 1500 = 7500$ when request #1 is allocated. In this way, the average time of a cycle made by the round-robin is kept short.

The maximum burst length w_i is designed to satisfy the following conditions:

$$w_i > \mathrm{Max}\{\mathrm{RTD} \times \text{transmission rate} + \text{OLT processing time} + \text{ONU processing time}, N \times \text{threshold}\}$$

In this case, when the ONU number with the longest maximum burst length w_i is j, the maximum length Rmax of a round-robin cycle, i.e. the maximum delay from request transmission to the start of transmitting the requested data is given by the following formula:

$$\mathrm{Rmax} = N \times (\mathrm{BOH} + \mathrm{Wreq}) + w_j + (\mathrm{Fmax} - 1) + \sum_{i \neq j} \mathrm{Th}_i \tag{5.6}$$

Here, N is the number of ONUs, and Th_i is the threshold of the ith ONU.

In addition, the efficiency Emax (or peak rate), when there is only one ONU transmitting upstream traffic, is given by the following formula:

$$\mathrm{Emax} = \frac{w_i}{N \times (\mathrm{BOH} + \mathrm{Wreq}) + w_i} \tag{5.7}$$

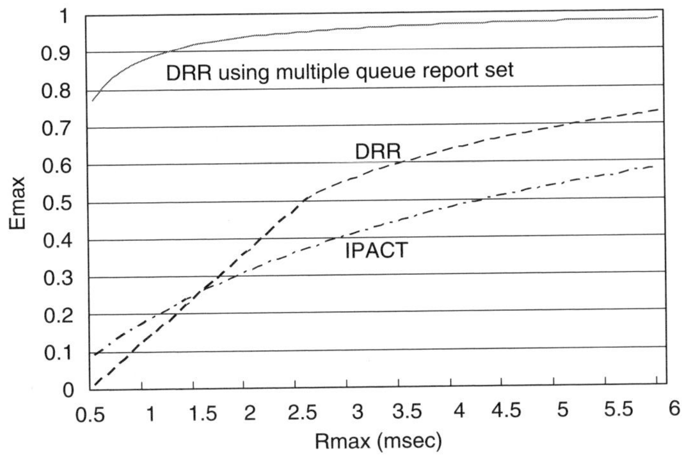

Figure 5.14 Comparison of bandwidth efficiency (Emax) among the three DBAs covered in this chapter. Here we have chosen the number of ONU $N = 32$, Burst overhead = 1.4 μsec, Max RTD = 100 μsec, $w_i = w$, $\mathrm{Th}_i = w/N$, OLT processing time = 16 μsec, and ONU processing time = 16 μsec.

Figure 5.14 shows the relationship between Rmax and Emax calculated from Eqs. (5.1), (5.2), (5.4), (5.5), (5.6), and (5.7). As seen in Fig. 5.14, DRR using Multiple Queue Report Set can achieve high efficiency together with short round-robin cycle.

In summary, we have given a DBA example which meets the three performance requirements mentioned in Sect. 5.2.3:

A. Fairness: Allocates the bandwidth among the users fairly.
B. Low delay: Can bound the maximum delay time within a designed value.
C. High efficiency: Can increase the efficiency of the bandwidth and increase the peak rate as much as possible.

REFERENCES

[1] ITU-T Rec. G.983.1(01/2005).
[2] ITU-T Rec. G.984.3(02/2004).
[3] IEEE Std. 802.3–2005.
[4] Marek Hajduczenia, "EPON versus APON and G-PON: a detailed performance comparison," in *Journal of Optical Networking*, vol.5, no.4, pp298–319, Apr., 2006.
[5] Available from http://internet.watch.impress.co.jp/cda/news/2006/03/23/11358.html.
[6] Glen Kramer et al., "Supporting differentiated classes of service in Ethernet passive optical networks," *Journal of Optical Networking*, vol.1, no.8/9, pp280–298, Aug., 2002.

[7] Chadi M. Assi et al., "Dynamic bandwidth allocation for quality-of-service over Ethernet PONs," *IEEE Journal on Selected Areas in Communications*, vol.21, no.9, pp1467–1477, Nov., 2003.
[8] IETF RFC1323.
[9] ITU-T Rec. Y.1541.
[10] A. Parekh and R. Gallerger, "A generalized processor sharing approach to flow control in integrated service network: the single node case," IEEE/ACM Transaction on Networking, vol.1, no.3, June, 1993.
[11] M. Shreedhar and G. Varghese, "Efficient fair queuing using deficit round robin," Proc. ACM SIGCOMM'95, pp231–242, Sept., 1995.
[12] S. Golestani, "A self-clocked fair queuing scheme for broadband applications," Proc. INFOCOM'94, pp636–646, 1994.
[13] L. Zhang, "Virtual clock a new traffic control; algorithm for packet switching networks," ACM Trans. Computer systems, vol.9, pp101–124, May, 1991.
[14] Glen Kramer et al., "IPACT: a dynamic protocol for an Ethernet PON (EPON)," *IEEE Communications Magazine*, pp74–80, Feb., 2002.
[15] Noriki Miki et al., "Optical Access Systems Suitable for Computer Communication," GLOBECOM'98, Access Network Mini Conference, 8.4, Nov., 1998.
[16] Osamu Yoshihara et al., "Dynamic Bandwidth Allocation Algorithm for GE-PON," COIN 2002, COIN.MoA1, pp22–24, July, 2002.

Chapter 6

Protection Architectures for Passive Optical Networks

Calvin C. K. Chan

The Chinese University of Hong Kong

6.1 INTRODUCTION

Over the last decade, passive optical networks (PONs) have emerged as an attractive and promising approach to deliver broadband services to a large number of subscribers. In a typical PON, services are originated from the optical line terminal (OLT) at a head end or central office (CO) and carried along an optical-fiber feeder for about 10–15 km, before the optical power is split into multiple output distribution fibers, via an optical power splitter located at the remote node (RN). Each distribution fiber, usually less than 5 km in length, then forwards the services toward the destined optical network unit (ONU), where the optical signal is terminated before being further distributed to all the subscribers attached to this ONU via other media, such as copper wire, etc. With such high fiber penetration into the access arena, the system cost of PONs has to be kept low in order to make them economically viable and competitive. PON features its passive remote node, which greatly relaxes the cost of managing optical elements in outside fiber plants. Besides, both the RN and the ONUs have to be kept simple and low-cost. Traditionally, the downstream services delivered over a PON are mostly distributive videos. Nevertheless, an upstream channel is often provisioned to carry subscribers' requests back to the OLT. Thus this upstream channel is usually of low data rate. Due to such distributive nature of the traffic as well as the low-cost concern of PONs, not much effort has been paid on the issue of network survivability.

With the recent huge popularity of the Internet and the multimedia services it carries, conventional PONs are evolving to carry two-way broadband interactive data signals. Bandwidth demands in both enterprise and residential broadband access networks have been increasing drastically and the services carried on

PONs are becoming more and more data-centric. Several variants of PONs such as B-PON (ITU-T G.983) [1–4], E-PON (IEEE 802.3ah) [5], and G-PON (ITU-T G.984) [6, 7] have been widely deployed recently to cope with the increasing demand for access bandwidths. With the recent availability of low-cost optical components, PONs using the wavelength division multiplexing (WDM) technique have been emerging as the next generation optical access networks. In a WDM-PON, each ONU is served by a dedicated set of wavelength channels to communicate with the OLT. Each individual ONU can enjoy a dedicated bandwidth, which is also scalable according to its own need. Thus, the system capacity and the network flexibility could be greatly enhanced. However, with conventional PON architectures, which have limited protection feature, any component or fiber failure would lead to huge loss of data or even business. Therefore, the issue of network survivability [8, 9] has aroused more attention over the recent years. Subscribers are now requesting high-availability services and connections. Thus, protection measures to enhance the network survivability are highly desirable to provide resilience against failures, for instance, in the cases of possible catastrophic events such as fires or flooding.

Fault management is one of the well-known crucial aspects in network management. Most of the conventional approaches of fault management rely on diagnosis in higher layers [10, 11], based on the status reports collected from various checkpoints on the managed optical network. However, such high-level fault diagnosis would impose excessive overhead in network signaling as well as in the network management system (NMS). Yet there is no guarantee that higher layers can provide recovery from faults in the physical layer. Therefore, in order to facilitate effective and prompt network protection and restoration, it is highly desirable to perform network survivability measures in the optical layer. This can be achieved by simple fiber link or equipment duplication with protection switching or some other intelligent schemes with minimal resource duplication or reservation for protection. For PON applications, equipment failure at either OLT or ONU can be easily remedied by having a backup unit in the controlled environment. However, for any fiber cut, it would take a relatively long time to perform the repair. Therefore, it is highly desirable to have survivable PON architectures with protection switching against any fiber cut.

In this chapter, several considerations in designing survivable network architectures for PONs will be discussed. Besides, several feasible protection schemes and architectures for both the conventional PONs and WDM-PONs will be reviewed.

6.2 CONSIDERATIONS OF PROTECTION IN PASSIVE OPTICAL NETWORKS

The following considerations are useful when designing survivable network architectures or other protection schemes for PONs.

6.2.1 Protection or Restoration

In fault management of optical networks, there are two common approaches to assure the network survivability, namely preplanned protection and dynamic restoration. Preplanned protection aims to incorporate protection lightpath at the network design stage and automatic protection switching can be readily performed to the designated protection lightpath, once any network failure is detected. The traffic restoration time can be kept very short. On the contrary, dynamic restoration would search for the spare lightpaths or network resources for restoring the disrupted traffic only after the occurrence of network failures. Although the amount of spare resources can be preallocated, the traffic restoration time may be long as it depends on the dynamic rerouting of the disrupted traffic. For most of the PON applications, as the network topology is quite regular, such as tree or ring, preplanned protection schemes are usually adopted to enhance the network survivability with short traffic restoration time. Besides, protections on the optical layer should have quick restoration time in order to minimize the disturbance to the upper layers, which usually rely on some time-out features for fault detection. Upper-layer restoration often involves routing table topology recalculation and slow convergence time. Improper handing of restoration time among different layers can cause network instability.

6.2.2 Network Topology

There are two common network topologies for PONs, namely tree and ring. In a tree topology, the optical signal sent from the OLT is split at the RN and delivered to the designated ONUs via the respective distribution fibers. For ring topologies, the OLT is connected to multiple access nodes (ANs) via single or double fiber rings. Each AN comprises an optical add-drop multiplexer (OADM) or simply an optical power splitter, to which multiple ONUs are further connected, either in a star or ring topology. In addition to ring and tree topologies, there are some other topologies such as bus and star-ring. When the protection network architecture is designed, the network topology definitely determines the paths or connections between the OLT and the ONUs and thus influences how the protection lightpath or fiber should be duplicated or incorporated, when appropriate. For instance, in a PON with tree topology, any fiber cut in the distribution fiber will make the affected ONU unreachable from the OLT. However, in a PON with ring topology, any fiber cut in the fiber ring will isolate all of the downstream ANs beyond the cut position along that ring, if no additional protection ring is incorporated.

6.2.3 Network Type

Conventional PONs usually support one (for video) or two (for data and video) downstream optical carriers, as well as one upstream optical carrier. These optical carriers are time-shared among all ONUs. In order to enable the delivery of higher capacity services to the subscribers and enhance the flexibility for network upgrade, PONs using WDM techniques, namely WDM-PONs, have recently emerged and have aroused much research interests. In WDM-PONs, each ONU is served by a set of dedicated and distinct wavelength channels to communicate with the OLT. The ranging problem in conventional time-shared PONs is also alleviated, as all upstream wavelengths are multiplexed at the RN without any signal collision. Each individual ONU enjoys dedicated bandwidth, which is readily scalable according to its own needs. Protection in a PON is usually achieved by lightpath diversity with fiber link duplication and switching. As WDM-PONs offer one more dimension for the optical channels, lightpath diversity can be much more flexibly realized by adopting wavelength routing on existing network architectures. Hence, excessive fiber-link duplications could be avoided.

6.2.4 Resources To Be Protected

In general, there are two major kinds of network resources which have to be protected in PONs, namely fiber links and active component or equipment. Both of them affect the availability of the wavelength channels running on the network. As fiber links are laid outside plant, they are more vulnerable to environmental conditions and it takes much more time to repair the fiber, whenever a fiber cut occurs. Moreover, any fiber cut may stop all wavelength channels traversing over it. Thus, alternate lightpath or redundant protection fibers in either $1 + 1$ or 1:1 protection scheme are usually preplanned to achieve lightpath diversity. For active component or equipment, such as the optical transceivers and optical switches residing at OLT or ONUs, they are easily protected by colocated backup units in $1{:}N$ or even $1 + 1$ protection scheme. In addition to fiber links and active components, the passive wavelength multiplexer at the RN in WDM-PONs may also need to be protected. If the spectral property of the WDM multiplexer is temperature-sensitive, such as conventional array waveguide grating (AWG), changes in the environmental temperature at the RN may possibly induce misalignment between the optical carriers and the transmission passband of the WDM multiplexer. This may induce excessive filtering and power loss to the optical carriers. Therefore, athermal WDM multiplexer, such as athermal AWG or thin-film-based WDM multiplexers, should be employed at the RN.

6.2.5 Single or Multiple Failures

Most of the existing efforts in network protection focus on the scenarios of a single failure at a time. In most cases, the occurrence of failure in a fiber link or a piece of network equipment is statistically independent in a network. Moreover, the mean time between failures is generally much longer than the mean time to repair a single failure. However, a single failure, such as a fiber cut, may sometimes lead to failure in multiple logical connections. Recently, there are also some research efforts on studying network protection under multiple failures.

6.2.6 Automatic Protection Switching

Automatic protection switching (APS) can be realized by either centralized or distributed control. In centralized control, all protection switching are performed at the OLT, after the fault alarms are collected. The ONUs still stay connected with the OLT after the APS. On the contrary, protection switching can be performed at individual ONUs instead, to realize distributed control. In this case, protection switches are incorporated in the individual ONUs, which continuously monitor the status of their attached fiber links or components. APS will be triggered only at the affected ONU when any fault is detected. In case of any failure, the OLT does not need to perform any remedy and is transparent to such APS. However, this approach increases ONU complexity and costs.

6.2.7 Operation, Administration, and Management

When incorporating APS into a network, fault monitoring units have to be installed at strategic checkpoints to gather network status information. A monitoring unit can be as simple as mere optical power level monitoring to identify the possible loss of signal at a low detected optical power level. Some other novel techniques could also be employed to detect other parameters such as the presence of a particular wavelength, and the identification of a specific faulty fiber branch in a PON with tree topology [12]. The collected monitoring information has to be delivered to APS units for appropriate remedies. In some cases, a signaling channel may be needed to carry the monitoring information.

6.2.8 Traffic Restoration Time

Once a network failure is detected, the time period required to perform protection switching and restore the affected traffic is known as the traffic restoration time. This time period should be kept small, say less than a few tens of milliseconds [8], to reduce the amount of data loss during service disruptions. In most cases, such traffic restoration time greatly depends on the intrinsic response of the optoelectronic detection and the optical switching devices used in the APS, as well as the possible induced additional latency of the protection lightpath. Sometimes, higher-layer protocols are employed to retransmit the lost data.

6.2.9 Complexity

The design of survivable PON architectures should require the least amount of additional fiber link or equipment duplication to keep the complexity and cost low. For example, a backup transceiver may be employed at the OLT for 1:N protection of all other working transceivers, instead of 1 + 1 protection. This backup transceiver can be of fixed wavelength or tunable wavelength, depending on the network type and requirement. In the case of WDM-PONs, alternate lightpath routing of wavelength channels may be adopted to bypass the failed fiber links and minimize the required additional fiber links for protection.

6.3 PROTECTION ARCHITECTURES

In this section, several survivable network architectures for both conventional PONs as well as WDM-PONs, in both tree and ring topologies, will be reviewed. As discussed in the previous sections, fiber link failure is the most frequent and critical in PONs, as it may interrupt many wavelength channels in a network. Therefore, the following architectures mainly focus on the protection of fiber links in PONs.

6.3.1 Conventional Passive Optical Networks

6.3.1.1 Tree Topology

Most of the conventional PONs, such as A-PON (ITU-T G.983.1) [1, 2], B-PON (ITU-T G.983.3) [1–4], EPON (IEEE 802.3ah) [5], and G-PON (ITU-T G.984) [6, 7], adopt a tree topology to provide point-to-multipoint connections. An optical power splitter is employed to split the received optical signal at the RN to all outgoing distribution fibers. Figure 6.1 shows the four protection architectures with different levels of protection, as suggested by ITU-T G..983.1 [1]. Basically, they are duplicating the fiber links and/or the components for protection.

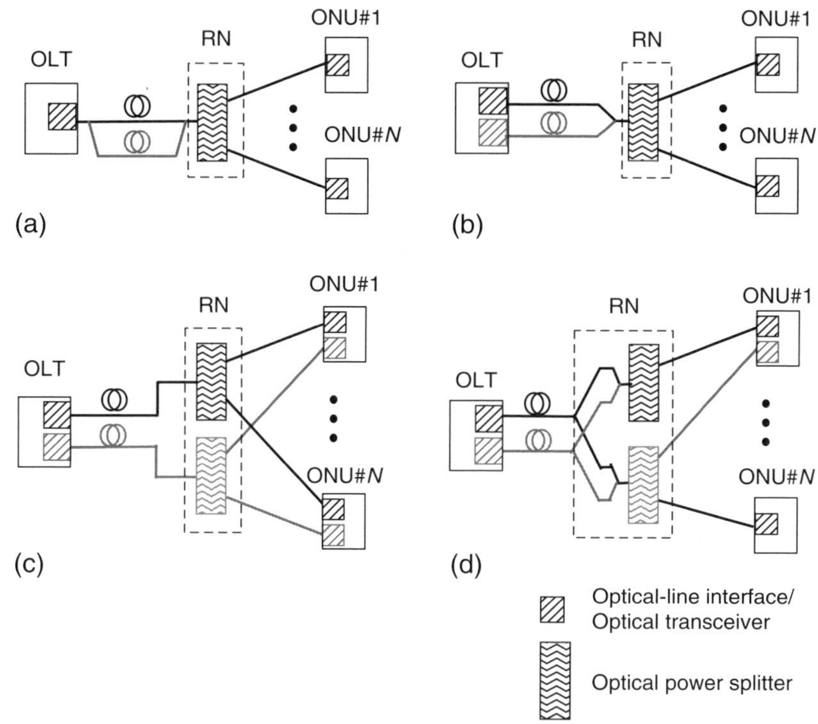

Figure 6.1 Protection switching architectures suggested by ITU-T G.983.1.

Figure 6.1 (a) duplicates the feeder fiber between the OLT and the RN only. Only the feeder fiber is protected. Figure 6.1 (b) doubles the optical transceivers at the OLT and also duplicates the feeder fiber between the OLT and the RN. Protection switching is done by switching the data to the backup optical transceiver at the OLT. This can reduce the cost of duplicating the optical transceivers at the ONUs. Figure 6.1 (c) doubles the optical transceivers not only at the OLT but also at the ONUs. In addition, the optical power splitter at the RN is also duplicated. Hot standby circuits are placed at the OLT and the ONUs, enabling hitless switching by switching to the backup facilities in the event of a failure of fiber or equipment. Figure 6.1 (d) is similar to Fig. 6.1 (c) except that an additional optical power splitter circuit is incorporated at the RN to cope with the case that not all ONUs have duplicated optical transceivers, due to some system constraints.

In addition to these four suggested protection switching architectures, several 1:N protection schemes for OLT [13–16] were also proposed, where one backup OLT interface was employed to protect all or a group of the working OLT interfaces. Figure 6.2 shows an example [14] of using an optical-splitting circuit at the RN to enable the backup OLT interface to perform the 1:N protection. This could alternatively be done by employing a switching circuit inside the OLT [16], instead.

In [17], protection of the distribution fibers was achieved by serially interconnecting the ONUs, as shown in Fig. 6.3. At each ONU, an optical switch (OSW) was employed for protection switching purpose. Half of the received downstream signal at the ONU (say ONU_2 in Fig. 6.3) was tapped off and fed to the OSW residing at the previous connected ONU (say ONU_1 in Fig. 6.3), via a piece of interconnection fiber. Therefore, if the distribution fiber for ONU_1 was broken, ONU_1 could still receive the downstream signal via ONU_2 and OSW_1 in protection state. Thus, the distribution fiber for ONU_2 and the interconnection fiber for both ONU_1 and ONU_2 served as the protection path for ONU_1, at the expense of installing the additional inter-ONU interconnection fibers.

In [18], two distribution fibers, one for working and the other one for protection, were designated for each ONU. The protection distribution fiber was looped back to one of the input ports of the optical star coupler at the RN. A protection feeder fiber was also installed, connecting the OLT and one of the output ports of the optical star coupler at the RN. In case of any cut at the distribution fiber, the optical switch at the ONU would switch to the protection distribution fiber, while the OLT would switch to the protection feeder fiber, as the same time. Thus, both the feeder and the distribution fibers could be protected at the expense of two distribution fibers per ONU and simultaneous protection switching at both the OLT and the affected ONU (Fig. 6.4). Moreover, the switching at the OLT would induce a hit to all the ONUs on the same PON.

In [19], the coarse WDM (CWDM) technology was employed to provide protection of the feeder fiber among two PONs, as shown in Fig. 6.5. The feeder

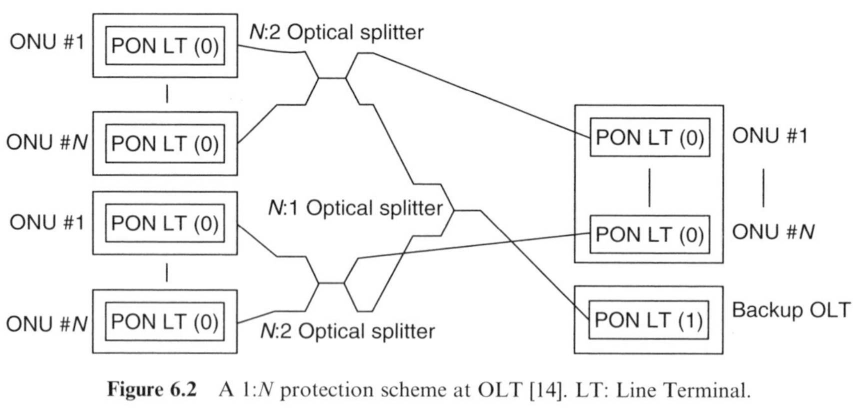

Figure 6.2 A 1:N protection scheme at OLT [14]. LT: Line Terminal.

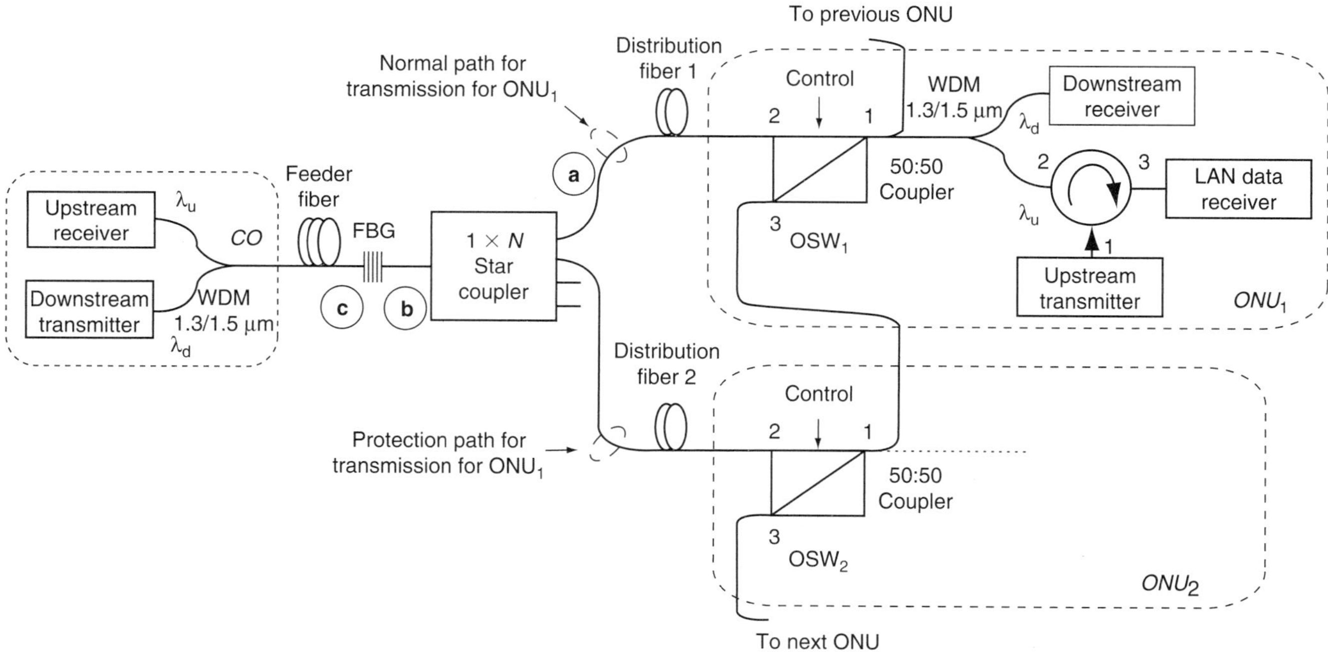

Figure 6.3 A PON architecture with protection of distribution fibers by interconnecting ONUs with protection switching [17]. OSW: optical switch, FBG: fiber Bragg gratings, λ_d: downstream wavelength, λ_u: upstream wavelength.

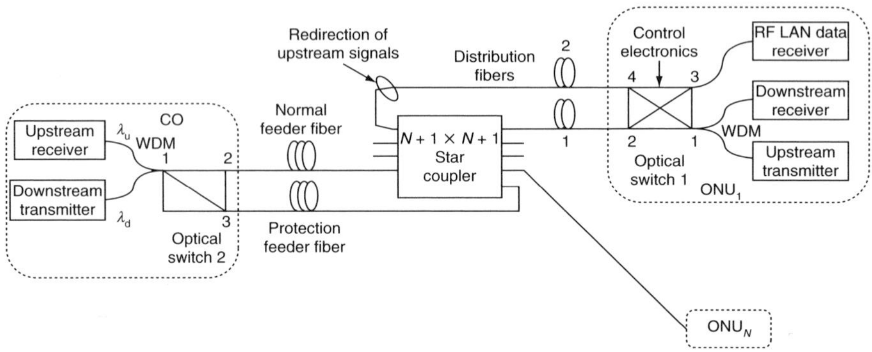

Figure 6.4 A PON architecture with protection of both feeder and distribution fibers by an additional loopback distribution fiber with protection switching [18]. λ_d: downstream wavelength, λ_u: upstream wavelength.

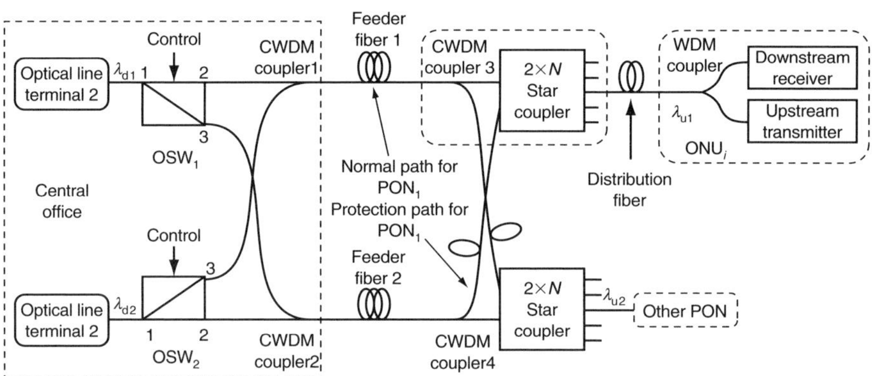

Figure 6.5 Protection of feeder fibers in two PONs using CWDM technology [19]. OSW: optical switch, λ_d: downstream wavelength, λ_u: upstream wavelength.

fibers of both PONs were cross-connected via the optical switches attached to each OLT, while the two RNs were also cross-connected as shown in Fig. 6.5. By employing CWDM technology, PON_1 could offer its feeder fiber as the protection path for PON_2, and vice versa, by means of optical multiplexing of one's working signals and the protected signals from the other PON, on the same feeder fiber, without any interference. The protection switching was performed at the OLT only.

6.3.1.2 Ring Topology

Conventionally, network protection schemes for ring-type PONs are similar to the SONET self-healing rings (SHRs) [20], in which duplicated protection fibers are employed to provide redundant paths, and line or path protection switching is incorporated at both the OLT and the access nodes. There are three common approaches [21–25] for optical ring protection, namely two-fiber unidirectional path-switched ring (UPSR), four-fiber bidirectional line-switched ring (BLSR/4), and two-fiber bidirectional line-switched ring (BLSR/2). UPSR is essentially a $1 + 1$ dedicated protection ring; while BLSR/2 and BLSR/4 are 1:1 shared protection rings. Figure 6.6 illustrates these protection switching operations of shared protection rings [25].

In [26], a star-ring architecture for a subcarrier-multiplexed PON was proposed. As shown in Fig. 6.7, multiple feeder fibers were employed in a star topology and each of them was connected to an RN. Every two adjacent RNs were interconnected via a ring network, on which multiple ONUs were connected. Protection switches and circuits were incorporated in the RN, as shown in the inset of Fig. 6.7, such that the two attached rings on both its sides could be combined to form a larger ring, when the feeder fiber connected to the RN failed. Thus, the failed feeder fiber could be isolated. In this scenario, the network was reconfigured such that the affected ONUs would still be in connection with the CO via the two neighboring working RNs.

6.3.2 WDM Passive Optical Networks

6.3.2.1 Tree Topology

In a WDM-PON with a tree topology, each ONU is designated with a dedicated set of wavelengths for both the downstream and the upstream channels. Therefore, a wavelength multiplexer is employed at the RN for wavelength routing, and different distribution fibers are carrying the respective set of wavelengths destined for their attached ONUs. Due to the similar tree topology as in conventional PONs, the protection switching architectures proposed in ITU-T G.983.1, as shown in Fig. 6.1, could also be employed for WDM-PONs, except that the optical power splitter at the RN has to be replaced by a wavelength multiplexer. However, these approaches incur excessive fiber duplication and protection switching for network protection. Figure 6.8 shows an example of WDM-PON with feeder fiber duplication [27], where a separate pair of protection feeder fibers were needed.

On the other hand, by utilizing the additional feature of wavelength routing in the AWGs employed in the WDM-PONs, lightpath diversity could be realized with higher flexibility and the amount of fiber duplication could also be reduced.

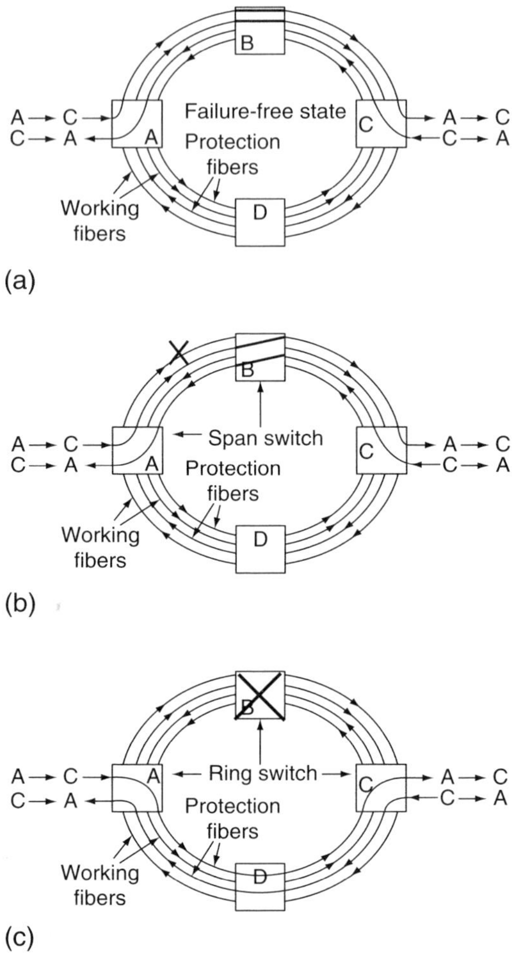

Figure 6.6 An example of a four-fiber shared rings in (a) normal operation; (b) span switching; and (c) ring switching [25].

With the recent much aroused research interests in the enabling technologies for WDM-PONs, several interesting and feasible survivable architectures for WDM-PONs in tree topology have been proposed.

In [28], a self-protected WDM-PON architecture with the idea of group protection of ONUs was proposed to protect against any failure at the distribution fibers. Two adjacent ONUs were grouped and their corresponding downstream and upstream wavelengths were connected to the OLT via the same output port of the AWG at the RN. This was achieved by utilizing the

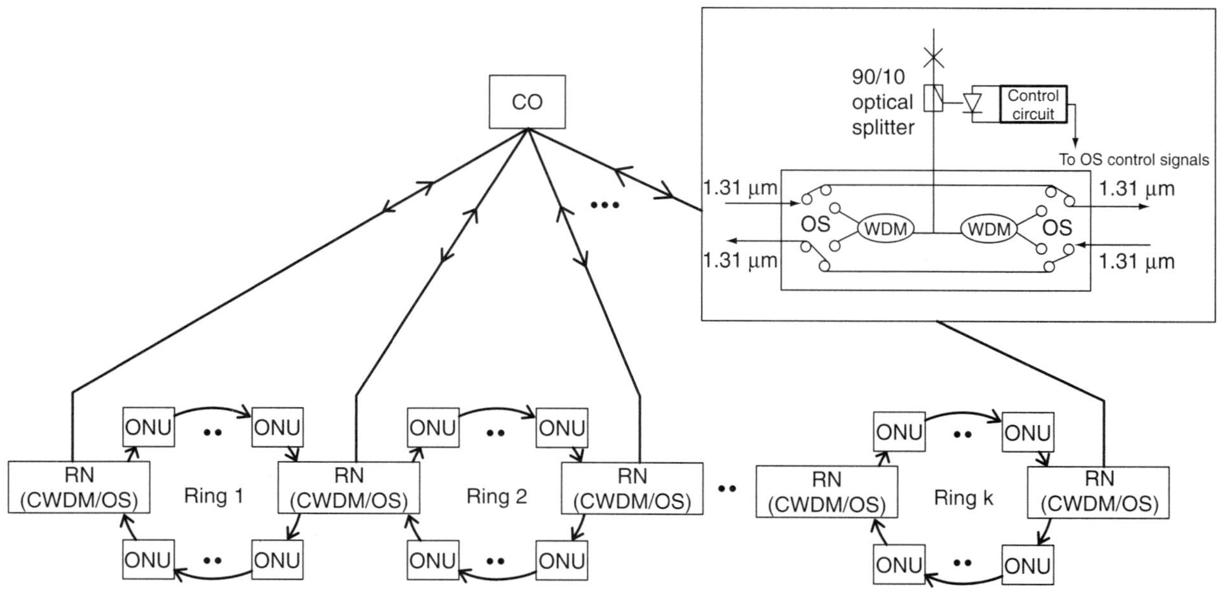

Figure 6.7 A modified star-ring PON with protection capability [26]. Inset shows the structure of the RN to illustrate the protection mode. OS: optical switch.

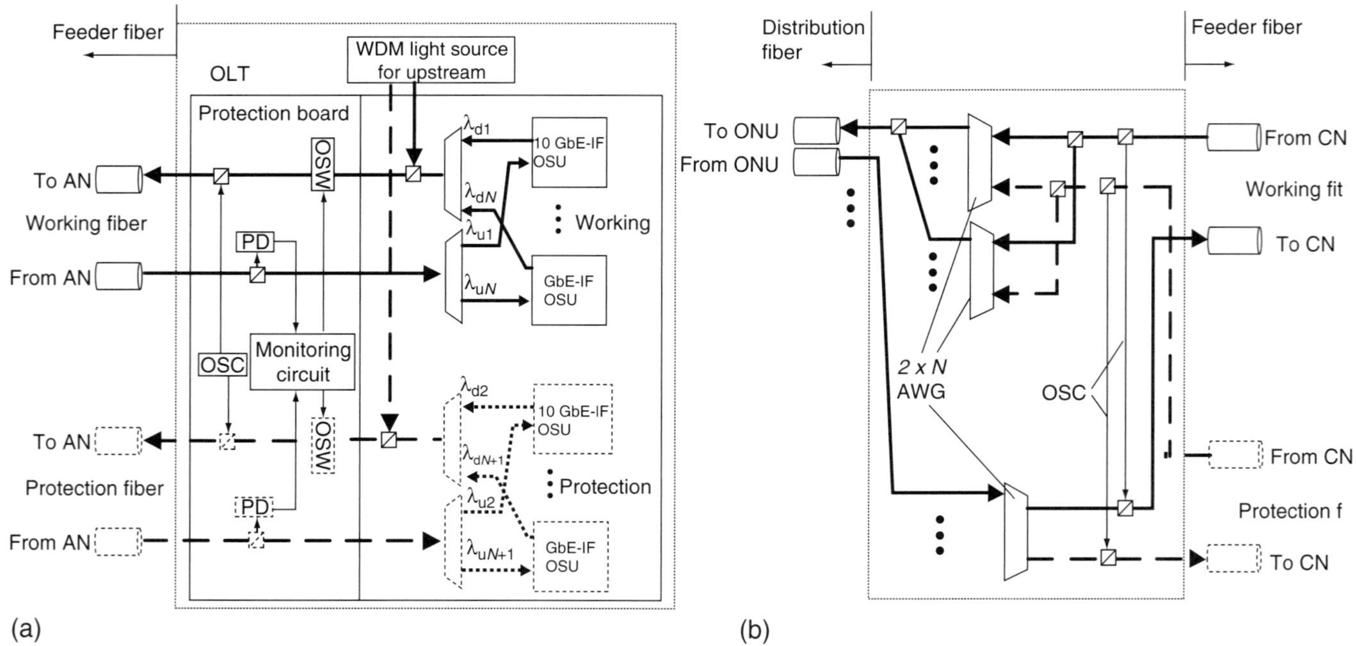

Figure 6.8 The structure of (a) the OLT and (b) RN of a WDM-PON with duplicated fiber feeders for protection [27]. AN: access node, CN: central node, OSC: optical supervisory channel, PD: photodiode, OSW: optical switch.

periodic spectral property of the AWG and with proper wavelength assignment. The two ONUs in the same group were connected by a piece of protection fiber and a pair of protection switches were incorporated into each ONU for signal rerouting. In case of a fiber cut between a particular ONU and the RN, the protection switches in the ONUs in the same group would be triggered via simple power monitoring at the modules M1 and M2, as illustrated in Fig. 6.9. Both the affected downstream and upstream wavelengths would be rerouted to its adjacent ONU before being routed back to the OLT via the same AWG output port. The normal traffic on the adjacent ONU was not affected while the OLT could still keep its connection with the affected ONU. In this way, the two ONUs in the same group protected each other and the OLT was transparent to such fiber failure. An improved version [29] which greatly reduced the number of optical couplers needed at the RN by means of a novel wavelength assignment was also proposed. These two architectures performed protection switching at the ONU side. This might increase the complexity at the ONUs. In [30, 31], centrally controlled WDM-PON survivable architectures were proposed to have all the protection switching performed at the OLT. This could greatly facilitate the control and management of all the protection switching and help to keep the ONUs simple. The protection switching in [30] was performed by a 2×2 OSW connecting the fiber feeders, thus having the possibility of inducing service interruption to all in-service wavelength channels during the short switching period. In [31], as shown in Fig. 6.10, multiple 2×2 OSWs were employed at the OLT to perform protection switching at the wavelength channel level. The wavelength assignment plan for both the downstream and the upstream wavelength channels was depicted in Fig. 6.10 (c), utilizing the cyclic spectral properties of the AWGs employed at the OLT and the RN. Therefore, any toggling of the switching state of any OSW at the OLT would activate the protection path by alternate wavelength routing between the OLT and the respective ONU. Whenever the feeder fiber or the distribution fiber connected to a particular ONU was broken and caused service outage, the respective protection switches for those affected ONUs at the OLT would have their switching states automatically toggled. Therefore, the affected wavelengths would be routed via their designated protection paths, without affecting any other in-service wavelength channels. All protection switching were performed at the OLT only. In [32], a novel WDM-PON architecture with protection of both the feeder fibers and distribution fibers were proposed, as shown in Fig. 6.11. Both the RN and the distribution fibers were duplicated. The wavelength channels for a destined ONU were copied and routed simultaneously in two different lightpaths, one for normal operation and the other for protection purpose, to achieve lightpath diversity. When a fiber cut occurred, the OSW at the ONU would be triggered to redirect the disrupted signals to the protection path.

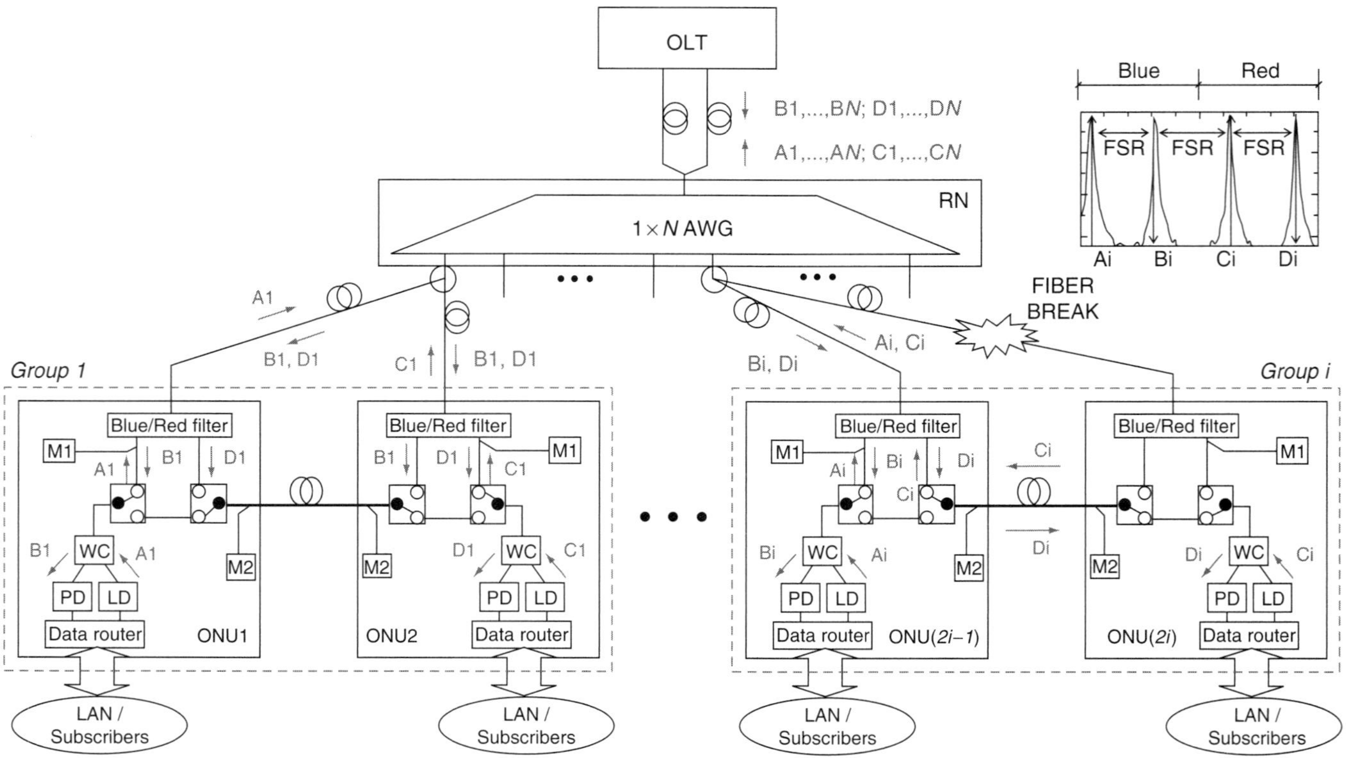

Figure 6.9 Self-protected network architecture for WDM-PON. LD1–4: laser diode; PD1–4: PIN photodiode; WC: WDM coupler; M1&M2: optical power monitors; {A*i*, C*i*} for $i \in \{1, \ldots, N\}$: upstream wavelengths; {B*i*, D*i*} for $i \in \{1, \ldots, N\}$: downstream wavelengths. Inset shows the spectral response of one of the output ports of the AWG. FSR: free-spectral range of the AWG [28].

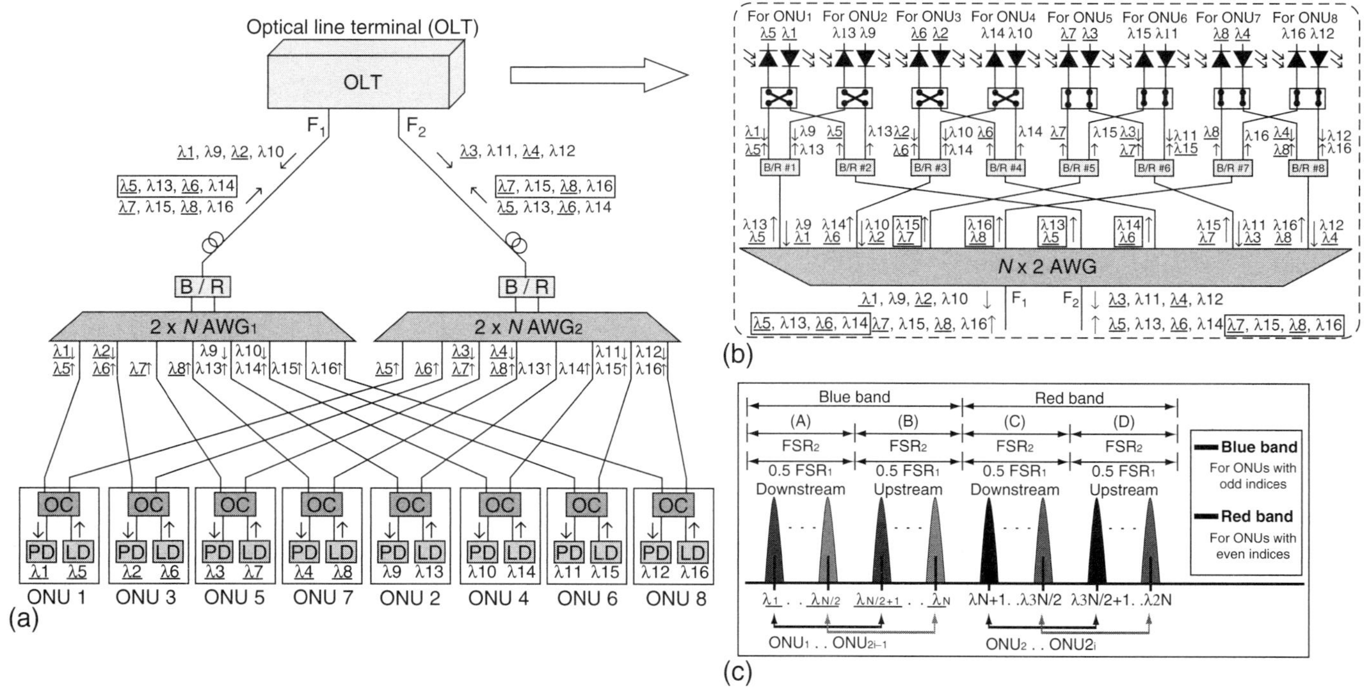

Figure 6.10 (a) A WDM-PON survivable architecture with eight ONUs and centralized protection switching control; (b) OLT configuration under normal operation; (c) wavelength assignment plan. B/R: blue/red filter; OC: 3-dB fiber coupler; LD: laser diode; PD: photodiode. Note: FSR1 stands for free-spectral range of the $N \times 2$ ($N = 8$) AWG at the OLT while FSR2 stands for that of both AWG1 and AWG2 at the RN. The wavelengths quoted in boxes are the working upstream wavelengths. The wavelengths in blue band are underlined but those in red band are not [31].

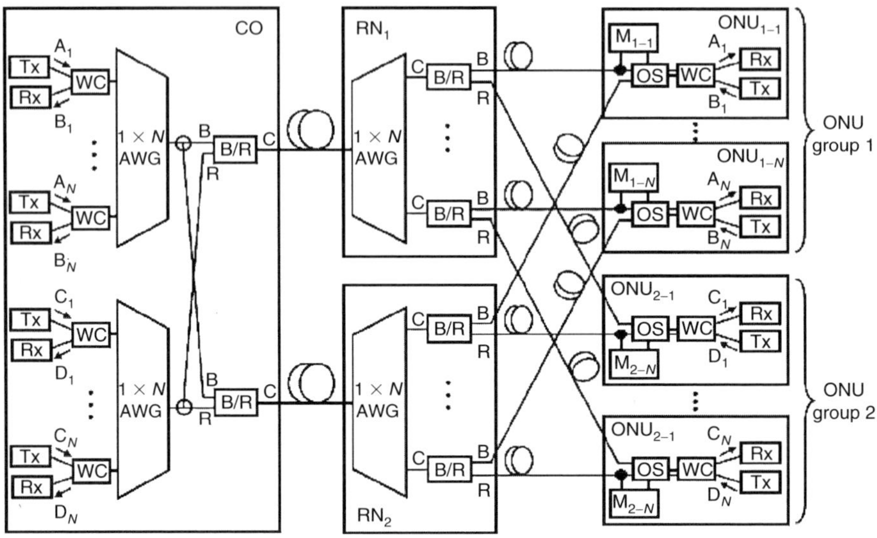

Figure 6.11 A WDM-PON architecture with self-protection capability against both feeder fiber and distribution fiber failures [32]. WC: wavelength coupler, B/R: blue/red filter, OS: optical switch, M: power monitoring modules.

Besides, there were some other survivable network architectures proposed, such as broadcast-and-select WDM-PON with an ONU group protection [33], star-ring WDM-PON with interconnected ONUs via an additional fiber ring [34], WDM-PON with 1:*N* protected OLT with tunable transceivers [35], and WDM-PON with waveband filters for protection [36].

6.3.2.2 Ring Topology

In WDM-PONs with ring topology, the OLT is connected to multiple ANs via single or double fiber rings. Each AN comprises an OADM or simply an optical power splitter, to which multiple ONUs are further connected either in star or ring topologies [37, 38]. Similar to conventional PONs with the ring topology, protection is achieved by means of self-healing rings [20]. Duplicated protection fiber rings are employed to provide redundant paths; and line or path protection switching is incorporated at both the OLT and the ANs. Several interesting WDM self-healing access ring networks have been recently reported [39–46]. In [39], a dense-WDM self-healing ring network, with a unidirectional

OADM, which was based on acousto-optic switches, at each network node, was proposed. In [40], optical filters and OSWs were employed at ANs for wavelength dropping and protection switching respectively. In [40, 41], an AWG add-drop filter was employed as the OADM and a loopback circuit was implemented to provide protection switching at each AN. However, these approaches still required two working fiber paths to support both protection as well as bidirectional transmission. In [42–46], single-fiber bidirectional SHR (Self Healing Ring) networks were also reported. This could further reduce the system cost and increase the fiber efficiency in SHR networks. In [47], an interesting star-ring-bus dense-WDM subcarrier-multiplexed network, with simple protection switching at the ANs was proposed. However, extensive fiber connections were required in the proposed network architecture.

In [45], a simple single-fiber CWDM optical access ring network with unidirectional OADMs incorporated in the ANs was demonstrated, as shown in Fig. 6.12. The downstream wavelength channels were power-split and sent along the ring network in both counter-propagating directions. A 2×2 OSW and a pair of low-cost CWDM OADM was employed at each AN, as illustrated in Fig. 6.12 (b). When a fiber cut along the ring network was detected, by means of signal-loss detection, the OSWs at the affected ANs would be triggered to toggle their switching states such that they could still communicate with the OLT via the input fiber in another propagation direction. In [46], a single-fiber bidirectional self-healing optical access ring networks with bidirectional OADM was demonstrated. The idea was to apply the same OLT architecture and alternate path switching scheme as in [31] to achieve self-healing function in a single-fiber optical access ring.

In [47, 48], an interesting protected optical star-shaped ring network was proposed, as shown in Fig. 6.13. The physical network topology was star-shaped, but the logical connections of all nodes, in form of wavelength paths, were actually in a ring topology, as shown in Fig. 6.13 (b). It was realized by the optical foldback at the AWG employed at the OLT and the wavelength routing properties of the AWG device. Another set of backup wavelength paths are also designated for protection against any fiber link failure. This set of backup wavelength paths can be activated by switching the fiber connections at the designated input ports of the AWG. When a fiber or a network node failed, the protection switch at the OLT would be reconfigured to activate the backup wavelength paths, so as to bypass the failed nodes or link.

6.4 SUMMARY

With the fast-growing deployments of PONs in both enterprise and residential areas, high network availability should be assured to all subscribers. Therefore,

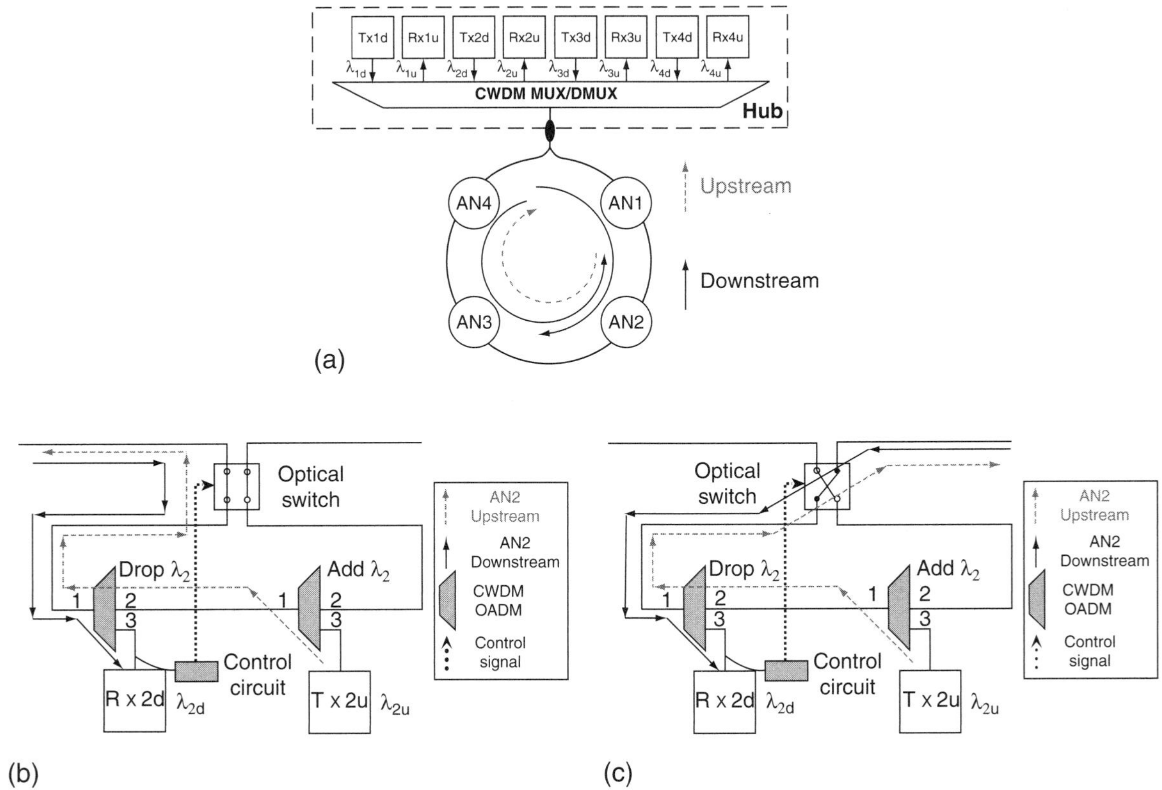

Figure 6.12 (a) A single-fiber CWDM optical access ring network; (b) & (c) the structure of AN2 in (b) normal state and (c) protection state when there was a fiber cut between AN2 and AN3 [45]. AN: access node, Tx: transmitter, Rx: receiver.

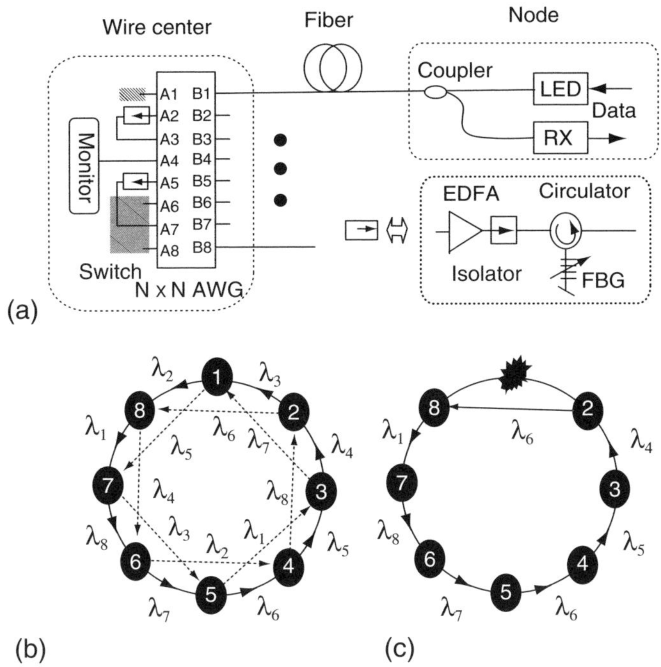

Figure 6.13 (a) A protected optical star-shaped ring network; (b) lightpath diagram, dotted lines were the designated protection paths; (c) the protection lightpath was adopted when node 1 failed [47]. FBG: fiber Bragg gratings.

the issue of network survivability has aroused much attention recently. It is highly desirable to have flexible and robust survivable network architectures as well as protection switching schemes in the optical layer to provide good resilience against any failure of fibers or network equipment. Several important design considerations of protection in PONs have been discussed and some of the recently proposed survivable network architectures for both conventional PONs and WDM-PONs have been reviewed. In most of these schemes, the critical optical components are the OSWs to enhance the protection switching. Hence, optical-switching technologies [49] with low-cost and fast-switching responses would be very crucial and of high practical importance to enable wide deployments of the survivable PONs. Typical protection switching time should be kept below a few tens of milliseconds, say 50 ms, to reduce the amount of data loss during traffic restoration.

ACKNOWLEDGMENTS

The author would like to thank his colleagues and graduate students at the Lightwave Communications Laboratory, The Chinese University of Hong Kong: Professor Chinlon Lin, Professor Lian K. Chen, Dr. Zhaoxin Wang, Mr. Tsan-Jim Chan, Mr. Chi-Man Lee, Mr. Xiaofeng Sun, and Mr. Qiguang Zhao, for research collaborations and stimulating discussions on the topic of optical access networks and technologies. The author would also like to thank Dr. Cedric Lam for his valuable comments on this chapter.

REFERENCES

[1] ITU-T Recommendation G.983.1, *Broadband optical access systems based on Passive Optical Networks (PON)*, 1998.
[2] ITU-T Recommendation G.983.2, *The ONT management and control interface specification for ATM PON*, 2000.
[3] ITU-T Recommendation G.983.3, *A broadband optical access system with increased service capability by wavelength allocation*, 2001.
[4] ITU-T Recommendation G.983.5, *A broadband optical access system with enhanced survivability*, 2002.
[5] IEEE 802.3 Ethernet in the First Mile Study Group, *Ethernet Passive Optical Networks (EPONs)*, 2001.
[6] ITU-T Recommendation G.984.1, *Gigabit-capable passive optical networks (GPON): general characteristics*, 2003.
[7] ITU-T Recommendation G.984.2, *Gigabit-capable passive optical networks (GPON): physical media dependent (PMD) layer specification*, 2003.
[8] D. Zhou, and S. Subramaniam, "Survivability in optical networks," *IEEE Network*, vol.14, no.6, pp16–23. Nov./Dec., 2000.
[9] T. H. Wu, "Emerging technologies for fiber network survivability," *IEEE Communication Magazine*, vol.33, no.2, pp62–74, Feb., 1995.
[10] D. Colle, S. De Maesschalck, C. Develder, P. Van Heuven, A. Groebbens, J. Cheyns, I. Lievens, M. Pickavet, P. Lagasse, and P. Demeester, "Data-centric optical networks and their survivability," *IEEE Journal on Selected Areas in Communications*, vol.20, no.1, pp6–20, 2002.
[11] L. Sahasrabuddhe, S. Ramamurthy, and B. Mukherjee, "Fault management in IP-over-WDM networks: WDM protection versus IP restoration," *IEEE Journal on Selected Areas in Communications*, vol.20, no.1, pp21–33, 2002.
[12] C.K. Chan, F. Tong, L.K. Chen, J. Song, and D. Lam, "A practical passive surveillance scheme for optically-amplified passive branched optical networks," *IEEE Phot. Technol. Lett.*, vol.9, no.4, pp526–528, Apr., 1997.
[13] D.J. Xu, W. Yen, and E. Ho, "Proposal of a new protection mechanism for ATM PON interface," *International Conference on Communications (ICC)*, vol.7, pp2160–2165, Helsinki, Finland, 2001.
[14] W.T. P'ng, S. Khatun, S. Shaari, and M.K. Abdullah, "A novel protection scheme for Ethernet PON FTTH access network," *International Conference on Networks*, vol.1, Malaysia, Nov., 2005.

[15] M.K. Abdullah, W.T. P'ng, P.W. Lau, and E.R. Tee, "FTTH access network protection using a switch," *Asia Pacific Conference on Communications (APCC)*, vol.3, pp1219–1222, Penang, Malaysia, 2003.

[16] A.J. Philips, J.M. Senior, R. Mercinelli, M. Valvo, P.J. Vetter, C.M. Martin, M.O. van Deventer, P. Vaes, and X.Z. Qiu, "Redundancy strategies for a high splitting optically amplified passive optical network," *IEEE/OSA J. Lightwave Technol.*, vol.19, no.2, pp137–149, 2001.

[17] N. Nadarajah, E. Wong, M. Attygalle, and A. Nirmalathas, "Protection switching and local area network emulation in passive optical networks," *IEEE/OSA J. Lightwave Technol.*, vol.24, no.5, pp1955–1967, 2006.

[18] N. Nadarajah, E. Wong, and A. Nirmalathas, "Automatic protection switching and LAN emulation in passive optical networks," *IEE Elect. Lett.*, vol.42, no.3, pp171–173, 2006.

[19] N. Nadarajah, A. Nirmalathas, and E. Wong, "Self-protected Ethernet passive optical networks using coarse wavelength division multiplexed transmission," *IEE Elect. Lett.*, vol.41, no.15, pp866–867, 2005.

[20] T.H. Wu and W.I. Way, "A novel passive protected SONET bidirectional self-healing ring architecture," *IEEE/OSA J. Lightwave Technol.*, vol.10, no.9, pp1314–1322, 1992.

[21] R. Ramaswami and K.N. Sivarajan, *Optical Networks: A Practical Perspective*, 2/e, Chap. 10: Network Survivability, Morgan Kaufmann, 2002.

[22] X. J. Fang, R. Iraschko, and R. Sharma, "All-optical four-fiber bidirectional line-switched ring," *IEEE/OSA J. Lightwave Technol.*, vol.17, no.8, pp1302–1308, 1999.

[23] H. Hayashi, K. Ohara, H. Tanaka, M. Daikoku, T. Ontani, and M. Suzuki, "Highly reliable optical bidirectional path switched ring applicable to photonic IP networks," *IEEE/OSA J. Lightwave Technol.*, vol.21, no.2, pp1356–1364, 2003.

[24] M.J. Li, M.J. Soulliere, D.J. Tebben, L. Nederlof, M.D. Vaughn, and R.E. Wagner, "Transparent optical protection ring architectures and applications," *IEEE/OSA J. Lightwave Technol.*, vol.23, no.10, pp3388–3403, Oct., 2005.

[25] J. Manchester, P. Bonenfant, and C. Newton, "The evolution of transport network survivability," *IEEE Communication Magazine*, vol.37, no.8, pp44–51, Aug., 1999.

[26] W.P. Lin, M.S. Kao, and S. Chi, "The modified star-ring architecture for high capacity subcarrier multiplexed passive optical networks," *IEEE/OSA J. Lightwave Technol.*, vol.19, no.1, pp32–40, Jan., 2001.

[27] H. Nakamura, H. Suzuki, J. Kani, and K. Iwatsuki, "Reliable wide-area wavelength division multiplexing passive optical network accommodating gigabit Ethernet and 10-Gb Ethernet services," *IEEE/OSA J. Lightwave Technol.*, vol.24, no.5, pp2045–2051, 2006.

[28] T.J. Chan, C.K. Chan, L.K. Chen, and F. Tong, "A self-protected architecture for wavelength division multiplexed passive optical networks," *IEEE Phot. Technol. Lett.*, vol.15, no.11, pp1660–1662, Nov., 2003.

[29] C.M. Lee, T.J. Chan, C.K. Chan, L.K. Chen, and C.L. Lin, "A group protection architecture (GPA) for traffic restoration in multi-wavelength passive optical networks," *European Conference on Optical Communications (ECOC)*, Paper Th.2.4.2, Rimini, Italy, Sep., 2003.

[30] Z.X. Wang, X.F. Sun, C.L. Lin, C.K. Chan, and L.K. Chen, "A novel centrally controlled protection scheme for traffic restoration in WDM passive optical networks," *IEEE Phot. Technol. Lett.*, vol.17, no.3, pp717–719, Mar., 2005.

[31] X.F. Sun, C.K. Chan, and L.K. Chen, "A survivable WDM PON architecture with centralized alternate-path protection switching for traffic restoration," *IEEE Phot. Technol. Lett.*, vol.18, no.4, pp631–633, Feb., 2006.

[32] E.S. Son, K.H. Han, J.H. Lee, and Y.C. Chung, "Survivable network architectures for wavelength-division-multiplexed passive optical networks," *Photonic Network Communications*, vol.12, no.1, pp111–115, 2006.

[33] Z.X. Wang, B. Zhang, C.L. Lin, and C.K. Chan, "A broadcast and select WDM-PON and its protection," *European Conference on Optical Communications (ECOC)*, Paper We4.P.24, Glasgow, United Kingdom, 2005.

[34] X.F. Sun, Z.X. Wang, C.K. Chan, and L.K. Chen, "A novel star-ring protection architecture scheme for WDM passive optical access networks," *IEEE/OSA Optical Fiber Communication Conference (OFC)*, Paper JWA53, Anaheim, CA, USA, Mar., 2005.

[35] Y.M. Kim, T.H. Kim, J.H. Bae, B.W. Kim, and H.S. Park, "A novel protection architecture for WDM-PON," *International Conference on Optical Internet (COIN)*, Paper MoB2–4, Jeju, Korea, 2006.

[36] L. Zong, T. Wang, P. Ji, L. Xu, and M. Cvijetic, "A novel protection scheme for WDM-PONs using waveband filters," *European Conference on Optical Communications (ECOC)*, Paper We3.P.173, Cannes, France, 2006.

[37] P.P. Iannone, K.C. Reichmann, A. Smiljanic, N.J. Frigo, A.H. Gnauck, L.H. Spiekman, and R.M. Derosier, "A transparent WDM network featuring shared virtual rings," *IEEE/OSA J. Lightwave Technol.*, vol.18, no.12, pp1955–1963, Dec., 2000.

[38] M. Kuznetsov, N.M. Froberg, S.R. Henion, H.G. Rao, J. Korn, K.A. Rauschenbach, E.H. Modiano, and V.W.S. Chan, "A next-generation optical regional access network," *IEEE Communication Magazine*, vol.38, no.1, pp66–72, Jan., 2000.

[39] A.F. Elrefaie, "Multiwavelength survivable ring network architectures," *IEEE International Conference on Communications (ICC)*, vol.2, pp1245–1251, Geneva, May, 1993.

[40] B. Glance, C.R. Doerr, I.P. Kaminow, and R. Montagne, "Optically restorable WDM ring networks with simple add/drop circuitry," *IEEE/OSA J. Lightwave Technol.*, vol.14, no.11, pp2453–2456, 1996.

[41] H. Toba, K. Oda, K. Inoue, K. Nosu, and T. Kitoh, "An optical FDM-based self-healing ring network employing arrayed waveguide grating filters and EDFA's with level equalizers," *IEEE Journal of Selected Areas in Communications*, vol.14, no.5, pp800–813, June, 1996.

[42] C.H. Kim, C.-H. Lee, and Y.C. Chung, "Bidirectional WDM self-healing ring network based on simple bidirectional add–drop amplifier modules," *IEEE Phot. Technol. Lett.*, vol.10, pp1340–1342, Sept., 1998.

[43] Y.H. Joo, G.W. Lee, R.K. Kim, S.H. Park, K.W. Song, J. Koh, S.T. Hwang, Y. Oh, and C. Shim, "1-fiber WDM self-healing ring with bidirectional optical add/drop multiplexers," *IEEE Phot. Technol. Lett.*, vol.16, pp683–685, Feb., 2004.

[44] S.B. Park, C.H. Lee, S.G. Kang, and S.B. Lee, "Bidirectional WDM self-healing ring network for hub/remote nodes," *IEEE Phot. Technol. Lett.*, vol.15, pp1657–1659, Nov., 2003.

[45] Z.X. Wang, C.L. Lin, and C.K. Chan, "Demonstration of a single-fiber self-healing CWDM metro access ring network with uni-directional OADM," *IEEE Phot. Technol. Lett.*, vol.18, no.1, Jan., 2006.

[46] X.F. Sun, Z.W. Wang, C.K. Chan, C.L. Lin, and L.K. Chen, "A single-fiber bi-directional WDM self-healing ring network with bi-directional OADM for metro-access applications," *IEEE Journal on Selected Areas in Communications*, vol.25, no.4, pp18–24, Apr., 2007.

[47] C.J. Chae and R.S. Tucker, "A protected optical star-shaped ring network using an N×N array waveguide grating and incoherent light source," *IEEE Phot. Technol. Lett.*, vol.13, no.8, pp878–880, Aug., 2001.

[48] C.J. Chae and R.S. Tucker, "A flexible and protected virtual optical ring network," *IEEE Phot. Technol. Lett.*, vol.14 no.11, pp1626–1628, Nov., 2002.

[49] R. Appelman and Z. Zalevsky, "All-optical switching technologies for protection applications," *IEEE Communication Magazine*, vol.42, no.11, ppS35–S40, Nov., 2004.

Chapter 7

Optical Characterization, Diagnosis, and Performance Monitoring for PON

Alan E. Willner[1] *and Zhongqi Pan*[2]

[1] *University of Southern California*
[2] *University of Louisiana at Lafayette*

7.1 INTRODUCTION

Optical diagnosis, performance monitoring, and characterization are essential for ensuring the high quality operation of any lightwave systems. In fact, an efficient and reliable optical network, such as a passive optical network (PON), depends on appropriate testing and measurement. During the construction phase, proper testing is the only way to guarantee that all the required transmission specifications are met, the network is ready for actual traffic, and subscribers are provided with the expected service quality. During initial commissioning and subscriber activation, testing and diagnosis can ensure that the whole system operates within the acceptable specifications. When the network is activated and operation begins, the quality of service (QoS) must be tested and monitored in order to meet service-level agreements with subscribers. When problems are detected and diagnosed (e.g. low signal or no signal), troubleshooting networks help to minimize network downtime, rapidly restore failed services, and efficiently manage network performance.

A PON is a point-to-multipoint, fiber-to-the-premises network architecture in which unpowered optical splitters (either splitting in optical power or wavelength) are used to enable a single optical fiber to serve multiple premises. Figure 7.1 shows the generic PON architecture. A PON does not use any active electronic components (devices consuming power), from the central office (CO) to the consumers' premises. The network carries a single strand of fiber, which

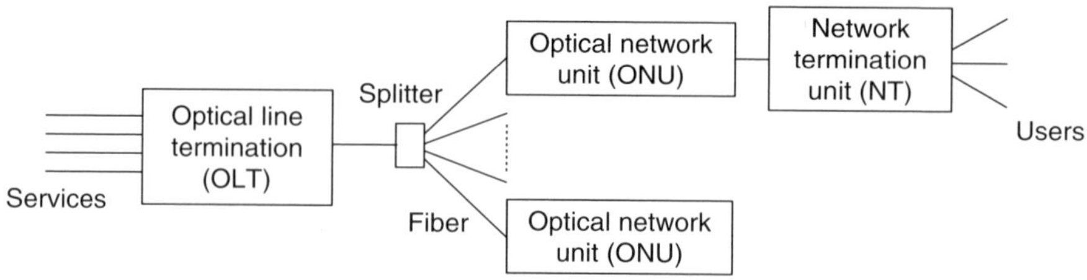

Figure 7.1 The architecture of a passive optical-access network.

undergoes multiple splits to serve many consumer installations. This splitting is achieved by means of passive splitters. On the side of the local exchange there is an optical line termination (OLT), on the user side there is an optical network unit (ONU). Using a passive point-to-multipoint fiber network, consisting of optical fibers and one or more splitters (in cascade), a number of ONUs are connected to an OLT in a tree topology. An ONU can be combined with a network termination unit (NT). This produces an optical network termination (ONT). The OLT has the interfaces with the backbone network that provide the services to the users. Hence a PON's passive part consists of splitters and fibers situated in the field. Reasonably complex active components are required in the local exchange (the OLT) and on the side of users (the ONU/ONTs).

Unlike the point-to-point terrestrial and undersea amplified wavelength-division-multiplexed (WDM) fiber systems, the point-to-multipoint nature of PON has made the optical diagnosis, performance monitoring, and characterization a challenge. The key tests performed during a PON's construction include total link loss measurement (optical power budget), optical return loss (ORL) measurement—especially when cable TV (CATV) services are provided, link characterization using an optical time-domain reflectometer (OTDR). During a PON's operation, network operators need to detect signal presence, measure them, and verify that they are within acceptable power ranges. Thorough performance assessment, accurate bit-error-rate (BER) measurement can help to define competitive, customer-retaining service-level agreements, and, more importantly, to assure and maintain them.

In this chapter, we first will introduce fiber's key properties briefly and then discuss the challenges in PON characterization and monitoring. Second, we will discuss the testing requirements for PON, the state-of-the-art measurement techniques, and network management functions. We will focus on the optical monitoring technologies, including basic monitoring techniques (optical channel monitoring: power, wavelength, and optical signal-to-noise ratio), and advanced monitoring techniques [chromatic dispersion (CD) and polarization mode dispersion (PMD) monitoring]. At the end of the chapter, we will give a brief conclusion on the future trends.

7.2 NETWORK TESTING, CHARACTERIZATION, AND MONITORING CHALLENGES FOR PON

7.2.1 Fiber's Key Characteristics

The key characteristic of a fiber link is the power loss that is caused by fiber attenuation and fiber connections. Attenuation, defined as the ratio of the input power to the output power, is the loss of optical power as light travels along the fiber. Attenuation in an optical fiber is caused by absorption, scattering, and bending losses. The fundamental physical limits imposed on the fiber attenuation are due to scattering of the silica atoms at shorter wavelengths and the material absorption at longer wavelengths. There are two minima in the loss curve, one near 1.3 μm and an even lower one near 1.55 μm. Fiber bending can also induce power loss because radiation escapes through its bends. The bending loss is inversely proportional to the bend radius and is wavelength-dependent. It is important to mention that there may be many fiber macro bends during a PON installation, and macro bends may have more significant losses at longer wavelengths.

Power loss is also present at fiber connections, such as connectors, splices, and couplers. Coupling of light into and out of a small-core fiber is much more difficult to achieve than coupling electrical signals in copper wires since: (i) photons are weakly confined to the waveguide whereas electrons are tightly bound to the wire, and (ii) the core of a fiber is typically much smaller than that of an electrical wire. First, light must be coupled into the fiber from a diverging laser beam, and two fibers must be connected to each other. Second, connecting two different fibers in a system must be performed with great care due to the small size of the cores. One wishes to achieve connections exhibiting: (i) low loss; (ii) low back reflection; (iii) repeatability; and (iv) reliability. Two popular methods are the permanent splice and the mechanical connector. The permanent "fusion" splice can be accomplished by placing two fiber ends near each other, generating a high-voltage electric arc which melts the fiber ends, and "fusing" the fibers together. Losses and back reflection are extremely low being <0.1 dB and <-60 dB respectively. Disadvantages are that: (i) a splice is delicate and must be protected and (ii) it is permanent. Alternatively, there are several types of mechanical connectors, such as ST, and FC/PC. Losses and back reflection are still fairly good, and are typically <0.3 dB and <-45 dB respectively.

Low loss is extremely important since a light pulse must contain a minimum amount of power in order to be detected such that "0" or "1" data bit can be unambiguously detected; thus optical power measurement (or fiber-link loss measurement) is the most basic testing in any optical-fiber network. Note that ORL characterization is related to optical power measurement, and is required for many digital and analog fiber systems, such as PONs. ORL along a fiber span is a combination of Rayleigh scattering and Fresnel reflections, which can reduce fiber system performance and increase BER by degrading transmitter stability.

Other fiber-based effects that limit the maximum transmission distance and the bit rate include chromatic dispersion, polarization mode dispersion, and nonlinearities. These are described briefly as follows:

1. Chromatic dispersion (CD): In any medium other than vacuum and in any waveguide structure (other than ideal infinite free space), different electromagnetic frequencies travel at different speeds. This is the essence of chromatic dispersion. The velocity in fiber of a single monochromatic wavelength is constant. However, data modulation causes a broadening of the spectrum of even the most monochromatic laser pulse. Thus, all modulated data has a nonzero spectral width which spans several wavelengths, and the different spectral components of modulated data travel at different speeds. In particular, for digital data intensity modulated on an optical carrier, chromatic dispersion leads to pulse broadening—which in turn leads to chromatic dispersion limiting the maximum data rate that can be transmitted through optical fiber. The unit of chromatic dispersion is (ps/nm)/km; thus, shorter time pulses, wider frequency spread due to data modulation, and longer fiber lengths will each contribute to temporal dispersion.
2. Polarization mode dispersion (PMD): Single-mode fibers actually support two perpendicular polarizations of the original transmitted signal (fundamental modes). In an ideal fiber (perfect) these two modes are indistinguishable, and have the same propagation constant owing to the cylindrical symmetry of the waveguide. However, the core of an optical fiber may not be perfectly circular, and the resultant ellipse has two orthogonal axes. The index-of-refraction of a waveguide, which determines the speed of light, depends on the shape of the waveguide as well as the glass material itself. Therefore, light polarized along one axis travels at a different speed than does the light polarized along the orthogonal axis. Fiber asymmetry may be inherent in the fiber from the manufacturing process, or it may be a result of mechanical stress on the deployed fiber. The inherent asymmetries of the fiber are fairly constant over time, while the mechanical stress due to movement of the fiber can vary, resulting in a dynamic aspect to PMD. Since the light in the two orthogonal axes travel with different group velocities, to first order, this differential light speed will cause a temporal spreading of signals, which is termed the differential group delay (DGD).

Chromatic dispersion and polarization mode dispersion are considered key degradation effects in long-haul transmission systems that operate at 10-Gbit/s and beyond.

3. Nonlinearities: The index of refraction of optical fiber is slightly dependent on the optical power it carries, with higher intensities experiencing a higher index. Moreover, the electric fields of several WDM channels will mix with

each other and produce sum and difference frequency products. Nonlinearities can be controlled by carefully introducing and balancing chromatic dispersion, perhaps with fixed and/or tunable dispersion compensation. The following nonlinear effects tend to significantly degrade the signal integrity: (i) Self-phase modulation (SPM) occurs because the intensity profile of an optical pulse on a single channel causes a time-varying index-of-refraction profile and, thus, the higher intensity center of a pulse travels slower than the lower-intensity pulse wings. (ii) When considering many WDM channels co-propagating in a fiber, photons from channels 2 through N can distort the index profile that is experienced by channel 1. This cross-phase modulation (XPM) index distortion translates into a lightwave speed distortion. (iii) The optical intensity propagating through the fiber is the square of the electric field. When squaring the sum of different fields, products emerge that are beat terms at various sum and difference frequencies to the original signals. If a WDM channel exists at one of the four-wave-mixing (FWM) beat-term frequencies, the beat term will interfere coherently with this overlapping WDM channel and potentially destroy the data.

In this section, we briefly introduced fiber's key properties that could affect the performances of transmission systems. The reader can find more in-depth information about the fiber's impairments and their effects from the following references [32, 44, 62, 63, 64]. It is important to mention that the considerable impairments in current deployed PON are the optical power loss and optical return loss. The reason is PON's transmission speed in general is limited to less than 10 Gbps and the transmission distance usually is less than 50 km. CD and PMD are not considered as the restrictive effects at this level of bit rate and distance.

7.2.2 Characterization and Monitoring Challenges for PON

Perhaps not all the above-mentioned fiber-based effects need to be characterized or monitored in a PON. In this section, we will discuss the requirements and challenges in testing and monitoring PONs.

A PON needs to be tested during the installation for the following reasons: (i) to ensure the fiber-optic cable is properly installed to specified industry standards (IEEE, ITU-T, Telcodia, and TIA/EIA); (ii) to ensure the equipment intended for use on the cable plant will operate properly; (iii) to ensure the communications equipment is working to specifications; and (4) to document the network for reference in case of future problems [1].

A PON's transmission speed in general is limited to less than 10 Gbit/s (e.g. GPONs are running at up to 2.488 Gbit/s downstream or upstream). The transmission distance usually is less than 50 km. Compared to high-speed long-haul

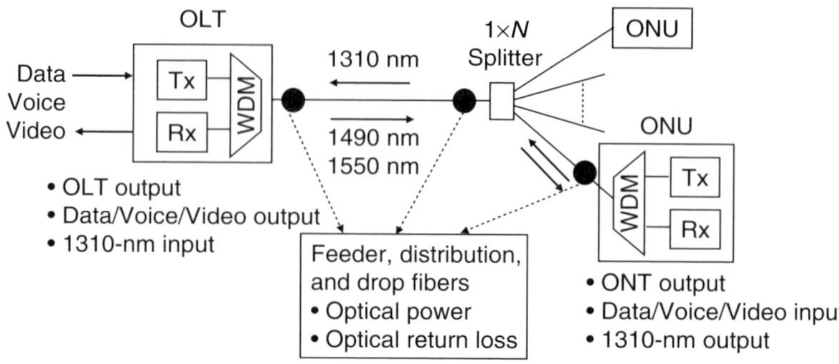

Figure 7.2 Points where the test required for a WDM, point-to-multipoint, and bidirectional PON.

WDM networks, a PON has some unique properties, such as point-to-multipoint, WDM, and bidirectional network architecture in which unpowered optical splitters are used to enable a single optical fiber to serve multiple premises. Downstream signals are broadcast to each premise sharing a fiber. Upstream signals are combined using a multiple access protocol, invariably time division multiple access (TDMA). Moreover, several standards [2, 3] proposed different data rates and protocols. As shown in Fig. 7.2. a typical PON has less optical components than a long-haul optical WDM network. The major components in a PON include: fiber-optic cables, optical splitters, drop terminals (such as patch panels), and connectors. Due to the nature of these passive optical components, the main testing and characterization parameters for a PON are the optical power level at different points. The proper power level (or end-to-end power loss) will ensure proper data transmission. In addition, optical back reflection must be reduced to a minimum. This is measured by optical return loss (ORL) that is defined as the ratio of incident optical power to reflected optical power. ORL can be measured at the input of a device and is commonly measured in dB. The higher the ORL, the lower the back reflection is, thus the better the system will perform. Keeping the back reflection low is especially important for PON systems employing analog video overlay at the 1.55-μm wavelength.

Due to its unique properties, characterization, diagnosis, and monitoring of a PON are different from other optical networks and have some challenges. We will summarize the main challenges in the following subsections.

7.2.2.1 New Wavelengths

The International Telecommunication Union Telecommunication Standardization Sector (ITU-T) has proposed the G.983.3 standard that uses one wavelength for downstream traffic and another for upstream traffic [2]. This recommendation designates the 1490-nm wavelength for downstream voice and data signals, and the 1310-nm wavelength for upstream voice and data signals. The 1550-nm band is purposely left open in case the service provider wishes to share the PON fiber with a hybrid fiber-coax (HFC) network, which is the traditional Cable TV architecture.

Traditional optical networks operate largely in the 1550-nm and 1300-nm bands; thus most of the test equipment covers these wavelengths very well. It is generally true that 1550-nm testing is enough to cover the 1490-nm region for newer fiber that was installed after the late 1990s. However, it is questionable for older fibers deployed in the early 1990s, prior to the existence of G.652C when the water peak (E band) was not attracting so much interest. The fiber-optic test equipment must be able to test at all three wavelengths. Even though people are quite familiar with both the 1310-nm and 1550-nm transmission wavelengths, the 1490-nm transmission wavelength presents new challenges for OTDR.

7.2.2.2 Bidirectional and WDM Network

A PON takes the advantage of WDM, using two wavelengths (1490 nm, 1550 nm) for downstream traffic and another (1330 nm) for upstream traffic. This allows for two-way traffic on a single fiber-optic cable. Therefore, bidirectional testing is important, because it allows the averaging of loss values and because several events, such as mismatched core size, generate different loss levels depending on whether the light comes from one direction or the other [4]. Multi-wavelength testing is also required because some events such as macro bends can induce more significant losses at higher wavelengths (1550 nm) than at lower ones (1310 nm). For the same reason, monitoring should also be performed bidirectionally for different wavelengths.

7.2.2.3 Point-to-Multipoint

Point-to-multipoint systems can be much more complex than point-to-point systems. The loss budget requirement for a PON, based on the ITU Recommendation [2] could be very tight, especially when high-port-count splitters are used in the design. The splitters in a PON cause an inherent loss because the input power is divided among several outputs. Splitting loss depends on the split ratio and is about 3 dB for a 1×2 splitter, increasing by 3 dB each time the number of outputs is doubled. A 1×32 splitter has a splitting loss of at least 15 dB. This loss is seen by both downstream and upstream signals. Combining the losses of the WDM coupler, splices, connectors, and fiber itself, it is therefore critical to perform bidirectional optical loss and ORL measurements on each transmission link of a PON during installation.

After splicing the coupler to the fiber from the CO, it is recommended to characterize the coupler for loss and back reflection to make sure the measurements comply with manufacturers' specifications. Characterization is performed using an optical time domain reflectometer (OTDR) combined with a bare-fiber adapter and pulse-suppressor box. This combination of tools is used to test fibers from coupler output ports toward the OLT in the CO and characterize the losses of each splitter port at 1310 nm, 1490 nm, and 1550 nm. The OTDR involves sending an optical signal along the transmission path and detecting the back-scattered light from the fiber to determine the state of the transmission path. Therefore, only the use of a dead-zone remover will allow the splitter's measurement; otherwise, the losses of the splitter would be within the OTDR dead zone and therefore uncharacterizable.

It is recommended to test from the CO toward the splitter(s), all the way to the ONTs, which can be accomplished by performing a point-to-multipoint OTDR test. However, the point-to-multipoint test is only possible if drop fibers from the splitter to the ONTs are all different from one another. Otherwise, the test is useless because reflections and back-scattering from each ONT branch will appear together in the trace [5]. Moreover, using OTDR for testing and monitoring is difficult in some events. For example, monitoring or troubleshooting one end user using 1330/1550-nm OTDR may interrupt the traffic for other users or OLTs, or OTDR cannot work properly.

7.2.2.4 Sparse Signaling

A traditional power-splitting PON is a shared network, in that the OLT sends a single stream of downstream traffic that is seen by all ONTs. Each ONT only reads the content of those packets that are addressed to it. Encryption may be used to prevent unauthorized snooping of downstream traffic. The OLT also communicates with each ONT in order to allocate upstream bandwidth to each node. When an ONT has traffic to send, the OLT assigns a time slot in which the ONT can send its packets. Because bandwidth is not explicitly reserved for each ONT but allocated dynamically, a PON allows statistical multiplexing and over-subscription of both upstream and downstream bandwidths. Therefore, not only the fiber but also the bandwidth can be shared across a large group of users. However, in the upstream direction, OLT receiver sensitivity and BER differs from regular point-to-point links. Conventional links can be measured regardless of the higher-layer protocol due to their continuous transmission mode. In the upstream direction of PON architectures, one has to take into account the bursty nature of the data and the distinguished distances of each ONU. Thus, when characterizing the OLT receiver sensitivity one has to do so as part of a PON in conjunction with several ONUs and not only as a stand-alone link. The OLT receiver has to capture a weak upstream burst that follows a strong one. This gives a challenge on PON testing and monitoring.

In summary, PON architectures raise the need for PON-specific test methods that differ from conventional optical test methods, to characterize the parameters of their subsystems and modules.

7.3 METHODS FOR CHARACTERIZATION, DIAGNOSIS, AND MONITORING

In this section, we will discuss the required testing and characterization during the PON installation, equipment testing, troubleshooting, and diagnosis guidelines in case of any traffic problems, and a few examples of in-service performance monitoring to ensure the quality of service (QoS) to end users. The readers can find more in-depth discussions in related literatures such as [6, 7].

7.3.1 Required Physical-Layer Measurement

The three main optical tests to be performed during a PON installation are: bidirectional ORL measurement, bidirectional PON-element optical loss measurement, and bidirectional end-to-end link characterization.

Ideally, a PON should be characterized and tested after each segment is installed. For example, once a cabled fiber is installed, end-to-end tests should be performed between the fiber end and OLT in the central office (CO). It is important to characterize each segment because it includes many closely spaced events such as connectors, tie cables, mechanical splices or fusion splices. All the tests should be performed bidirectionally.

Link losses and ORL measurements need to be performed at different wavelengths, such as 1490/1550 nm in the downstream direction and 1310 nm in the upstream direction. End-to-end link characterization includes parameters such as the attenuations of individual fiber segments, locations and losses of splices, connectors, optical splitters, WDM couplers, and link anomalies.

7.3.2 Basic Test Equipment

The following tools are needed to test and troubleshoot the fiber-optic cable plant, system, or link properly:

1. Optical loss test set (OLTS) or power meter and test source with optical ratings matching the specifications of the installed system (fiber type, transmitter type, and wavelength) and proper connector adapters. An OLTS that

merely tests cable plant losses may not include a calibrated power meter needed for testing transmitter and receiver power, so a calibrated power meter and a calibrated source are better choices for link or system testing.

2. OTDR involves sending an optical signal along the transmission path and detecting the back-scattered light from the fiber to determine the state of the transmission path. The faults that can be detected by OTDR include: fiber misalignment/mismatch, fiber breaks, macro bends, dirt on connector ferrules, and angular faults. For PON testing, the OTDR should be able to test at three wavelengths (1310 nm, 1490 nm, and 1550 nm) and possibly a fourth one (1625 nm or 1650 nm for macro-bending-sensitive detection due to the fact that macro bends have more significant losses at longer wavelengths). The dead zone of the OTDR is another important consideration and should be as short as possible. Note that using a special length of fiber called an optical pulse suppressor (OPS) can move the dead zone from the beginning of the fiber under test to this special fiber, thus reducing the dead zone to about 1 m.
3. ORL test meter includes a source and an optical power meter to measure reflected power. Bidirectional ORL measurement should be performed for each fiber section, each splitter branch, and overall end-to-end paths between each drop and the OLT.
4. Visual fiber tracer and/or visual fault locator (VFL) uses visible laser source to locate events such as fiber breaks, overly tight bends, or poorly mated connectors. The visible light allows the user to see a fiber fault or high-loss point as a glowing or blinking light, thus enabling a quick identification of a fault in an optical-fiber link.

Other test equipment includes PON wavelength-isolating power meter that can measure the burst optical power of upstream traffic; live fiber detector (LFD) that can detect live fibers with transmitting traffic; optical spectrum analyzer (OSA) that measures optical power as a function of wavelength; and BER test equipment that uses standard eye-pattern masks to evaluate the data-handling ability of an optical fiber link.

One of the active sectors in PON testing is expected to be the development of a diagnostic analysis software engine that will be integrated into the test equipment. The intelligent test equipment will be able to make a diagnosis and provide the results in plain-language text form, graph form, or values that will be easily understood by all technicians. With the addition of the FTTx software automation, the concept of a one-button FTTx modular test instrument, featuring physical- and protocol-layer test modules, with automated test software based on industry and company standards, would greatly contribute to meeting the aggressive mass deployments of new FTTx subscribers that operators are targeting [8].

7.3.3 Network Testing, Characterization, and Diagnosis Guidelines

All fiber cable plants require certain basic tests to assure they were installed correctly and will meet expected performance values. After a PON network is installed and put into operation, the inevitable optical-fiber cable plant and communication electronic degradation and failures will occur. The operational anomalies or failures that might occur at an ONT, an OLT, or distribution cabinet include: (1) degradation of fiber connectors resulting from moisture, dirt, damage, or misalignment; (2) OLT or ONT circuit card failure; (3) performance degradation in an ONT resulting from other customer-connected equipment; and (4) malicious and accidental damage to cables, distribution cabinet, or access terminals. These conditions may lead to the loss of optical signal at one or more ONTs, received optical power that is below its specified value, an increased BER or degraded signal. We will discuss briefly the guidelines for installation, testing, and network troubleshooting in this section [1, 6–8]. Note that the most valuable data one can have for troubleshooting is the installation testing documentation.

7.3.3.1 Testing and Troubleshooting Installed Fiber Plant

There are a number of possible problems with fiber-optic cable installations that are caused by installation practice, such as improper pulling techniques, improper splicing procedures or termination, and high-loss connectors. Before installation, it is advisable to test the continuity of all the cables as received on the reel using a visual tracer or fault locator. After installation, splicing, and termination, all cables should be tested for insertion loss using an OLTS and OTDR. In general, cables are tested individually and then a complete concatenated cable plant is tested end-to-end bidirectionally. The test is used to compare with the link power budget and communications equipment power budget to ensure proper operation. Note that the testing should be done at the intended operation wavelengths.

The guidelines for troubleshooting a fiber plant problem are:

1. Determine if the problem is with one or all the fibers in the cable. If all the fibers have a problem, there is a likelihood of a severe cable installation issue. If all fibers are broken or have higher than expected loss, an OTDR will show the location of the problem on longer cables but premises cables may be too short and need physical inspection of the cable run. If the problem is caused by kinking or too tight a bend, the cable will have to be repaired or replaced.
2. Bad splices or terminations are the most likely causes for high-loss fibers.

3. Testing for high loss should start with microscope inspection of terminations for proper polish, dirt, scratches, or damage.
4. If the reason for high loss is not obvious and the connectors are adhesive/polish style, the problem may be a fiber break in the back of the connector. A VFL may help in finding fiber breaks, depending on the connector style and the opacity of the cable jacket.
5. Splice-loss problems can be pinpointed during OTDR testing. Confirmation with a VFL should be done if the length from the end of the cable is short enough (~2–3 km) where a VFL is usable. The VFL can find high-loss splices or cracks in fibers caused by handling problems in the splice tray.

7.3.3.2 Testing and Troubleshooting Communication Equipment

After the cable plant has been tested, the communication equipment should be properly connected using matching patch cords. If the cable plant loss is within the loss budget of the equipment (including the loss of the patch cords), the communication links should work properly. If a link does not work, most likely potential problems are:

1. Improper connections: The system requires a transmitter be connected to a receiver, of course, so it is important to verify this for each link. Even if the cable plant is properly documented, fibers may have been crossed at intermediate connections, so using a visual tracer or visual fault locator will allow quick confirmation of the connection.
2. Malfunctions of the communication equipment: If it is connected to the cable plant but not operating properly, begin by checking the power at the receiver on one end of the link. Make sure the equipment is trying to transmit a signal and the power is within the operating specifications. Some equipment have a testing mode to force transmission of a test signal or the equipment may simply keep transmitting to try to complete a connection. If the receiver power is within the specifications, the receiver or electronics beyond the link may be the problem. Use equipment diagnostics or consult the manufacturer for assistance.

If the transmitter tests are good but receiver power is low, the problem is probably in the cable plant. Thus the troubleshooting of the fiber plant is required.

7.3.4 Network Performance Monitoring

Live-fiber testing and monitoring is defined as a nonintrusive test on a given fiber, whether it is one section or multiple sections carrying live data, voice, or video signals, for assessing the quality of an optical-fiber link. It is becoming clear that optical performance monitoring, featuring both physical and protocol layer, can be quite advantageous for the maintenance and management of a PON, and would greatly contribute to meeting the aggressive mass connectivity of new FTTx subscribers that operators are targeting.

7.3.4.1 Basic Monitoring Techniques for Optical Network

Fundamentally, optical performance monitoring (OPM) is a potential mechanism to improve the control of transmission and physical layer fault management. It must be able to diagnose the health of the system. In the unlikely event of a failure, it must determine the location and nature of the failure so that an appropriate repair can be planned and executed quickly. OPM should isolate the specific cause and location of the problem rather than simply sound an alarm; thus, monitors should probably be deployed ubiquitously around the network. Furthermore, it can be quite advantageous to determine when a data signal is beginning to degrade, so that the network operator can take preventive actions to correct the problem (i.e. change a laser wavelength, tune a compensator) or route the traffic around the degraded area [9].

OPM is essential for managing complicated traditional high-capacity WDM transmission and switching systems. Examples of the functions that require OPM include amplifier control, channel identification, and signal quality assessment. OPM can be broken down into three layers [10], as shown in Fig. 7.3. First, WDM channel management layer monitoring involves a determination of the optical domain characteristics essential for transport and channel management at the WDM layer, such as real-time measurements of channel presence, power levels and optical signal-to-noise ratio (OSNR). Second, optical signal

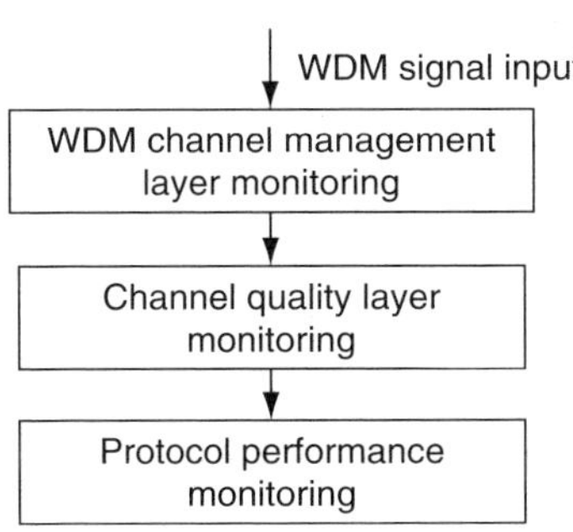

Figure 7.3 Three layers of optical performance monitoring: transport monitoring, signal quality monitoring, and protocol monitoring.

quality monitoring involves a single wavelength and performs signal transition-sensitive measurements such as the eye statistics, Q-factor, the electronic signal-to-noise ratio (SNR), and distortions that occur within the eye due to dispersion and nonlinear effects. Finally, the data protocol monitoring involves protocol performance information, such as the BER.

Another issue related to OPM is the troubleshooting that is often a trial-and-error process and the cost is likely to scale rapidly with the number of sites visited and components tested. Troubleshooting will be especially difficult if the maintenance team cannot obtain accurate information concerning signal routes and channel configurations. This reality of managing a large network places a greater premium on accurate and automated performance information about a network.

In addition, an effective monitor must be able to detect faults for network fault management, i.e. the identification, diagnosis, resolution, and tracking of faults in a network. A fault is recognized when a component or monitor alarm is triggered or when a customer report is filed. Note that the monitors must have the sensitivity equal to or better than that of the end terminal receiver, where sensitivity is defined relative to a particular impairment. For example, noise builds throughout the network and waveform distortion follows the dispersion map. The combination of both noise and distortion at a given location will determine the sensitivity requirements at that location.

7.3.4.2 Fiber Network Impairments Monitoring

In general, optical impairments can be classified into catastrophic and non-catastrophic problems. Catastrophic problems such as component faults include individual or multiple component malfunctions, improperly installed or configured equipment, and damage or intrusion to the network. Impairments due to such faults are as diverse as the components and network designs deployed in the field. In addition to faulty operation, there are many other well-known effects that are always present and must be minimized or controlled. For a reconfigurable network, all of these effects should be controlled dynamically through the network design. The most common effects and the measurement parameters are given in the following list [10].

Notable transmission impairments:

1. Amplifier noise [amplified spontaneous emission (ASE) noise];
2. Amplifier distortion and power transients;
3. Chromatic dispersion;
4. Polarization-mode dispersion;
5. Polarization effects;
6. Fiber nonlinearity induced distortion and cross talk;
7. Timing jitter;

8. Interference effects;
9. Pump laser relative-intensity-noise transfer;
10. Optical-filter distortion;
11. Cross talk.

7.3.4.3 Basic Monitoring Techniques

The general requirements of OPM include in-line (in-service), fast, sensitive, and relative low cost. High-performance OPM should be ubiquitous around a network to enable proper diagnosis, such that the monitor should isolate the specific cause and location of the problem, not just simply sound an alarm. It should be noted that high-data-rate optical networks are quite susceptible to "noncatastrophic" problems, in which there is sufficient optical signal power but the data bits themselves are unrecoverable due to various linear and nonlinear dispersive effects. Exhaustive monitoring might be possible with an unlimited budget. Although the future of optical networks is difficult to predict, the value of monitoring increases with increasing transparency. It might be used to realize new methods of traffic managements. For instance, routing decisions based upon performance monitoring is one possibility. By monitoring the channel quality and link security and updating the routing look-up tables continually, high capacity and priority traffic can be dynamically tuned to high-performance optical channels [11], thus ensuring that the data channels achieve acceptable BER and the whole network achieves sufficient transmission and protection capacity.

In general, OPM is physical-layer monitoring and therefore the required OPM depends strongly on the physical network design. Different OPM parameters often require different monitors. Therefore OPM is highly constrained by the available optical-monitoring technology. The possible physical-layer measurement parameters include:

1. Average power (per wavelength or aggregate);
2. Peak power;
3. Pulse/bit shape;
4. Eye diagram;
5. Intensity/field autocorrelation (including higher order);
6. Amplitude power spectrum (RF spectrum);
7. Polarization state;
8. Optical spectrum (wavelength);
9. Amplitude histogram (synchronous and asynchronous);
10. Q-factor/BER (V-curve measurement);
11. Polarization-mode dispersion (DGD, including higher order);
12. Chromatic dispersion;
13. Phase/optical carrier characteristics;

In general, monitoring techniques can be either analog or digital. Digital techniques use high-speed logic to process digital information encoded in the optical waveform. Measurements on the digital signal are used to infer the characteristics of the optical signal. Digital methods have the strongest correlation with the BER, but are usually less effective at isolating the effects of individual impairments. Analog-measurement techniques treat the optical signal as an analog waveform and attempt to measure specific characteristics of this waveform. These measurements are typically protocol-independent and can be subdivided further into either time domain methods or spectral methods. Time domain monitoring includes eye diagram measurements and auto- or cross-correlation measurements. Spectral methods must be broken down into optical spectrum and amplitude power spectrum (RF spectrum) measurements. The optical spectrum is conveniently measured using highly sensitive optical techniques which can provide optical noise information. The amplitude power spectrum is a better measure of signal quality because it measures the spectrum of the signal that is imposed on the optical carrier. Noise and distortion on the amplitude power spectrum will usually directly translate to impairments on the signal.

7.3.4.3.1 Power and Wavelength Monitoring

The optical power at a given wavelength is the most important parameter in any fiber network. Other parameters or impairment monitoring will depend upon the optical power measurement. Therefore, power monitoring is the basic requirement for any optical network. We actually can categorize the power and wavelength monitoring to optical channel monitoring that is becoming more common in WDM systems and may be well adapted to standardization. Frequently suggested OPM parameters include optical power, wavelength, and wavelength range/spacing, and OSNR [10, 12]. For example, one optical performance monitor uses proprietary thin-film filter technology combined with a micro-actuator and high-speed electronics [12]. It is used to measure critical information on optical transmission signals in DWDM networks for monitoring signal dynamics, determining system functionality, identifying performance change, and providing feedbacks for controlling network elements so as to optimize operational performance. It can also automatically scan the C-, L-, and/or C + L-band wavelength range and precisely measure channel wavelength, power, and OSNR with a 60-dB dynamic range.

Most of proposed PON monitoring actually is the monitoring of the optical power. We will discuss a few examples in this section to provide the readers with more in-depth PON power-monitoring techniques.

There are several reports on WDM-PON monitoring and fault location [13–17, 60]. In general an OTDR has been used to localize fiber failures. However, the conventional OTDR is not suitable for use in WDM-PONs due to the wavelength selective component placed at the remote node. It has been proposed to localize the fiber failures in WDM-PONs by using a wavelength-tunable OTDR or implementing additional paths at remote node (RN) to bypass arrayed-waveguide grating (AWG). One method added an AWG operated at 1550 nm to the star coupler of a PON in the remote node and separates the original PON (operating at 1310 nm) and the overlay via 1310/1550-nm splitters, which guarantees the maintenance function [13]. The overlay in the RN can be employed for the maintenance function of localization of losses and fiber breaks in the individual paths. Another in-service individual line monitoring technique demonstrated for WDM-PON can compensate for the temperature-dependent channel drift of the arrayed-waveguide grating router (AWGR) remote node [14]. These methods require either an expensive OTDR or a complicated network configuration.

Recently, one demonstration shows the detection and localization of the fiber failures in WDM-PON by monitoring the status of the upstream signals [15]. Figure 7.4 shows the proposed experimental setup. When a failure occurred in the feeder fiber, the optical powers of every upstream channel were reduced. In this case, the control unit randomly selected one of the downstream light sources and used it to transmit OTDR pulses instead of data for the localization of fiber fault. When the failures occurred in several drop fibers simultaneously, the control unit identified the failed channels and then analyzed the locations of fiber failures channel by channel according to the predetermined priorities. The measurement has shown that the dynamic range was dependent only on the peak power of the OTDR pulse, and not sensitive to the noise characteristics of

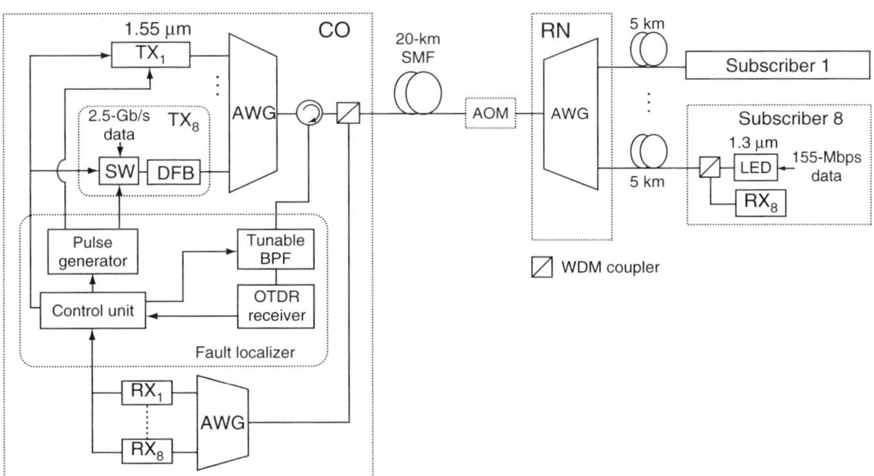

Figure 7.4 Example on the detection and localization of fiber failures in a bidirectional WDM-PON [15].

the light sources. The time for the upstream receiver to detect the failure is the propagation delay time. After the detection, the control unit immediately triggered the electrical switch to stop the transmission of downstream signal and send OTDR pulses. The advantage of this technique is that failures in drop fibers and the feeder fiber can be located without using a wavelength-tunable OTDR. Since this technique reuses the downstream light source of the failed channel to transmit the OTDR pulse, it does not require additional light sources for the localization of fiber failure. Moreover, the proposed approach can detect and localize the fiber failure in the failed channel without affecting the operations of other channels.

A WDM-PON based on wavelength-locked Fabry–Perot (FP) laser diodes using spectrum-sliced amplified spontaneous emission (ASE) injection can be low cost compared to conventional WDM transmitters [16]. A recent report has demonstrated a simple fiber-link monitoring method for bidirectional WDM-PON using the FP lasers [17]. As shown in Fig. 7.5, the system consists of a wide-band super-luminescent light emitting diode (E-SLD) and an optical reflector as a feedback element at each ONU. The scheme uses the cyclic transmission property of the AWG and the addition of an out-of-band ASE signal and optical reflectors. Each port of the AWG supports transmission and routing of three different wavelength signals (L-band upstream: 1583–1609 nm, C-band downstream: 1540–1565 nm, and E-band link monitor signal: 1423–1447 nm). Thus, three different bands of light are launched into the feeder fiber. At each ONU, an optical reflector reflects the E-band monitoring light and transmits the communication signal. The optical reflector consists of a WDM filter and a mirror. The link status of each channel can be monitored by checking the reflected power level of the corresponding wavelength. All reflected monitor signals are detected at the tap coupler (up-tap) by use of a tunable filter or optical-spectrum analyzer (OSA). If the power level at a certain wavelength degrades after a certain time or is below

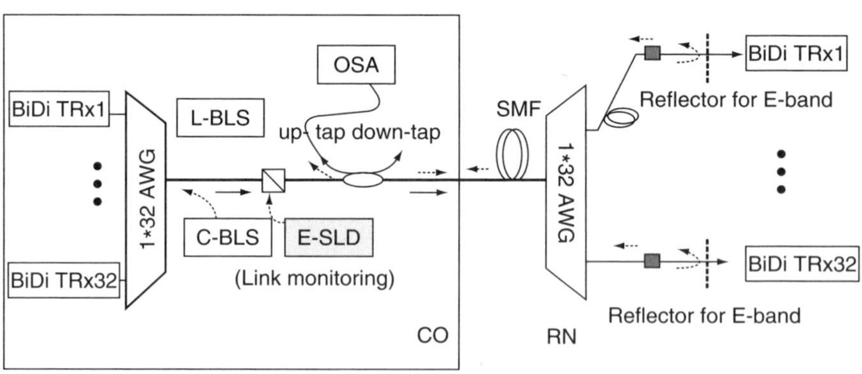

Figure 7.5 Fiber-link loss monitoring example in a bidirectional WDM-PON using ASE-injected FP-LD [17].

the detection limit, this indicates that there might be a progressive degradation or a fault at the corresponding channel. This technique enables one to identify the difference between a feeder fiber fault and a distribution fiber fault without affecting the in-service subscribers. Moreover, the proposed scheme can provide the additional functionality of in situ channel monitoring, allowing continuous supervision of communication channels for quality and performance.

Since a PON using optical power splitter mixes many almost identical paths together, OTDR analysis can sometimes be erroneous. To differentiate among 32 or more optical paths, an optical wavelength domain reflectometry (OWDR) method has been recently proposed [18]. The idea is to embed a permanent passive component at each ONT. The embedded component was called a wavelength coded tag (WCT) that has the following features: (1) having unique wavelength characteristics that can be distinguished by a spectrum analysis remotely; (2) transparent to the entire signal traffics that share the network; (3) wavelength and optical-loss stable over a severe environmental condition; and (4) low-cost structure and manufacturing process to justify the FTTH vast deployment. Figure 7.6 depicts the schematic diagram of an OWDR-embedded FTTH optical network from OLT to ONT. As an example to illustrate its functionality, the 1575–1600-nm band is used as the fiber diagnostic band. Each end user is assigned a unique wavelength code through a WDM tag or WCT inside the OLT. One can even embed additional WCTs into different sections of the network so long as each embedded one has a unique wavelength code. To make the OWDR work, an additional FWDM device, a single thin-film filters (TFF) based band separator, is

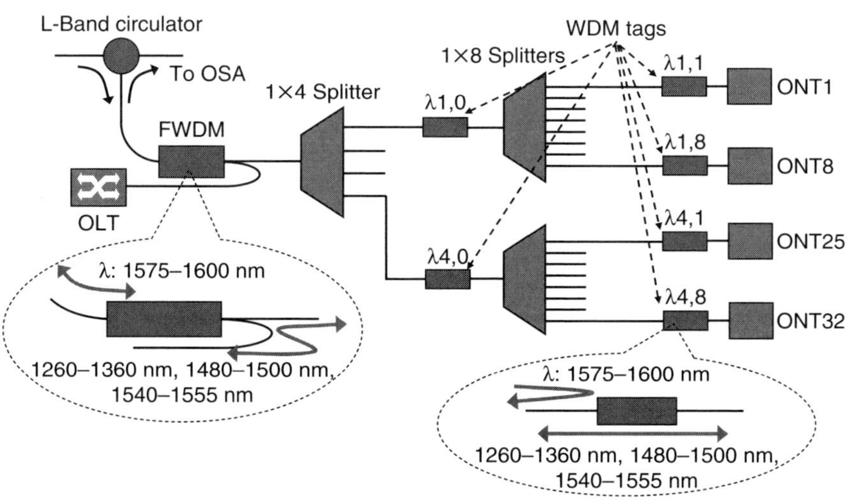

Figure 7.6 Schematic of an OWDR-embedded FTTH network from OLT to ONT [18].

inserted to serve as a MUX for the downstream to combine the diagnostic band with FTTH bands, and as a DEMUX for upstream signals. The monitoring band can be fed from a tunable laser or an SLED-based broadband light source covering wavelengths from 1575 nm to 1600 nm. Each WCT can retroreflect a unique narrowband WDM signal within the monitoring band of 1575–1600 nm. All other wavelengths should pass this WCT with minimum power loss. Thus, any fiber cut can be remotely detected through sending a probe signal while analyzing its echoes generated from all the WCTs. Any missing WCT indicates something happened to that section of the fibers. Then an additional OTDR can pinpoint the exact location of the fiber fault.

To meet the challenge of remote monitoring and diagnosis in a PON, the monitoring tool generally should be based on reflectometry that can send a signal out while detecting its reflected echoes, such as an OTDR or an OWDR [61, 18]. The use of an OTDR/OWDR for detecting and locating degradations is more valuable on live fibers than on dark ones because one can measure, pinpoint, and eventually correlate problems raised by other systems dedicated to measuring signal quality such as BER, Q-factor, or other more sophisticated parameters.

7.3.4.3.2 OSNR Monitoring

Common technologies for channel OSNR monitoring are optical spectrum analyzer method that is based on analyzing the optical spectrum of the detected signal [12]. Wavelength tunable devices for channel OSNR monitoring include fiber FP filters, fiber Bragg grating filters, free-space and micro-electro-mechanical systems (MEMS) diffractive optics, and dielectric thin film filters. Several techniques have been developed that do not directly measure the optical spectrum or focus on wavelength monitoring [19–22]. Modulation tone techniques have also been used as a low-cost alternative to spectral measurements. In principle, the techniques that measure signal power can also be used to obtain the optical-noise power, which is extrapolated from the power level adjacent to the channel. This approach works well if the optical noise can in fact be obtained from the power level adjacent to the channel. This condition is not true for many important types of optical noise including multipath interference effects, amplifier pump laser relative intensity noise (RIN) transfer noise, and four-wave mixing.

To monitor/measure the noise power within the individual channel optical bandwidth, the challenge is to discriminate between the noise and the signal (see Fig. 7.7). For example, when channels pass through different number of fiber amplifiers in long-haul network with optical add/drop multiplexers (OADMs), even adjacent channels, therefore, may experience a significantly different noise level as shown in Figure 7.7(a). Moreover, when channels pass through filters or MUX/DEMUX that suppress the spectral regions next to those channels, very low noise may be measured out-of-band (between channels) regardless of the

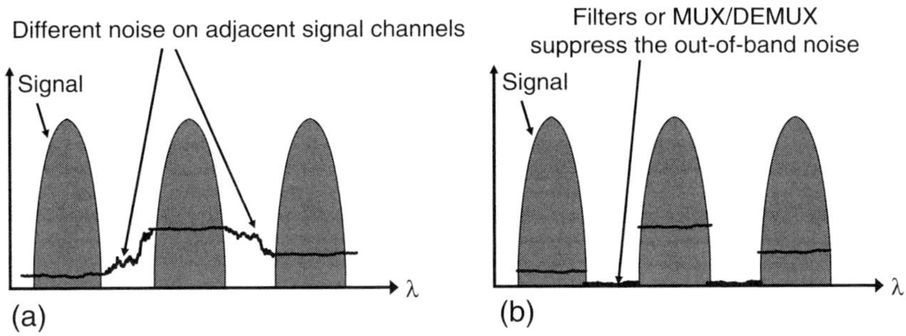

Figure 7.7 (a) Undermined noise level: different noise on adjacent signal channels; (b) "Missing" noise between channels: optical filters or MUX/DEMUX may remove the out of band noise.

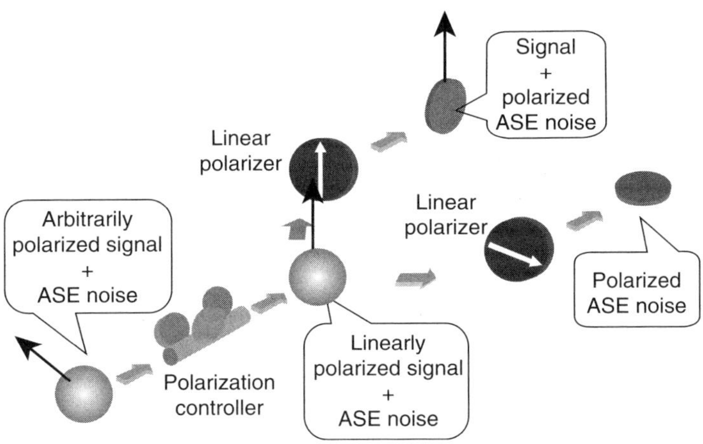

Figure 7.8 Operating principle of the polarization technique: the polarized noise (i.e. half of the total noises) can be measured by using the second linear polarizer, which is aligned to be orthogonal to the signal's polarization [25].

true OSNR level, as shown in Figure 7.7(b) [65]. To overcome this problem, one possible approach is to make use of the optical polarization [23, 24]. In principle, an optical signal will have a well-defined polarization, whereas optical noise will be unpolarized, as shown in Fig. 7.8. Therefore, the polarization extinction ratio is a measure of the optical SNR. In this technique, the received signal (together with noise) is split into two orthogonal polarization components in which one component consists of signal and polarized noise, while the other has polarized noise only (assuming that the signal is highly polarized, and the noise is completely unpolarized). Thus, it is possible to measure the signal and noise powers right at the signal's wavelength since the powers of the polarized noises measured in these polarization components should be the same (i.e. one half of total noise power) [25]. The performance of this technique could be affected by various polarization effects in the transmission link. For example, it could be seriously deteriorated if the signal is depolarized by PMD and nonlinear birefringence. The accuracy of this technique could be degraded significantly if the signal is depolarized by PMD and nonlinear birefringence, and/or the ASE noise is partially polarized due to polarization-dependent loss.

Another method to in-channel OSNR monitoring is to use the amplitude power spectrum of the data and monitor at spectral locations where the signal is not present. This can involve monitoring at low frequencies, high frequencies, or at special null locations within the spectrum [26]. Optical subcarrier monitoring has been used to directly measure the OSNR and correlate the optically measured value to the electrical SNR seen by the receiver [27]. This method has the advantage of monitoring the actual data signal itself as it propagates along the impairment path.

In fact, the pilot-tone-based monitoring technique can monitor many other physical parameters cost-effectively in the optical domain. This technique typically utilizes a small amplitude-modulated pilot tone dedicated to each WDM channel, as shown in Fig. 7.9. Because the tone is at a single low frequency, it is easy to generate and process using conventional electronics. More importantly, since the added tone is within the signal bandwidth, it will experience the same degradations as the signal does through the fiber network. Therefore, signal monitoring can be extrapolated from the measurement of the phase, power, or SNR of the added RF tone. In addition, RF tone can monitor the optical parameters without using expensive demultiplexing filters such as tunable optical filters and diffraction gratings. A commonly used RF tone monitoring method is to assign each WDM channel a different RF frequency tone. The average power in these tones will be proportional to the average optical power in the channel. Thus, the aggregate WDM optical signal on the line can be detected and the tones of all the channels will appear in the RF power spectrum in much the same way they would appear in the optical spectrum [28–31]. Moreover, the monitoring of RF tones can be used for measuring the accumulation of

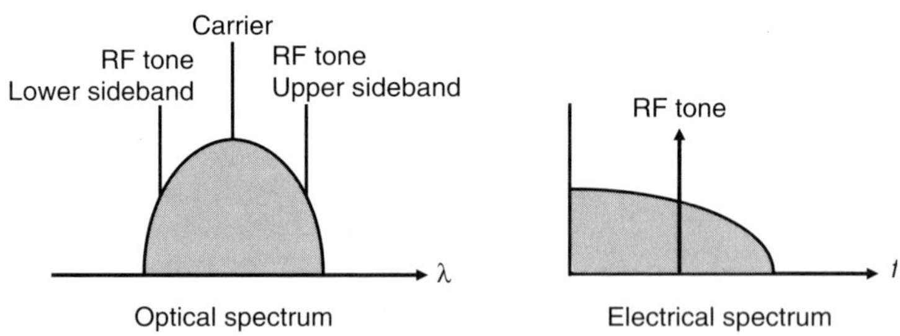

Figure 7.9 RF pilot tone added to the channel bandwidth as the signal quality/degradation monitor.

chromatic dispersion and PMD on a digital signal [32–36]. The spectral techniques have the advantage that they can be implemented with narrowband electronics. Even though high frequencies might be used, the narrow bandwidths will reduce the cost of the electronics. Furthermore, if the detection bandwidth can be narrowed as well, then the sensitivity can be increased. It is worth mentioning that one of the major drawbacks of this technique is that the amplitude-modulated tone and data could interfere with each other and cause deleterious effects. Thus, the amplitude of the pilot tone should be large enough to discern the tone signals from the noiselike random data, but small enough to not induce a significant degradation in the receiver sensitivity for data.

Note that OSNR monitoring is an excellent measure of optical amplifier performance and is frequently used for troubleshooting optical-power induced or amplifiers-induced network faults.

7.3.4.3.3 BER/Q-Factor Monitoring

BER is the ultimate measurement of transmission system performance, thus it is the preferred parameter to use for fault management. In fact, this is precisely the parameter used in electronic networks. However, one difficulty for BER monitoring in optical networks is that the signal is typically error-free within the network. For OPM at an amplifier site the signal is amplified only and not regenerated; therefore, noise will pass through and continue to accumulate. Measurement of the BER at a fault location would result in an error-free measurement. When the signal reaches the end terminal, however, due to accumulated noise it is not error-free and the performance degradation on the BER is observed. In order to detect the degradation within the network, one solution is to use noise loading. In this case, noise is intentionally added to the signal in order to bring the BER to a measurable level and then the additional noise caused by the impairment can be detected.

The common method to the low sensitivity of BER monitoring is to use Q-factor monitoring [37, 38]. The Q-factor is obtained by adjusting the decision threshold voltage of the monitoring receiver away from the optimum level so that errors are recorded. Figure 7.10 (a) shows typical measured data in the logarithm of the BER versus the decision threshold in the decision circuit. Once an error rate is generated, changes to that rate can be monitored and small degradations become visible. Several such techniques have been developed for measuring the Q-factor [39, 40]. Note that the Q-factor is essentially the SNR. It is defined as the difference between the average values of the marks (ones) and the spaces (zeros) divided by the sum of the standard deviations of the noise distributions around each. Figure 7.10 (b) shows the BER as a function of the received optical power [41]. If the Q-factor is measured using a receiver, then it is precisely the electronic SNR. If measured by other means such as optical sampling, then it is the in-band optical SNR.

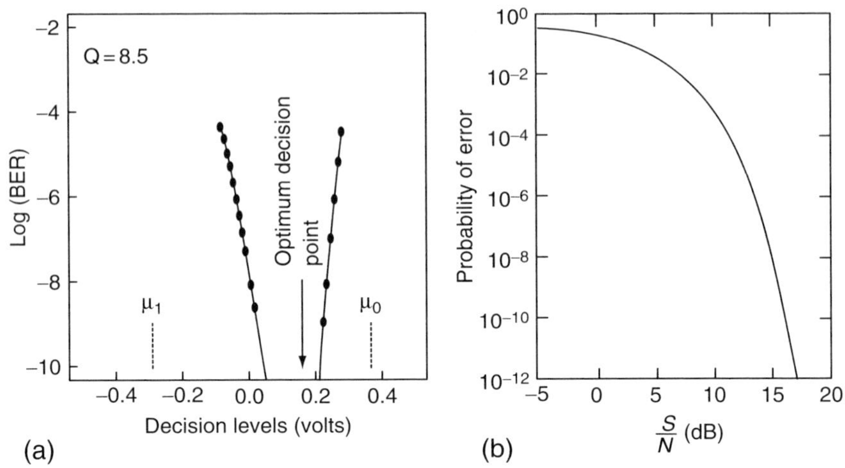

Figure 7.10 (a) Typical measured data for the logarithm of the BER versus the decision threshold [37]; (b) the BER as a function of the received optical SNR [41].

Due to the strong correlation between the Q-factor and BER, Q-factor measurement is highly effective for fault management. The Q-factor is sensitive to the same impairments that impact the end terminal receiver with the appropriate sensitivity. Although the cost of this approach may be high for many embedded network-monitoring applications, a portable unit can be a valuable tool in troubleshooting faults, particularly targeting the rare complication that is not identified by embedded optical channel monitors (OCMs).

7.3.4.4 In-Service Testing Techniques Using Different Wavelengths

Instead of monitoring the signal power or quality directly, there are other proposed methods that test the fiber link at other wavelengths. Due to the fact that the U-band (ultralong wavelength band, 1625–1675 nm) light has been reserved for standard PON monitoring, the following two proposed approaches use U-band light to perform maintenance and in-service testing.

An OCDM solution for in-service testing and management is illustrated in Fig. 7.11 [42]. Every network leg is terminated by a standard passive wavelength selector, widely used in PONs, isolating the standard-monitoring U-band from the other data bands at the front of every ONU and at the CO as well. The transmission section of the OCDM monitoring equipment at the CO consists of a U-band pulse laser driven by a processor to transmit 1 ns (or shorter) pulses with a predetermined low-frequency rate (few megahertz or lower). Every pulse

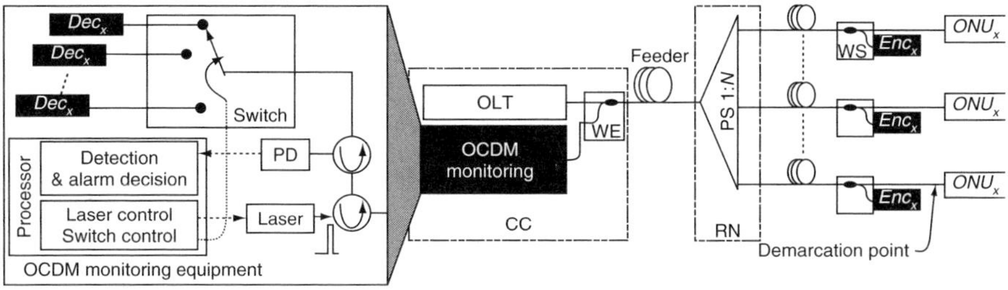

Figure 7.11 OCDM-based PON-monitoring system where every network leg is assigned an encoder. One tunable decoder is employed at the CO [42].

propagates through the tree network is coded by the encoders and is then reflected back to the CO. Every encoder at the leg termination implements a unique code. The coupler combines the upstream encoded pulses together as in standard OCDM. In the CO, a tunable direct sequences (DS)-decoder, consisting of an optical switch and a bank of fixed decoders, (similar to encoders but introducing delays in reverse order) discriminates responses coming from different branches of the tree network. Every healthy branch in the network contributes an autocorrelation peak. A missing autocorrelation peak indicates the corresponding network leg is broken or exhibits abnormal power loss.

Another method also uses a test light at the U-band that is different from the communication band [43]. The filters are installed in front of the ONTs to cut off only the test light, and in front of the OTDR to cut off only the communication light without degradation in the OTDR trace. In fact, U-band in-service line monitoring is an attractive way of maintaining optical-fiber networks cost effectively. Figure 7.12 depicts the system configuration of optical-fiber line testing system using the 1650-nm testing window. The system consists of a control terminal, a test equipment module (TEM) that contains test equipment (TE) such as an OTDR, a test control (TC) unit and a frame and test equipment selector (FTES), optical-fiber selectors that select fibers to be tested, test access modules for introducing test lights into an optical-fiber line, and a termination cable with an optical filter. Termination cables with filters are positioned in front of the termination equipment (e.g. ONT) on users' premises. The filter is a chirped FBG with a center wavelength of 1650 nm and a bandwidth of $\pm$ 5nm. This filtering technology enables monitoring an in-service line using a narrow test band window. The test access modules were installed for the U-band. Inasmuch as the wavelength of the test light is 1650 nm in the U-band, the system can perform maintenance tests on in-service fibers with no degradation to the transmission quality and be applied to a wide communication system covering the entire 1260–1600-nm band.

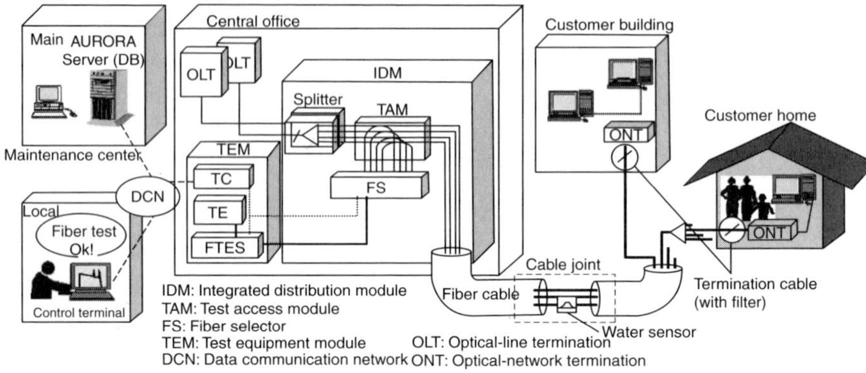

Figure 7.12 System configuration of optical fiber line testing system using a 1650-nm testing window [43].

7.3.4.5 Advanced Monitoring Techniques (Dispersion Monitoring)

In addition to power/wavelength, OSNR, and BER, there are many other physical-layer advanced parameters required to be monitored in high-speed reconfigurable WDM networks, including CD, PMD, polarization-dependent loss, and so on. These effects will become important in superPON systems and next generation high-speed PONs covering bigger service areas.

High-bit-rate transmission systems are susceptible to deleterious optical-fiber-based effects, such as CD and PMD. These impairments can create significant signal distortion and lead to BER floors; thus, isolating the accumulation of these effects may be valuable for network management and control. In today's networks, fibers and other dispersive components are precharacterized for CD and PMD and the transmission link engineered to accommodate or correct for these values. Errors in the network record failures in dispersive components, and improperly installed network elements are potential sources of dispersion-related faults. These problems are often handled in static networks through offline testing during installation. As bit rates increase and optical switching is employed, more emphasis is placed on embedded dispersion monitoring. Increased sensitivity to environmental effects has driven the development of active compensation devices. In reconfigurable optical networks different signals can traverse different paths at different times and, thus, the target dispersion becomes time-dependent. Keeping track of dispersion in a dynamically changing environment can become prohibitively difficult without embedded monitoring. The routing of individual wavelengths requires wavelength selective components, which are often dispersive. The use of these and other dispersive elements introduce a new set of component failure modes that might benefit from monitoring.

Using a dispersion monitor together with noise monitoring provides a highly effective combination for both fault localization and diagnosis (i.e. determining the root cause). Indeed, it may be desirable for management systems to mitigate or compensate these degrading effects separately. In a system that utilizes OCM and active dispersion compensation for control of transmission, this OPM combination might be implemented with little additional hardware. Real-time dispersion monitors that can measure the amount of distortion can be used to trigger alarms or feedback to active dispersion compensators [32, 44]. Here we provide a brief overview of several advanced monitoring techniques that are of interest.

7.3.4.5.1 Chromatic Dispersion (CD) Monitoring

Chromatic dispersion is a well-understood effect that arises from the frequency-dependent nature of the index of refraction in an optical fiber, and is one of the main impairments that limit the performance of optical-fiber systems. For robust high-bit-rate systems, it is essential that dispersion be compensated to within tight tolerances. Several techniques have been demonstrated for real-time CD monitoring to enable dynamic dispersion compensation and may be applied more generally as OPM techniques. One method is to detect the conversion of a phase-modulated signal into an amplitude-modulated signal due to chromatic dispersion [45]. A second method is inserting a subcarrier (RF tone) at the transmitter. The subcarrier approach measures the resulting delay of the subcarrier sidebands relative to the baseband and can be used to measure the accumulated dispersion with fine and medium accuracy without the knowledge of the signal transport history [27, 29, 33, 46, 47]. These two methods are simple and applicable to WDM systems, but require modifications to the transmitter. Based on the dispersion-induced RF power-fading effect, an alterative technique is to extract the bit-rate frequency component (clock) from photo-detected data and monitor its RF power [48, 49]. This technique does not require modification of the transmitter, but is bit-rate and modulation-format-dependent. Although this approach cannot isolate CD, like other tone fading techniques it is sensitive to a variety of distortion effects including PMD and pulse carver misalignment, which is advantageous for fault localization.

Another powerful technique is to detect the relative group delay between the upper and lower vestigial-sideband (VSB) signals in transmitted data [50]: the lower and upper vestigial sidebands are obtained by tuning an optical filter away from the optical-spectrum center of the double-sideband data, as shown in Fig. 7.13. Since the two optical sidebands occupy different wavelength ranges, fiber CD induces a relative group delay between the lower and upper VSB signals.

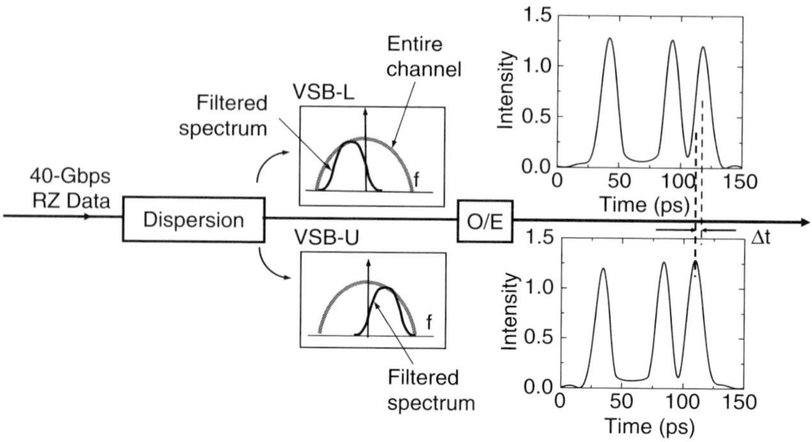

Figure 7.13 Conceptual diagram for monitoring chromatic dispersion using optical vestigial-sideband (VSB) filtering: the recovered bits from either part of the spectrum arrive at slightly different times depending on the chromatic dispersion [50].

This group delay can be measured through clock recovery and phase-sensitive detection. This technique requires no modification at the transmitter, is highly sensitive, is unaffected by PMD, fiber nonlinearity and transmitter chirp, and can be applied to WDM signals by sweeping the optical filter.

7.3.4.5.2 Polarization-Mode-Dispersion Monitoring

PMD is a limitation in optical-communication systems due to either the PMD of the fiber plant, particularly for high-PMD legacy fiber, or of in-line components [44]. PMD is based on the concept that the same spectral component of optical data splits on two orthogonal states of polarization [i.e. principal states of polarization (PSPs)] within a fiber and these two spectral copies travel down the fiber at slightly different speeds. Deleterious PMD effects are stochastic, time-varying, temperature-dependent, and worsen as the bit rate rises. Moreover, the instantaneous first-order PMD [i.e. differential group delay (DGD)] follows a Maxwellian probability distribution, always with some finite possibility of a network outage.

A number of monitoring techniques have been demonstrated to provide appropriate control signals for PMD mitigation. Several techniques are based on spectral analysis such as RF tones [34, 35, 51–53]. A given optical-frequency component splits on two orthogonal PSPs and each replica travels down the fiber with a different speed that dephases these replicas. This effect reduces the corresponding spectral component in the detected RF power spectrum through destructive interference. Another technique involves measuring the phase difference between two optical-frequency components for the two orthogonal PSPs [54]. This technique requires polarization tracking at the receiver to be able to find the PSPs so the phase can be measured. It is well known that any lightwave can be represented by a superposition of a polarized and unpolarized component. The degree of polarization (DOP) is the percentage of polarized component, and is PMD dependent. Its value is in between 0% to 100%. Monitoring techniques based on measuring the DOP of the signal [55, 56] have the advantage of not requiring high-speed circuits and are insensitive to the other degrading effects [57]. Note that the DOP is pulse-width-dependent [52], as shown in Fig. 7.14 (a). However, DOP-based techniques suffer from the following disadvantages: (i) a small DGD monitoring range for short pulse return to zero (RZ) signals; (ii) a lack of sensitivity for non-return to zero (NRZ) signals; and (iii) they are affected by higher-order PMD. Higher-orders of PMD decrease the signal's maximum DOP at the receiver to less than unity [58], as shown in Fig. 7.14 (b). These limitations can potentially be overcome by centering a narrowband optical filter at either the optical–central frequency or one of the signal's sidebands [59].

To summarize this section, there are many impairment effects in fiber networks that could affect system performances and may need to be monitored. However, it is impossible for exhaustive monitoring with reasonable cost and budget. Meanwhile, depending upon the specific network, it is not necessary to monitor all the previously mentioned parameters. For example, a PON's transmission speed in general is limited to 10 Gbps. The transmission distance

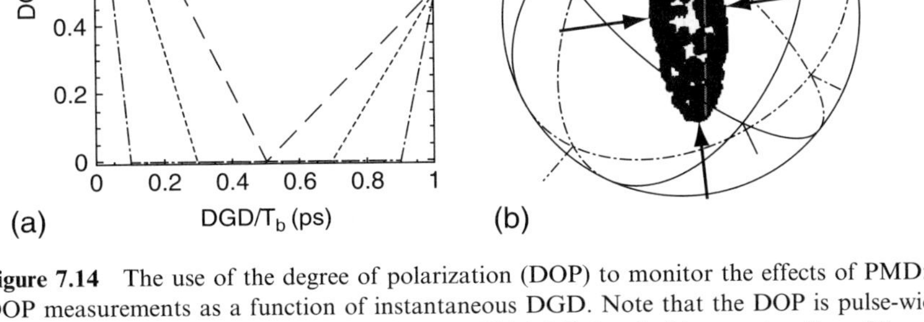

Figure 7.14 The use of the degree of polarization (DOP) to monitor the effects of PMD: (a) DOP measurements as a function of instantaneous DGD. Note that the DOP is pulse-width-dependent [52]; and (b) measured DOP for a 40-Gbit/s RZ signal with concatenation of 6-ps and 4-ps DGD sections [58]. Note that higher-orders of PMD decrease the signal's maximum DOP at the receiver to less than unity.

usually is less than 50 km. Since CD and PMD are not considered as the restrictive effects at this level of bit rate and distance, monitoring of CD and PMD in general is not required in PONs. However, this may not be true for future PON systems with larger transmission distances and higher bit rates, such as superPON and 100-Gbps Ethernet.

7.4 CONCLUSION

The progress in deployed PON has already achieved an astounding rate of growth over the past few years, and is expected to grow exponentially over the next few years. In order to enable robust and cost-effective automated operation, PON systems should probably be able to intelligently monitor the physical state of the network as well as the quality of propagating data signals, isolate specific impairments, and locate faults, and automatically diagnose and repair the network. It is certain that the value of monitoring increases with increasing intelligence.

In this chapter, we have covered a variety of topics on testing and diagnosis, and performance monitoring of PONs. Compared to traditional optical networks, the main parameter for PON testing and monitoring is the loss or power budget. Focus on 100-Gbps Ethernet is growing and high-speed optical components and transmission technologies are being explored. Other physical-layer parameters and signal quality monitoring may be applicable to future PONs, thus there will be many research challenges and therefore opportunities. Since exhaustive monitoring is possible with an unlimited budget, to apply the techniques with the right balance between monitoring coverage, sensitivity, and cost is always the biggest challenge.

ACKNOWLEDGMENTS

The authors wish to thank the kind and critical help from Mr. Hunter Boudreaux at UL Lafayette, Mr. Vahidreza Arbab and Jeng-Yuan Yang at USC's Optical Communications Laboratory.

REFERENCES

[1] The Fiber Optic Association, Inc. (FOA) website: www.thefoa.org
[2] ITU-T G.983 (B-PON) and G.984 standards (G-PON).
[3] IEEE 802.3 Ethernet standard (E-PON).
[4] Available from http://exfo.com/en/support/WaveReview/2004-June/WRarticle1.asp
[5] Available from http://documents.exfo.com/appnotes/anote110-ang.pdf

[6] G. Keiser, *FTTx Concepts and Applications*, John Wiley & Sons Inc., Hoboken, New Jersey, 2006, ISBN 0-471-70420-2.
[7] A. Girard, FTTx PON Technology and Testing, 2005 EXFO Electro-Optical Engineering Inc., ISBN1-5542-006-3.
[8] N. Gagnon and S. Legault, "FTTx testing will require an equipment evolution," *Lightwave*, Dec., 2005.
[9] A.E. Willner, "The optical network of the future: can optical performance monitoring enable automated, intelligent and robust systems?" *OPN*, Mar., 2006.
[10] D.C. Kilper, R. Bach, D.J. Blumenthal, D. Einstein, T. Landolsi, L. Ostar, M. Preiss, and A.E. Willner, "Optical performance monitoring," *IEEE/OSA J. Lightwave Technol.*, vol.22, no.1, pp294–304, 2004.
[11] R. Friskney, K. Warbrick, S. Poliakoff, and R. Heath, "Link-based photonic path performance prediction and control," in Europ. Conf. Optical Commun. (ECOC), 2002.
[12] Available from http://www.optoplex.com/Optical_Channel_Monitor.htm
[13] U. Hilbk, M. Burmeister, B. Hoen, T. Hermes, J. Saniter, and F.J. Westphal, "Selective OTDR measurements at the central office of individual fiber link in a PON," in Optical Fiber Commun. (OFC'97), Paper Tuk3, p54, 1997.
[14] K. Tanaka, H. Izumita, N. Tomita, and Y. Inoue, "In-service individual line monitoring and a method for compensating for the temperature-dependent channel drift of aWDM-PON containing an AWGR using a 1.6 μm tunable OTDR," in Europ. Conf. Optical Commun. (ECOC'97), Paper 448, pp295–298, Sept., 1997.
[15] K.W. Lim, E.S. Son, K.H. Han, and Y.C. Chung, "Fault localization in WDM passive optical network by reusing downstream light sources," *IEEE Phot. Technol. Lett.*, vol.17, no.12, pp2691–2693, Dec., 2005.
[16] H.D. Kim, S.G. Kang, and C.H. Lee, "A low-costWDMsource with an ASE injected Fabry–Pérot semiconductor laser," *IEEE Phot. Technol. Lett.*, vol.12, no.8, pp1067–1069, Aug., 2000.
[17] K. Lee, S.B. Kang, D.S. Lim, H.K. Lee, and W.V. Sorin, "Fiber link loss monitoring scheme in bidirectional WDM transmission using ASE-injected FP LD," *IEEE Phot. Technol. Lett.*, vol.18, no.3, pp523–525, 2006.
[18] Y. Li, D. Wang, and J. Li, "FTTH remote fiber monitoring using optical wavelength domain reflectometry (OWDR) and wavelength coded tag (WCT)," in Optical Fiber Commun. (OFC'06), 2006, Paper OThu3.
[19] C.J. Youn, S.K. Shin, K.J. Park, and Y.C. Chung, "Optical frequency monitoring technique using arrayed-waveguide grating and pilot tones," *Elect. Lett.*, vol.37, pp1032–1033, 2001.
[20] L.E. Nelson, S.T. Cundiff, and C.R. Giles, "Optical monitoring using data correlation for WDM systems," *IEEE Phot. Technol. Lett.*, vol.10, pp1030–1032, 1998.
[21] M. Teshima, M. Koga, and K.I. Sato, "Performance of multiwavelength simultaneous monitoring circuit employing arrayed-waveguide grating," *IEEE/OSA J. Lightwave Technol.*, vol.14, pp2277–2286, 1996.
[22] D.C. Kilper, S. Chandrasekhar, L. Buhl, A. Agarwal, and D. Maywar, "Spectral monitoring of OSNR in high speed networks," in Europ. Conf. Optical Commun. (ECOC), p7.4.4, 2002.
[23] S.K. Shin, K.J. Park, and Y.C. Chung, "A novel optical signal-to-noise ratio monitoring technique for WDM networks," *OSA Trends in Optics and Photonics* (*TOPS*), vol.54, pp182–184, 2000.
[24] J.H. Lee, D.K. Jung, C.H. Kim, and Y.C. Chung, "OSNR monitoring technique using polarization-nulling method," *IEEE Phot. Technol. Lett.*, vol.13, pp88–90, 2001.
[25] J.H. Lee, H.Y. Choi, S.K. Shin, and Y.C. Chung, "A review of the polarization-nulling technique for monitoring optical-signal-to-noise ratio in dynamic WDM networks," *IEEE/OSA J. Lightwave Technol.*, vol.24, no.11, pp4162–4171, Nov., 2006.

[26] X. Tian, Y. Su, W. Hu, L. Leng, P. Hu, L. Yi, Y. Dong, and H. He, "Nonlinearity-tolerant in-band OSNR monitoring for synchronous traffic using gated-signal RF spectral analysis," in Conf. Optical Fiber Commun. (OFC), Paper OThPl, 2006.
[27] M. Rohde, E.-J. Bachus, and F. Raub, "Monitoring of transmission impairments in long-haul transmission systems using the novel digital control modulation technique," in Europ. Conf. Optical Commun. (ECOC), 2002.
[28] G.R. Hill et al., "A transport network layer based on optical network elements," *IEEE/OSA J. Lightwave Technol.*, vol.11, pp667–679, 1993.
[29] G. Rossi, T.E. Dimmick, and D.J. Blumenthal, "Optical performance monitoring in reconfigurable WDM optical networks using subcarrier multiplexing," *IEEE/OSA J. Lightwave Technol.*, vol.18, pp1639–1648, 2000.
[30] D.C. Kilper and W. Weingartner, "Monitoring optical network performance degradation due to amplifier noise," *IEEE/OSA J. Lightwave Technol.*, vol.21, pp1171–1178, May, 2003.
[31] K.-P. Ho and J.M. Kahn, "Methods for crosstalk measurement and reduction in dense WDM systems," *IEEE/OSA J. Lightwave Technol.*, vol.14, pp1127–1135, 1996.
[32] A.E. Willner and B. Hoanca, "Fixed and tunable management of fiber chromatic dispersion," in *Optical Fiber Telecommunications IVB*, I. Kaminow and T. Li, Eds. San Diego, CA: Academic, pp642–724, 2002.
[33] M.N. Petersen, Z. Pan, S. Lee, S.A. Havstad, and A.E. Willner, "Online chromatic dispersion monitoring and compensation using a single inband subcarrier tone," *IEEE Phot. Technol. Lett.*, vol.14, pp570–572, 2002.
[34] G. Ishikawa and H. Ooi, "Polarization-mode dispersion sensitivity and monitoring in 40-Gbit/s OTDM and 10-Gbit/s NRZ transmission experiments," in Conf. Optical Fiber Commun. (OFC) 1998, pp117–119, 1998.
[35] T. Takahashi, T. Imai, and M. Aiki, "Automatic compensation technique for timewise fluctuating polarization mode dispersion in in-line amplifier systems," *Elect. Lett.*, vol.30, pp348–349, 1994.
[36] S. M. R. M.S. M. R. Motaghian Nezam, Y.-W. Song, A.B. Sahin, Z. Pan, and A.E. Willner, "PMD monitoring in WDM systems for NRZ data using a chromatic-dispersion-regenerated clock," in Conf. Optical Fiber Commun. (OFC), pp200–202, 2002.
[37] N.S. Bergano, F.W. Kerfoot, and C.R. Davidson, "Margin measurements in optical amplifier systems," *IEEE Phot. Technol. Lett.*, vol.5, pp304–306, 1993.
[38] ITU-T Recommendation O.201.
[39] R. Wiesmann, O. Bleck, and H. Heppner, "Cost effective performance monitoring in WDM systems," in Conf. Optical Fiber Commun. (OFC), pp171–173, 2000.
[40] J.D. Downie and D.J. Tebben, "Performance monitoring of optical networks with synchronous and asynchronous sampling," in Conf. Optical Fiber Commun. (OFC), 2001.
[41] W.B. Jones, Jr., *Introduction to Optical Fiber Communication Systems*, Holt, Rinehart, and Winston, New York, 1988.
[42] H. Fathallah, and L.A. Rusch, "Network management solution for PS/PON, WDM/PON and hybrid PS/WDM/PON using DS-OCDM," in Conf. Optical Fiber Commun. (OFC), Paper OThE2, 2007.
[43] N. Honda, H. Izumita, and M. Nakamura, "Spectral filtering criteria for u-band test light for in-service line monitoring in optical fiber networks," *IEEE/OSA J. Lightwave Technol.*, vol.24, no.6, pp2328–2335, June, 2006.
[44] H. Kogelink, R.M. Jopson, and L.E. Nelson, "Polarization-mode dispersion," in *Optical Fiber Telecommunications IVB*, I. Kaminow and T. Li, Eds. San Diego, CA: Academic, pp725–861, 2002.

[45] A. Chraplyvy, R. Tkach, L. Buhl, and R. Alferness, "Phase modulation to amplitude modulation conversion of CW laser light in optical fibers," *Elect. Lett.*, vol.22, pp409–411, 1986.
[46] M. Murakami, T. Imai, and M. Aoyama, "A remote supervisory system based on subcarrier overmodulation for submarine optical amplifier systems," *IEEE/OSA J. Lightwave Technol.*, vol.14, pp671–677, 1996.
[47] T.E. Dimmick, G. Rossi, and D.J. Blumenthal, "Optical dispersion monitoring technique using double sideband subcarriers," *IEEE Phot. Technol. Lett.*, vol.12, pp900–902, 2000.
[48] G. Ishikawa and H. Ooi, "Demonstration of automatic dispersion equalization in 40 Gb/s OTDM transmission," in Europ. Conf. Optical Commun. (ECOC), pp519–520, 1998.
[49] Z. Pan, Q. Yu, Y. Xie, S.A. Havstad, A.E. Willner, D.S. Starodubov, and J. Feinberg, "Real-time group-velocity dispersion monitoring and automated compensation without modifications of the transmitter," *Optics Communications*, vol.230, no.1–3, pp145–149, Jan. 15, 2004.
[50] Q. Yu, Z. Pan, L.-S. Yan, and A.E. Willner, "Chromatic dispersion monitoring technique using sideband optical filtering and clock phase-shift detection," *IEEE/OSA J. Lightwave Technol.*, vol.20, no.12, pp2267–2271, Dec., 2002.
[51] H.Y. Pua, K. Peddanarappagari, B. Zhu, C. Allen, K. Demarest, and R. Hui, "An adaptive first-order polarization-mode dispersion compensation system aided by polarization scrambling: theory and demonstration," *IEEE/OSA J. Lightwave Technol.*, vol.18, pp832–841, 2000.
[52] A.E. Willner, S.M.R. Motaghian, L.S. Yan, Z. Pan, and M. Hauer, "Monitoring and control of polarization-related impairments in optical fiber systems," Invited paper, *IEEE/OSA J. Lightwave Technol.*, vol.22, no.1, Jan., 2004.
[53] R. Noe, D. Sandel, M. Yoshida-Dierolf, S. Hinz, V. Mirvoda, A. Schopflin, C. Gungener, E. Gottwald, C. Scheerer, G. Fischer, T. Weyrauch, and W. Haase, "Polarization mode dispersion compensation at 10, 20, and 40 Gb/s with various optical equalizers," *IEEE/OSA J. Lightwave Technol.*, vol.17, pp1602–1616, 1999.
[54] B.W. Hakki, "Polarization mode dispersion compensation by phase diversity detection," *IEEE Phot. Technol. Lett.*, vol.9, pp121–123, 1997.
[55] F. Roy, C. Francia, F. Bruyere, and D. Penninckx, "Simple dynamic polarization mode dispersion compensator," in Conf. Optical Fiber Commun. and Int. Conf. Integrated Optics (OFC/IOOC), vol.1, pp275–278, 1999.
[56] S. Lanne,W. Idler, J.-P. Thiery, and J.-P. Hamaide, "Fully automatic PMD compensation at 40 Gbit/s," *Elect. Lett.*, vol.38, pp40–41, 2002.
[57] N. Kikuchi, "Analysis of signal degree of polarization degradation used as control signal for optical polarization mode dispersion compensation," *IEEE/OSA J. Lightwave Technol.*, vol.19, pp480–486, 2001.
[58] H. Rosenfeldt, C. Knothe, R. Ulrich, E. Brinkmeyer, U. Feiste, C. Schubert, J. Berger, R. Ludwig, H.G. Weber, and A. Ehrhardt, "Automatic PMD compensation at 40 Gbit/s and 80 Gbit/s using a 3-dimensional DOP evaluation for feedback," in Proc. Conf. Optical Fiber Commun., vol.4, ppPD27-1–PD27-3, 2001.
[59] S. M. R. M. S. M. R. Motaghian Nezam, L.-S.Yan, J.E. McGeehan,Y.Q. Shi, A.E. Willner, and S. Yao, "Wide-dynamic-range DGD monitoring by partial optical signal spectrum DOP measurement," in Conf. Optical Fiber Commun. (OFC), pp924–926, 2002.
[60] E. Wong, X. Zhao, and C.J. Chang-Hasnain, "Novel fault monitoring and localization scheme in WDM-PONs with upstream VCSEL transmitters," in Conf. Optical Fiber Commun. (OFC), Paper OThE3, 2007.
[61] W. Chen, B. De Mulder, J. Vandewege, X.Z. Qiu, J. Bauwelinck and B. Baekelandt, "A novel technique for low-cost embedded non-intrusive fiber monitoring of P2MP optical access networks," in Conf. Optical Fiber Commun. (OFC), Paper OThE4, 2007.

[62] G.P. Agrawal, Fiber-Optic Communication Systems, Wiley, New York, 2002.
[63] W.C. Young, "Optical fiber connectors, splices, and joining technology," in Optoelectronic Technology and Lightwave Communication Systems, C. Lin, ed., Van Nostrand Reinhold, 1989.
[64] F. Forghieri, R.W. Tkach, and A.R. Chraplyvy, "Fiber nonlinearities and their impact on transmission systems," Optical Fiber Telecommunications III, Ivan P. Kaminow and Thomas, L. Koch, editors, Academic Press, 1997.
[65] O. Shalitin, G. Shabtay, and Z. Zalevsky, "Future wave in optical network monitoring," Lightwave Web, June 2003.

Appendix I

G-PON PMD Characteristics

This section reproduces the gigabit-speed optical interface requirements in ITU-T G.984.2 standards. The reference points in the following architecture drawing are used in the specifications.

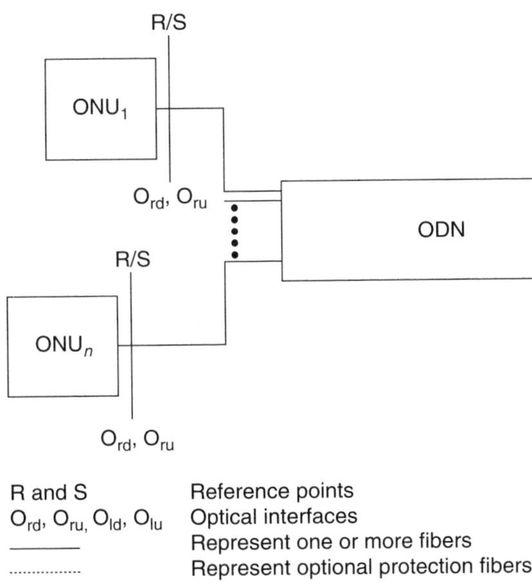

Figure A1.1 Generic physical configuration of the optical distribution network (reproduced from Figure 5/G.983.1, with kind permission from ITU).

Upstream PMD characteristics at 2488.32 Mbps are still under study and yet to be finalized.

Table A1.1

G.984.2—Physical medium dependant layer parameters of ODN (reproduced from Table 2a/G.984.2, with kind permission from ITU)

Items	Unit	Specification
Fibre type (Note 1)	—	ITU-T Rec. G.652
Attenuation range (ITU-T Rec. G.982)	dB	Class A: 5–20 Class B: 10–25 Class C: 15–30
Differential optical path loss	dB	15
Maximum optical path penalty	dB	1
Maximum logical reach	km	60 (Note 2)
Maximum differential logical reach	km	20
Maximum fiber distance between S/R and R/S points	km	20 (10 as option)
Minimum supported split ratio	—	Restricted by path loss PON with passive splitters (16-, 32-, or 64-way split)
Bidirectional transmission	—	1-fiber WDM or 2-fiber
Maintenance wavelength	nm	To be defined

NOTE 1—For future extended reach (>20 km), the use of different types of fiber is for further study, for a future PMD specification.
NOTE 2—This is the maximum distance managed by the higher layers of the system (MAC, TC, Ranging), in view of a future PMD specification.

Table A1.2

G.984.2—Optical interface parameters of 1244 Mbit/s downstream direction (reproduced from Table 2b/G.984.2, with kind permission from ITU)

Items	Unit	Single fiber			Dual fiber		
		OLT transmitter (optical interface O_{ld})					
Nominal bit rate	Mbit/s	1,244.16			1,244.16		
Operating wavelength	nm	1,480–1,500			1,260–1,360		
Line code	—	Scrambled NRZ			Scrambled NRZ		
Mask of the transmitter eye diagram	—	Figure 2			Figure 2		
Maximum reflectance of equipment, measured at transmitter wavelength	dB	NA			NA		
Minimum ORL of ODN at O_{lu} and O_{ld} (Notes 1 and 2)	dB	More than 32			More than 32		
ODN class		A	B	C	A	B	C
Mean-launched power MIN	dBm	−4	+1	+5	−4	+1	5
Mean-launched power MAX	dBm	+1	+6	+9	+1	+6	+9
Launched optical power without input to the transmitter	dBm	NA			NA		
Extinction ratio	dB	More than 10			More than 10		
Tolerance to the transmitter incident light power	dB	More than −15			More than −15		
If MLM laser—maximum RMS width	nm	NA			NA		
If SLM laser—maximum −20 dB width (Note 3)	nm	1			1		
If SLM laser—minimum side mode suppression ratio	dB	30			30		

Continued

Table A1.2

G.984.2—Optical interface parameters of 1244 Mbit/s downstream direction (reproduced from Table 2b/G.984.2, with kind permission from ITU)—Cont'd

Items	Unit	Single fiber			Dual fiber		
		ONU receiver (optical interface O_{rd})					
Maximum reflectance of equipment, measured at receiver wavelength	dB	Less than −20			Less than −20		
Bit error ratio	—	Less than 10^{-10}			Less than 10^{-10}		
ODN class		A	B	C	A	B	C
Minimum sensitivity	dBm	−25	−25	−26	−25	−25	−25
Minimum overload	dBm	−4	−4	−4 (Note 4)	−4	−4	−4
Consecutive identical digit immunity	bit	More than 72			More than 72		
Jitter tolerance	—	Figure 5			Figure 5		
Tolerance to the reflected optical power	dB	Less than 10			Less than 10		

NOTE 1—The value of "minimum ORL of ODN at point O_{ru} and O_{rd}, and O_{lu} and O_{ld}" should be more than 20 dB in optional cases which are described in Appendix I/G.983.1.

NOTE 2—The values on ONU transmitter reflectance for the case in which the value of "minimum ORL of ODN at point O_{ru} and O_{rd}, and O_{lu} and O_{ld}" is 20 dB are described in Appendix II/G.983.1.

NOTE 3—Values of maximum −20-dB width, and minimum side mode suppression ratio are referred to in ITU-T Rec. G.957.

NOTE 4—While only −6-dBm overload is required to support the class C ODN, a −4-dBm overload value has been chosen here for ONU receiver uniformity across all ODN classes.

Table A1.3

G.984.2—Optical interface parameters of 2488 Mbit/s downstream direction (reproduced from Table 2c/G.984.2, with kind permission from ITU)

Items	Unit	Single fiber			Dual fiber		
		OLT transmitter (optical interface O_{ld})					
Nominal bit rate	Mbit/s	2,488.32			2,488.32		
Operating wavelength	nm	1,480–1,500			1,260–1,360		
Line code	—	Scrambled NRZ			Scrambled NRZ		
Mask of the transmitter eye diagram	—	Figure 2			Figure 2		
Maximum reflectance of equipment, measured at transmitter wavelength	dB	NA			NA		
Minimum ORL of ODN at O_{lu} and O_{ld} (Notes 1 and 2)	dB	More than 32			More than 32		
ODN class		A	B	C	A	B	C
Mean-launched power MIN	dBm	0	+5	+3 (Note 4)	0	+5	+3 (Note 4)
Mean-launched power MAX	dBm	+4	+9	+7 (Note 4)	+4	+9	+7 (Note 4)
Launched optical power without input to the transmitter	dBm	NA			NA		
Extinction ratio	dB	More than 10			More than 10		
Tolerance to the transmitter incident light power	dB	More than −15			More than −15		
If MLM laser—maximum RMS width	nm	NA			NA		
If SLM laser—maximum −20-dB width (Note 3)	nm	1			1		
If SLM laser—minimum side mode suppression ratio	dB	30			30		

Continued

Table A1.3

G.984.2—Optical interface parameters of 2488 Mbit/s downstream direction (reproduced from Table 2c/G.984.2, with kind permission from ITU)—Cont'd

Items	Unit	Single fiber			Dual fiber		
		ONU receiver (optical interface O_{rd})					
Maximum reflectance of equipment, measured at receiver wavelength	dB	Less than −20			Less than −20		
Bit error ratio	—	Less than 10^{-10}			Less than 10^{-10}		
ODN class		A	B	C	A	B	C
Minimum sensitivity	dBm	−21	−21	−28 (Note 4)	−21	−21	−28 (Note 4)
Minimum overload	dBm	−1	−1	−8 (Note 4)	−1	−1	−8 (Note 4)
Consecutive identical digit immunity	bit	More than 72			More than 72		
Jitter tolerance	—	Figure 5			Figure 5		
Tolerance to the reflected optical power	dB	Less than 10			Less than 10		

NOTE 1—The value of "minimum ORL of ODN at point O_{ru} and O_{rd}, and O_{lu} and O_{ld}" should be more than 20 dB in optional cases, which are described in Appendix I/G.983.1.

NOTE 2—The value on ONU transmitter reflectance, for the case that the value of "minimum ORL of ODN at point O_{ru} and O_{rd}, and O_{lu} and O_{ld}" is 20 dB, which is described in Appendix II/G.983.1.

NOTE 3—Values of maximum −20-dB width, and minimum side mode suppression ratio are referred to in ITU-T Rec. G.957.

NOTE 4—These values assume the use of a high-power DFB laser for the OLT transmitter and of an APD-based receiver for the ONU. Taking future developments of SOA technology into account, a future alternative implementation could use a DFB laser + SOA, or a higher power laser diode, for the OLT transmitter, allowing a PIN-based receiver for the ONU. The assumed values would then be (conditional to eye-safety regulation and practice):

Mean-launched power MAX OLT transmitter: +12 dBm
Mean-launched power MIN OLT transmitter: +8 dBm
Minimum sensitivity ONU receiver: −23 dBm
Minimum overload ONU receiver: −3 dBm

Table A1.4

G.984.2—Optical interface parameters of 1244 Mbit/s upstream direction (reproduced from Table 2f-1/G.984.2, with kind permission from ITU)

Items	Unit	Single fiber			Dual fiber		
		ONU transmitter (optical interface O_{ru})					
		1,244.16			1,244.16		
Operating wavelength	nm	1,260–1,360			1,260–1,360		
Line code	—	Scrambled NRZ			Scrambled NRZ		
Mask of the transmitter eye diagram	—	Figure 3			Figure 3		
Maximum reflectance of equipment, measured at transmitter wavelength	dB	Less than −6			Less than −6		
Minimum ORL of ODN at O_{ru} and O_{rd} (Notes 1 and 2)	dB	More than 32			More than 32		
ODN class		A	B	C	A	B	C
Mean-launched power MIN	dBm	−3 (Note 5)	−2	+2	−3 (Note 5)	−2	+2
Mean-launched power MAX	dBm	+2 (Note 5)	+3	+7	+2 (Note 5)	+3	+7
Launched optical power without input to the transmitter	dBm	Less than min sensitivity −10			Less than min sensitivity −10		
Maximum Tx Enable (Note 3)	bits	16			16		
Maximum Tx Disable (Note 3)	bits	16			16		
Extinction ratio	dB	More than 10			More than 10		
Tolerance to transmitter incident light power	dB	More than −15			More than −15		
MLM laser—maximum RMS width	nm	(Note 5)			(Note 5)		
SLM laser—maximum −20 dB width (Note 4)	nm	1			1		

Continued

Table A1.4

G.984.2—Optical interface parameters of 1244 Mbit/s upstream direction (reproduced from Table 2f-1/G.984.2, with kind permission from ITU)—Cont'd

Items	Unit	Single fiber			Dual fiber		
If SLM laser—minimum side mode suppression ratio	dB	30			30		
Jitter transfer	—	Figure 4			Figure 4		
Jitter generation from 4.0 kHz to 10.0 MHz	UI p-p	0.33			0.33		
		OLT receiver (optical interface O_{lu})					
Maximum reflectance of equipment, measured at receiver wavelength	dB	Less than −20			Less than −20		
Bit error ratio	—	Less than 10^{-10}			Less than 10^{-10}		
ODN Class		A	B	C	A	B	C
Minimum sensitivity	dBm	−24 (Note 6)	−28	−29	−24 (Note 6)	−28	−29
Minimum overload	dBm	−3 (Note 6)	−7	−8	−3 (Note 6)	−7	−8
Consecutive identical digit immunity	Bit	More than 72			More than 72		
Jitter tolerance	—	NA			NA		
Tolerance to the reflected optical power	dB	Less than 10			Less than 10		

NOTE 1—The value of "minimum ORL of ODN at point O_{ru} and O_{rd}, and O_{lu} and O_{ld}" should be more than 20 dB in optional cases which are described in Appendix I/G.983.1.

NOTE 2—The values of ONU transmitter reflectance for the case that the value of "minimum ORL of ODN at point O_{ru} and O_{rd}, and O_{lu} and O_{ld}" is 20 dB are described in Appendix II/G.983.1.

NOTE 3—As defined in 8.2.6.3.1.

NOTE 4—Values of maximum −20-dB width, and minimum side mode suppression ratio are referred to in ITU-T Rec. G.957.

NOTE 5—While MLM laser types are not applicable to support the full ODN fiber distance of Table 2a, such lasers can be used if the maximum ODN fiber distance between R/S and S/R is restricted to 10 km. The MLM laser types of Table 2e can be employed to support this restricted fiber distance at 1244.16 Mbit/s. These laser types are subject to the same conditions as indicated in Note 5 of Table 2e.

NOTE 6—These values assume the use of a PIN-based receiver at the OLT for Class A. Depending on the amount of ONUs connected to the OLT, an alternative implementation from a cost point of view could be based on an APD-based receiver at the OLT, allowing it to use more economical lasers with lower fiber-coupled emitted power at the ONUs. In this case, the values for Class A would be:
Mean-launched power MIN ONU transmitter: −7 dBm
Mean-launched power MAX ONU transmitter: −2 dBm
Minimum sensitivity OLT receiver: −28 dBm
Minimum overload OLT receiver: −7 dBm

Table A1.5

G.984.2—Optical interface parameters of 1244 Mbit/s upstream direction using power-levelling mechanism at ONU Transmitter (reproduced from Table 2f-2/G.984.2, with kind permission from ITU)

Items	Unit	Single fiber			Dual fiber		
		ONU transmitter (optical interface O_{ru})					
ODN Class		A	B	C	A	B	C
Mean-launched power MIN	dBm	−2 (Note 2)	−2	+2	−2 (Note 2)	−2	+2
Mean-launched power MAX	dBm	+3 (Note 2)	+3	+7	+3 (Note 2)	+3	+7
		OLT receiver (optical interface O_{lu})					
ODN class		A	B	C	A	B	C
Minimum sensitivity	dBm	−23 (Note 2)	−28	−29	−23 (Note 2)	−28	−29
Minimum overload	dBm	−8 (Note 2)	−13	−14	−8 (Note 2)	−13	−14

NOTE 1—This table only indicates the parameters of Table 2f-1 that change due to the application of the power-leveling mechanism at ONU transmitter, namely the launched powers of the ONU transmitter and the sensitivity and overload of the OLT receiver. All other parameters and notes are identical to those in Table 2f-1.

NOTE 2—These values assume the use of a PIN-based receiver at the OLT for Class A. Depending on the amount of ONUs connected to the OLT, an alternative implementation from a cost point of view could be based on an APD-based receiver at the OLT, allowing it to use more economical lasers with lower fiber-coupled emitted power at the ONUs. In this case, the values for Class A would be:
Mean-launched power MIN ONU transmitter: −7 dBm
Mean-launched power MAX ONU transmitter: −2 dBm
Minimum sensitivity OLT receiver: −28 dBm
Minimum overload OLT receiver: −10 dBm
The impact of power-leveling is less, due to the restriction on the minimal power to be emitted for guaranteeing the eye-diagram.

Appendix II

EPON MPCPDU Formats

This section lists the EPON multipoint MAC protocol data units as specified in IEEE 802.3ah standard, Clause 64.

A2.1 Gate Frame

Octets	Field
6	Destination address
6	Source address
2	Length/Type (0 × 88–08)
2	Op-code (0 × 00–02)
4	Timestamp
1	Number of grants/Flags
0/4	Grant #1 start time
0/2	Grant #1 length
0/4	Grant #2 start time
0/2	Grant #2 length
0/4	Grant #3 start time
0/2	Grant #3 length
0/4	Grant #4 start time
0/2	Grant #4 length
2	Sync time
13–39	0 Padding/reserved
4	FCS

Bit	Flag field
0–2	Number of grants (0–4)
3	Discovery (0-normal, 1-discovery)
4	Force Report Grant 1
5	Force Report Grant 2
6	Force Report Grant 3
7(MSB)	Force Report Grant 4

When a force report grant bit is 0, no action is required. When it is 1, a Report frame should be issued at the corresponding transmission opportunity indicated in that Grant.

Sync time: Only present for discovery gate to indicate the synchronization time of OLT receiver.

Figure A2.1 Gate frame format.

A2.2 Report Frame

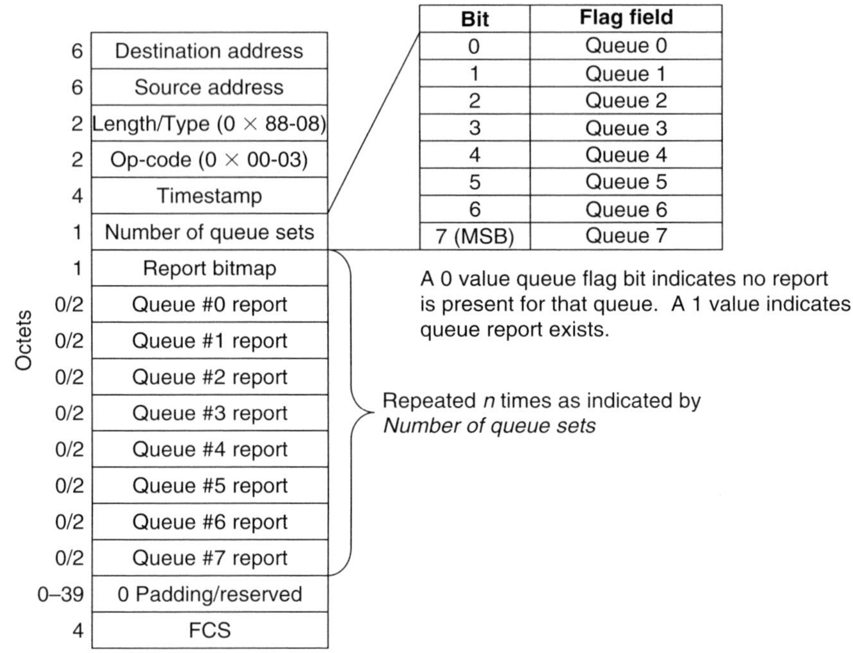

Figure A2.2 Report frame format.

A2.3 Register Request Frame

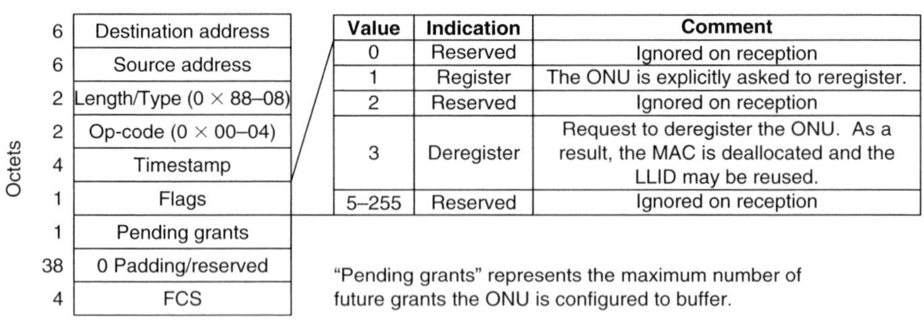

Figure A2.3 Register request frame format.

A2.4 Register Frame

Octets	Field
6	Destination address
6	Source address
2	Length/Type (0 × 88–08)
2	Op-code (0 × 00–05)
4	Timestamp
2	Assigned port (LLID)
1	Flags
2	Sync time
1	Echoed pending grants
34	0 Padding/reserved
4	FCS

Value	Indication	Comment
0	Reserved	Ignored on reception
1	Reregister	The ONU is explicitly asked to reregister.
2	Deregister	Request to deallocate the port and free the LLID
3	Ack	The requested registration is successful
4	Nack	The requested registration attempt is denied by the higher layer entity.
5–255	Reserved	Ignored on reception

Assigned Port represents the LLID assigned to ONU MAC following registration.

Figure A2.4 Register frame format.

A2.5 Register Acknowledgment Frame

Octets	Field
6	Destination address
6	Source address
2	Length/Type (0 × 88–08)
2	Op-code (0 × 00–06)
4	Timestamp
1	Flags
2	Echoed assigned port
2	Echoed sync time
35	0 Padding/reserved
4	FCS

Value	Indication	Comment
0	Nack	The requested registration attempt is denied by the higher layer entity.
1	Ack	The registration process is successfully acknowledged.
2–255	Reserved	Ignored on reception

Figure A2.5 Register acknowledgment frame format.

A2.6 Reference

IEEE 802.3 2005, Clause 64.

Index

2P1 Device. *See* Two-PONs-in-one device

A

Access nodes (AN), 245
Acousto-optic switches, 261
All-polymer AWG
  polymer waveguide materials, 103
  TE/TM polarization shift and temperature-dependent transmission spectrum of, 104
Amplified spontaneous emission (ASE), 74, 284
Amplitude modulation-vestigial side band (AM-VSB), 3
Analog CATV Overlay impact
  BER degradation, 205
  CNR degradation, 206–207
Analog TV distribution systems, 36
Analog voice frequency transmission, 2
Angled physical contact connectors, 89, 112
APDs. *See* Avalanche photodiodes
Arrayed waveguide grating router (AWGR), 72–73, 283
Arrayed waveguide grating, temperature and polarization insensitivity
  mechanically movable compensation plate for, 104
  polymer materials and overcladding, 103
Array waveguide grating (AWG), 246
  athermal, 99
  crosstalk of, 93–94
  cyclic, 94
  dual function, 96–99
  freepropagating regions or slabs, 93
  input and output Ports, 95–96
  for passive operation, 88
  propagation loss, 93
  schematic layout of $1\times N$, 93–94
  schematic layout of $N\times N$, 94–95
  as wavelength router and demultiplexer, 93
  WDM signal propagation, 93–94
Asymmetric DSL (ADSL), 3
Asynchronous transfer mode (ATM), 8
ATC method. *See* Automatic threshold control method
Athermal arrayed waveguide grating
  based on compensating plate, 104–106
  based on negative thermal coefficient materials, 103–104
  based on temperature-compensating rod, 106–107
  bimetal plate and temperature-dependent stress, 107–108
  channel spacing of, 99, 108
  diffraction loss and silicone resin-filled trenches, 101–103
  insertion losses and crosstalk level, 103–104
  insertion losses and triangular grooves, 99–100
  mechanically movable compensation plate in, 104
  polarization-dependent wavelength shift minimization, 103
  polymer materials and waveguide structure, 103–104
  polymer overcladding of, 103
  schematic diagram of, 100
  silicon resin-filled trenches in slab region, 102–103
  temperature and polarization insensitivity, 103–104
  temperature-dependent transmission spectrum, 99, 104
  TE/TM polarization shift, 104–105

Athermal AWG based on compensating plate
insertion loss, 106
optical spectra of, 105
Athermal AWG based on temperature-compensating rod
crosstalk levels and loss of, 106
temperature-dependent wavelength shift, 107
ATM adaptation layer (AAL), 40
ATM passive optical network (APON), 37
control and management, 40–41
services in, 40
system description, 37–40
ATM passive optical network (APON)/ G-PON
collision resolution in, 42
protection switching, 54
Automatic gain control (AGC) technique, 218
cell-AGC preamplifier configuration, 184–185
trans-impedance gain control, 184
Automatic protection switching (APS), 247
Automatic threshold control method, 183–184
Avalanche photodiodes
TO-CAN structure of, 165
V-I curve of, 164
Avalanche photo-diodes (APD), 44

B

Bandwidth allocation
dynamic bandwidth allocation
IPACT and its variants, 228–231
overview of, 221–222
requirements of, 224–226
target service, 222–224
traffic control fundamentals, 226–228
ranging
EPON, protocol of, 220–221
G-PON, protocol of, 218–220
procedures for, 216–218
purpose of, 215–216
Bendable SC connector and indoor installation, 110–111
Bend-insensitive single mode optical fiber
bend loss characteristics of, 113
for indoor fiber installation, 112
BER/Q-factor monitoring, 289–290
Bi-directional line-switched ring (BLSR), 253
Bidirectional optical loss, 273
Bidirectional optical subassembly
bulk-optic assembly technology
optical and electrical cross talk suppression, 168
optical system schematics of, 167
TO-CAN fabrication, 168
planar lightwave circuits
embedded-filter, 171
external-filter, 170–171
Bit error rate (BER), 30, 59, 268
BM Tx. *See* Burst-mode transmitter
BOSA. *See* Bidirectional optical subassembly
BPON
burst-mode timing specifications for, 158
power-leveling mechanism (PLM), 159
split ratios and power budgets for, 154–155
standards, 155–156
standards, physical-layer requirements of, 155–156
Broadband access networks
cable modem, 3–6
digital subscriber line (DSL), 2–3
Ethernet, 8–11
fiber access systems, 7–8
history of, 1
Broadband interactive data signals, 243
Broadband optical sources and ONU, 129–130
Broadcast emulation and point-to-point operation, 74
Burst-mode BER and sensitivity penalty, measurement of, 197, 199–201
Burst-mode protocol, 56

Burst-mode receiver
AC-coupled circuit, 185–186
DC-coupled implementations
automatic gain control technique, 184–185
automatic threshold control method, 183–184
feedback configuration
differential input/output trans-impedance amplifier, 181–182
feedback loop, 182–183
feed-forward implementation
DC-coupled pre-amplifier, 181
synchronization, 64
Burst-mode transmission, 33
Burst-mode transmitter
APC circuit, 178–179
automatic power control (APC) algorithm, 176–177
laser diode and laser diode driver IC (LDD), 175–176
optical and timing performance of, 179–180
Burst overhead (BOH), 218

C

Cable modem, 3–6
Cable modem termination system (CMTS), 5
Capital expenditure (CAPEX), 35
Carrier Sense Multiple Access with Collision Detection (CSMA/CD), 9–10, 215–216
Carrier-to-noise ratio (CNR), 25, 35
C-band analog CATV signal overlay, 35–36
Centralized supercontinuum broadband light source, 130
Central office (CO), 19–20, 243, 267, 275
Chromatic dispersion (CD), 23–24, 268, 293–294
Coarse wavelength division multiplexer (CWDM), 75, 250
Coarse WDM coupler
insertion loss, 98
pass bands, 97
signal isolation and, 97–98
Code division multiple access (CDMA), 215
Colorless ONUs
based on spectral-slicing techniques
broadband optical sources and, 129
transmission bit-rate, effect on, 130
optically injection-locked and wavelength-seeded
self injection-locked FP-LDs, 135–136
self-seeding reflective SOAs, 137
Community antenna TV (CATV) service, 1, 3, 5
Cross-phase modulation (XPM) index, 271
CWDM coupler. *See* Coarse WDM coupler
Cyclic redundancy check (CRC), 60

D

Data over cable service interface specification (DOCSIS), 5
DBR lasers. *See* Distributed Bragg reflector lasers
DDTC. *See* Digital diagnostics transceiver controller
Deficit round-robin (DRR) scheduling, 231
Degree of polarization (DOP), 295
DFB lasers. *See* Distributed feedback lasers
Differential group delay (DGD), 270, 294
Digital diagnostics transceiver controller, 188
Digital subscriber line (DSL) services, 2–3
Digital voice signals, 2
Discrete multitone modulation (DMT), 3
Dispersion tolerant transceivers, 71
Distributed Bragg reflector lasers, 123–124
Distributed feedback (DFB) lasers, 28, 123–124
L-I curve of, 164
TO-CAN package of, 165

Dual-function AWG
2P1 device, 97–99
CWDM, 97
insertion losses and transmission spectra of wavelength router, 97
multilayered dielectric filter, 97–98
schematic illustration of, 96
Dynamic bandwidth allocation (DBA)
applications of, 6
implementation of, 62
IPACT and its variants, 228–231
mechanisms of, 39
multiplexing capabilities and embedded OAM functions for, 47
overview of, 221–222
requirements of, 224–226
target service, 222–224
traffic control fundamentals, 226–228
Dynamic bandwidth report upstream (DBRu), 50

E

Electromagnetic waves, attenuation of, 3
End-to-end private-line Ethernet services, 10
EPON
encryption and protection mechanisms for, 67–68
ONU, autodiscovery of, 64
PMD layer, 56
point-to-point emulation in, 65
EPON MPCPDU
See EPON multipoint MAC control protocol data units
EPON multipoint MAC control protocol data units
gate frame, 311
register acknowledgement frame, 313
report frame and register request frame, 312
EPON standards, physical-layer requirements of, 156–157
EPON transceiver
burst-mode timing specifications for, 158
multipoint MAC control protocol data units
gate frame, 311
register acknowledgment frame, 313
report frame and register request frame, 312
PMD parameters of, 159–160
split ratios and power budgets, 154–155
standards, physical-layer requirements of, 156–157
system evaluation
continuous-mode and P2MP burst-mode BER, 202–203
dispersion penalties, 205
F-P laser diodes, 204
Erbium-doped fiber amplifier (EDFA), 24, 70
Ethernet
for connecting IP devices, 8–11
framing of, 59–60
MAC interfaces, 65
OAM sublayer, 68–69
Ethernet in the First Mile (EFM), 8

F

Fabry-Perot (FP) laser diodes, 11, 26, 123–124, 284
FA connector socket and plug. *See* Field assembly connector plug and socket
Fair Queuing, 226
FFTH installation. *See* Fiber-to-the-home installation
Fiber deep access network architecture, 14
Fiber-in-the-loop (FITL) systems, 7
Fiber network impairments, monitoring of, 280–281
Fiber-optic cable installations, 277
Fiber-optic communication systems, 22, 26
Fiber's key characteristics, 269
Fiber-to-the-curb (FTTC), 3, 20
Fiber-to-the-home (FTTH) system, 7, 15, 20, 223, 285
Fiber-to-the-home installation, 113
Fiber-to-the-premise (FTTP), 8

Fiber-to-the-x (FTTx) system, 7, 14–15, 19, 23
Field assembly connector plug and socket
components of, 109
connection loss and return loss trials, 110
Fixed bandwidth allocation (FBA), 221
Forward error correction (FEC), 53
Four-wave-mixing (FWM), 271
FP-LD. *See* Fabry-Perot laser diode
Frame and test equipment selector (FTES), 291
Frame check sequence (FCS), 60
Free spectral range
of athermal AWGs, 132
cyclic property of AWG, 94
Free spectral ranges (FSR), 73
Frequency division multiplexed (FDM) frequency, 3
Fresnel reflections, 269
FSR. *See* Free spectral range
Full service access network (FSAN), 7, 37

G
GEM frame synchronization, 53
Generic framing procedure (GFP), 8, 45
Gigabit capable PON (G-PON), 8
Gigabit media-independent interface (GMII), 55
GI-POF. *See* Graded index polymer optical fiber
G-PON and EPON, comparision between, 69–70
G-PON Architecture, 44
G-PON encapsulation mode (GEM), 8
G-PON transceiver
burst-mode timing specifications for, 158–159
PMD parameters of, 159
of ODN, 302
optical interface parameters for downstream direction, 303–306
optical interface parameters for upstream direction, 307–309
power-leveling mechanism (PLM), 159, 199
split ratios and power budgets for, 154–155
standards, physical-layer requirements of, 157
system evaluation
burst length and guard time, 197, 199
burst-mode penalty, 197, 199–200
performance parameters for upstream link, 201–202
G-PON transmission convergence (GTC) layer
GEM encapsulation, 52–53
GTC downstream framing, 47
GTC framing, 47
GTC upstream framing, 48
Graded index polymer optical fiber
bending loss and refractive index of, 119–120
and home networking, 119

H
Header error control (HEC), 53
Hybrid fiber-coax (HFC) network, 3, 273
Hybrid WDM/TDM-PON architecture, 76–80

I
Indoor fiber
bend-insensitive single mode optical fiber, 112–114
optical curl cord, 114–119
Integrated services digital network (ISDN) system, 2
Interframe gap (IFG), 59
Interleaved polling with adaptive cycle time (IPACT), 228, 231
Internet Engineering Task force (IETF), 10
Internet Protocol (IP), packets of, 223
Intersymbol interference, 28
ITU-T G.984.2 specification requirements, 201–202

L
Linear optical transmission system, 4
Live fiber detector (LFD), 276

Logical link identifier (LLID), 56
Low-loss optical components, 30

M

Management information bases (MIB), 42
Maximum transmission window (MTW), 228
Maxwellian probability distribution, 294
Media-independent interface (MII), 55
Medium access control (MAC), 5, 31
Metro Ethernet Forum (MEF), 10
MFD mismatch. *See* Mode field diameter mismatch
Mode field diameter mismatch, 116–117
Multilayered dielectric filter, 97–98
Multi-longitudinal-mode (MLM) lasers, 44
Multimode fibers (MMF), 22
Multiple service operators (MSO), 5
Multi-Point Control Protocol Data Unit (MPCPDU), 62–63
Multipoint control protocol data units (MPCPDU), 60
Multipoint control protocol (MPCP), 55, 60–65
Multipoint-to-multipoint (MP2MP) coaxial bus line, 9

N

Near-end cross talk (NEXT), 26
Network management system (NMS), 244
Network performance, monitoring of, 278–279
Noise funneling effect, 5
Nonlinear birefringence, 288
North American National Television System Committee (NTSC) standard, 3

O

ODN. *See* Optical distribution network
OIL-VCSELs. *See* Optically injection-locking vertical-cavity surface-emitting laser
OLT. *See* Optical line terminal
OLT burst mode receiver, 44
OLT receiver, 163
One-fiber single-wavelength full duplex, 26–27
ONT management and control channel (OMCC), 42
ONT management and control interface (OMCI), 42
ONU-embedded OTDR
  back-scattering, 195
  data bursts, 193
ONU transceiver
  components of, 172
  embedded OTDR functionality
    negative step response (NSR) approach, 193–195
    optic coupler and laser diode, 195–196
  fiber monitoring functionality of, 192–193
  PLM mode, 200–201
  triplexer transceiver module for, 172–173
ONU transmitter, 163
Open system interconnect (OSI), 54
Operation, administration, and management (OAM) sublayer, 10
Optical add-drop multiplexer (OADM), 245
Optical curl cord
  bending loss and bending radius, 115–116
  FTTH installation, 114
  laptop terminal and optical outlet, connection between, 115
  MFD mismatch, 116–117
Optical distribution network, 159
Optical distribution network (ODN), 38
Optical fiber link, 276
Optical fibers
  index of refraction of, 270
  multimode fibers (MMF), 22
  standard single-mode fiber (SMF), 22
Optical line terminal, 152
Optical line terminal (OLT), 6–7, 20, 215, 243, 268, 275
Optical loss test set (OLTS), 275

Optically injection-locking vertical-cavity surface-emitting laser, 125–128
Optical networks
basic monitoring techniques for, 279–280
fault management of, 245
Optical network termination (ONT), 268
Optical network units (ONU), 20, 88, 215, 243, 268
colorless ONUs, 129–137
source-free ONUs, 137–141
wavelength-specific ONUs, 123–129
Optical performance monitoring (OPM), 279
Optical power level monitoring, 247
Optical power splitter, 87
insertion losses in, 99
PLC-based, 91–93
Optical pulse suppressor (OPS), 276
Optical return loss (ORL), 268, 272
Optical signal
delay, 34
pulses, 23
Optical signal-to-noise ratio (OSNR), 279, 286
Optical spectral bandwidth, 74
Optical spectrum analyzer (OSA), 276, 284
Optical star coupler, 36
Optical switch (OSW), 250
Optical time-domain reflectometer (OTDR), 268, 274, 276
Optical time domain reflectometry system
components of, 192
fiber fault detection, 193
Optical to electrical (OE) conversion, 35
Optical transmission
classes of, 38
system
chromatic dispersion, 23–24
fiber loss, 25–26
optical fibers, 21–23
Optical transmit and receive devices
avalanche photodiodes, 164
Optical transmitter devices, 163
TO packaging techniques, 164–165
WDM coupler, 165
Optical transport network (OTN), 53
Optical wavelength domain reflectometry (OWDR), 285
OTDR system. *See* Optical time domain reflectometry system

P

Passive optical network (PON) , 87
affects of optical fiber loss, 25–26
characterization and monitoring challenges for, 271–272
components of, 152
commercial application of, 14
as fiber optic access network, 19
history of, 1
network testing for diagnosis guidelines in, 277–278
network topologies for, 245
optical diagnosis, performance monitoring, and characterization of, 267
optical line termination (OLT) interface in, 215
protection architectures for, 243
signal distribution for, 11
splitting ratio of, 25, 30
wavelength division multiplexed, 120
Payload length indicator (PLI), 52
Payload type indicator (PTI), 53
Phase alternate line (PAL) standard, 3
Physical coding sublayer (PCS), 59
Physical control block (PCBd), 47
Physical layer OAM (PLOAM) functions, 45
Physical layer operation, administration, and management upstream (PLOAMu), 50
Physical layer overhead upstream (PLOu), 48
Physical layer (PHY), 55
Physical-layer specifications
BPON standards, 155–156
data traffic, 155
of G-PON and EPON Standards, 156–157

Physical medium dependent (PMD) layer, 8, 155
Plain old telephone service (POTS), 2, 31
Planar lightwave circuits
  embedded-filter, WDM filtering technology, 171
  external-filter, 169
    passive alignment, 171
    WDM circuit and transmitter LD, 170
  technology, 87
PLC-based optical power splitter
  connector loss and splitting loss, 93
  insertion loss, 92
  network implementations, 91
  PLC chip and fiber array alignment, 91
PLM. *See* Power-leveling mechanism
P2MP power-splitting PON (PS-PON), 7
Point-to-point emulation (P2PE), 64
Polarization extinction ratio, 287
Polarization mode dispersion (PMD), 270, 292, 294
PON. *See* Passive optical network
PON burst-mode timing requirements
  for BPON, G-PON, and EPON, 158–159
  burst-mode chips and, 163
  ONU transmitter switching, 156–157
PON transceivers
  absolute temperature monitoring
    sensing voltage and temperature, 188–189
    signal stability, 190
  bidirectional optical subassembly
    bulk-optic assembly technology, 167–169
    planar lightwave circuits, 169–172
  burst-mode transmission
    for BPON, G-PON, and EPON, 158–159
    burst-mode data, 174–175
    burst-mode receiver, 181–186
    burst-mode transmitter, 175–181
    power drift, 157
  burst packet and burst-mode data, 174–175
  CATV overlay, impact of
    BER degradation, 205
    CNR degradation, 206–207
  components of, 152
  functional building blocks of, 162
    OLT side of, 162–163
    ONU side of, 162
  module multisource agreements, 172
  optical link budgets, 155
  physical-layer specifications, 154–156
  power-monitoring, 190
  split ratios and power budgets for, 154
  system evaluation
    burst-mode penalties, 197
    burst-mode test setup for, 197–198
    data bursts, 197
  types of
    BPON, GPON, and EPON, 154
    diplexer and triplexer, 161–162
  upstream and downstream traffic, 152
PON transceiver system evaluation
  data bursts, 197
Power-leveling mechanism, 159, 199
Power leveling sequence upstream (PLSu), 50
Power-splitting passive optical network, 87–88
Principal states of polarization (PSP), 294
Protocol data units (PDU), 60

Q

Quadrature amplitude modulation (QAM), 5
Quadrature phase shift keying (QPSK), 6
Quality of service (QoS), 35, 40, 223, 225, 267, 275

R

Radio frequency (RF) domain, 4
Raman amplification, 70
Rayleigh scattering, 25, 269
Receiver-power-monitoring, 190
Reflective semiconductor optical amplifiers (RSOA), 11, 79, 137
Regional Bell Operation Companies (RBOC), 14

Remote node (RN), 243
Remote terminal (RT), 15, 19–20
Request for proposal (RFP), 8, 70
Ring topology, 253
Round-trip delay (RTD), 215
Round-trip time (RTT), 34

S

Self-healing rings (SHR), 253, 261
Self-phase modulation (SPM), 271
Semiconductor optical amplifier (SOA), 70
Service Level Agreement (SLA), 10
Service network interface (SNI), 32, 43
Set-top box (STB), 35
Signal quality monitoring, 296
Silica optical fiber, 25
Single-copy broadcast (SCB), 67
Source-free ONUs
  amplitude shift keying, 139
  downstream DFB laser source, optical spectrum of, 140
  downstream signal encoding, 139–140
  round-trip loss, 140
  wavelength reuse scheme, 138–139
Spanning Tree Protocol (STP), 10
Sparse signaling, 274–275
Splitting strategies in a TDM-PON, 29
SRR. *See* Upstream signals to Rayleigh backscattering ratio
Start frame delimiter (SFD), 60
Start LLID delimiter (SLD), 66
Super PON, 70

T

Telecommunication networks
  applications of, 11–12
  IP data traffic in, 9
  policy and regulations for, 13–14
  standardization efforts for, 14
  structure of, 19
Temperature-dependent transmission spectrum, 99–100
Termination unit (NT), 268
Test equipment module (TEM), 291
Time division duplex, 27
Time division multiple access (TDMA), 215, 272
Time division multiplexing PON (TDMPON), 20
Time division multiplexing (TDM), 5, 84
Traffic restoration time, 247–248
Transceiver digital diagnostic monitoring
  absolute temperature, 188–189
  optical power, 190
  transmitter bias current, 190–191
  transmitter-power, 191
Transmitter bias current monitoring, 190–191
Transmitter power monitoring, 191
Tree topology, 248–252
Two-PONs-in-one device, 97–99

U

Unidirectional path-switched ring (UPSR), 253
Upstream signals to Rayleigh backscattering ratio
  downstream signals for, 128
  and Rayleigh backscattering effects, 127
User network interface (UNI), 10, 32, 43, 223

V

Vertical-cavity surface-emitting laser (VCSEL), 89–90
  BER measurements of, 127–128
  and colorless operation, 128–129
  modulation bandwidth, 126
  optical injection locking, 125–126
Very high data rate DSL (VDSL), 3
Vestigial-sideband (VSB) signals, 293
Video on-demand (VOD) services, 1
Video streaming services, 9
Virtual channel identifier (VCI), 40
Virtual circuits (VC), 40
Virtual path identifier (VPI), 40
Visual fault locator (VFL), 276
Voice-over IP (VoIP), 223, 225

W

Waveguide forming process
  polarization-dependent wavelength shift minimization, 103
  polymer overcladding, 103
Wavelength coded tag (WCT), 285
Wavelength division duplex method, 27–28
Wavelength division multiple access (WDMA), 215
Wavelength division multiplexed passive optical network, 120
  AWG as multiplexer, 93–94
  downstream/upstream bandwidths, 88
  OIL-VCSELs, implementation of, 126
  optical injection-locking of FP-LDs
    centralized broadband optical source, 131–133
    wavelength-seeded RSOAs deployment, 134–135
  self-injection-locked Fabry-Perot laser diode modulation, 136–137
Wavelength division multiplexed passive optical network (WDM-PON)
  advantages and challenges of, 72
  ring topology, 260–261
  tree topology, 253–260
Wavelength division multiplexing signal, 93, 97
Wavelength division multiplexing (WDM), 11, 19–20
Wavelength grating router (WGR), 73
Wavelength reuse schemes
  downstream signal encoding, 139–140
  and multichannel fiber Bragg grating, 141
Wavelength-specific ONUs
  DFB and DBR lasers
    feedback mechanism, 124
    FP etalons and Bragg filters, 124
    high-speed direct modulation, 123
  VCSELs
    BER measurements of, 127–128
    and colorless operation, 128–129
    modulation bandwidth, 126
    optical injection locking, 125–126
WDM. *See* Wavelength division multiplexing
WDM coupler, 15
  transmission wavelength spectrum of, 166
  upstream and downstream signal separation, 165–166
WDM Ethernet-PON (WE-PON), 79
WDM-on-WDM architecture, 75
WDM signal propagation, 93–94
WDM transmissions, 25